The Making of Measure and the Promise of Sameness

The Making of
Measure and the
Promise of Sameness

Emanuele Lugli

THE UNIVERSITY OF CHICAGO PRESS

Chicago and London

The University of Chicago Press, Chicago 60637
The University of Chicago Press, Ltd., London
© 2019 by The University of Chicago
All rights reserved. No part of this book may be used or reproduced in any manner whatsoever without written permission, except in the case of brief quotations in critical articles and reviews. For more information, contact the University of Chicago Press, 1427 East 60th Street, Chicago, IL 60637.
Published 2019
Printed in the United States of America

28 27 26 25 24 23 22 21 20 19   1 2 3 4 5

ISBN-13: 978-0-226-61249-2 (cloth)
ISBN-13: 978-0-226-61252-2 (e-book)
DOI: https://doi.org/10.7208/chicago/9780226612522.001.0001

This book has received the Weiss-Brown Publication Subvention Award from the Newberry Library. The award supports the publication of outstanding works of scholarship that cover European civilization before 1700 in the areas of music, theater, French or Italian literature, or cultural studies. It is made to commemorate the career of Howard Mayer Brown.

The University of Chicago Press gratefully acknowledges the generous support of Stanford University toward the publication of this book.

Library of Congress Cataloging-in-Publication Data

Names: Lugli, Emanuele, author.
Title: The making of measure and the promise of sameness / Emanuele Lugli.
Description: Chicago ; London : The University of Chicago Press, 2019. | Includes bibliographical references and index.
Identifiers: LCCN 2018054726 | ISBN 9780226612492 (cloth : alk. paper) | ISBN 9780226612522 (e-book)
Subjects: LCSH: Measurement—Italy—History—To 1500. | Measurement—Social aspects—Italy. | Metrology—Italy—History—To 1500. | Metrology—Social aspects—Italy. | Weights and measures—Social aspects—Italy. | Italy—Civilization—1268–1559. | Metrology—Italy—Historiography.
Classification: LCC QC85 .L84 2019 | DDC 389/.1509450902—dc23
LC record available at https://lccn.loc.gov/2018054726

♾ This paper meets the requirements of ANSI/NISO Z39.48-1992 (Permanence of Paper).

Contents

List of Illustrations —— vii
Preface: Written in Stone —— xi

I. SAFES

Thinking through History —— 3
Measurements, Epistemological Filters —— 9
Metrological Blurs —— 14
The Silent Maneuvering of the *Tavole di Ragguaglio* —— 25
Naturalizing Measurements —— 32
Metrologies —— 41
Measureless Art —— 49

II. SQUARES

The Pratissolo Deal —— 57
Opacifying the Invisible —— 60
The *Pietre di Paragone* —— 65
Measurements, Made and Remade —— 75
In the Open —— 80
Disciplining Standards —— 88
The Politics of Measurement —— 95
Measurements and the People —— 101

III. CITIES

Divine Measures — 109
From *Fratres Penitentie* to *Religiosi Viri* — 115
Cutting through Buildings — 121
Invisible Boundaries — 128
Imposing Self-Control — 134
The Ideology of Order — 139
The Height of Christ — 145
The Touch of Measurements — 153

IV. FIELDS

Dividing Up the Land — 161
The Origins of Medieval Measurements — 167
Geo-metria — 172
Thinking through Squares — 185
Frustrating Bodies — 194
Fibonacci's Standardizations — 201

Conclusion: The Metamorphoses of Measurements — 209
Acknowledgments — 219
Notes — 221
Index — 303

Illustrations

PLATES (*following page 94*)

1. The measurement standards of Bologna
2. The platinum meter (1799)
3. The length standards of Reggio Emilia
4. The *staio* of Venice (1263)
5. Ambrogio Lorenzetti, *Allegory of Good Government* (1338–39)
6. Ambrogio Lorenzetti, *Allegory of Good Government* (1338–39), detail
7. Monk Guido, treasurer of the commune of Siena (1280)
8. The *mensura Christi* oration (ca. 1293)
9. The *mensura Christi* oration (late fourteenth century)
10. Christ's footprints in Jerusalem
11. The *mensura Christi* canopy (fifteenth century)
12. A ribbon reproducing Saint Rose's height (eighteenth century)
13. Santo Stefano, Bologna
14. The length standard of Florence (1094)
15. The centuriation reticulum (sixth century)
16. Fibonacci's system for counting with fingers (thirteenth century)

FIGURES

2.1 Gabriel Mouton's universal standard of length (1670) — 12
3.1 Virgil's funerary site (1768) — 15
3.2 The plan of Virgil's tomb and the Neapolitan palm (1768) — 16
3.3 Giorgio Fossati, half a Vicenza foot and other measuring tools (1741) — 18
3.4 Maximilian Misson's records of Italian measurements of length (1739) — 19
3.5 The seven-palm *canna*, Naples Cathedral — 21
3.6 Nicola Carletti, fountain with measurement standards (1775) — 23
4.1 Conversion chart for the meter and the Milanese *braccio* (1803) — 26
5.1 Étienne Lenoir, prototype brass meter (1794) — 34
5.2 Maria Theresa's measurement standards (1756) — 35
6.1 Table of measurement standards (1760) — 43
7.1 The measuring of the *Piscina Mirabilis* (1768) — 51
9.1 The measurement standards of Modena — 61
9.2 The measurement standards of Pisa (1862) — 62
10.1 The length standards of Bergamo (doc. 1204) — 66
10.2 The length standards of Perugia — 68
10.3 The length standards of Ancona — 71
10.4 Bernardino Arrigoni, weight standard of Mantua in the shape of an amphora (mid-sixteenth century) — 74
12.1 Rome's volume standard for grain (fourteenth century), carved out of an ancient urn — 84
12.2 The measurement standards of Padua (1277) — 86
13.1 The measurement standards of Verona (1380) — 91
18.1 The measuring of the distance between Santa Maria della Scala and San Fermo in Verona (1327) — 126
19.1 The *terminatio* of Bologna's Piazza Maggiore (1286) — 131
19.2 Reconstruction of the "new" streets of Brescia (1237–49) — 133
22.1 Pompeo Ugonio, council hall of the Lateran (ca. 1588) — 151

23.1	The distance between the reproduction of the Holy Sepulcher and the Chapel of the Holy Cross in Bologna's Santo Stefano	157
25.1	The *canna* of Volterra	170
26.1	Florence's *campanile*	174
26.2	Andrea Pisano and workshop, *Invention of Agriculture* (1334–36)	175
26.3	The centuriation around Minturnae (820–40)	179
26.4	Extant centuriation reticulum near Bologna	180
26.5	Proportional scheme of the centuriation	181
26.6	Construction of the Roman centuriation	182
26.7	Diagram showing the diagonal of a square approximating 17/12 of the side	183
27.1	Filippo Rusuti, Pope Liberius recording the boundaries of the miraculous snowfall of 358 (ca. 1295)	187
27.2	The orientation of Lanfranc's cathedral in Modena	188
27.3	The proportions of the cathedral of Cremona	189
27.4	The proportions of the cathedral of Modena	191
28.1	The measuring of a tower using a rod stuck in the ground	199
29.1	Computation diagrams from a copy of Fibonacci's *Liber abaci* (early fourteenth century)	205

Preface

Written in Stone

In January 2009, red writing bled on the facade of the city hall of Bologna in northern Italy. It was a new slap in the face for a city council that for the previous two years had been cracking down on graffiti, working with the police to keep a register of vandals, and sending cleaning squads to each neighborhood while telling the press about the hefty fees taxpayers were paying for them.[1] Yet the attack in January was particularly chilling. Most delinquents targeted houses in deserted streets and the dark porticos of churches; they hardly ventured against a fifteenth-century palace overlooking the main square, where the traffic of four of the city's busiest arteries flows, day and night, under blaring lights and neon shop signs. But one vandal succeeded. It was shocking to imagine that he could have acted undisturbed right in the city center. But the most violent blow came from the triteness of his declaration, a statement so banal and so crassly executed that the word "graffito" would be a compliment: *"Daniela, Ercole ti ama"* (Ercole loves you, Daniela). Ercole wasn't the lionhearted activist in a black hood that the police imagined, but a sweet nobody who took a towering Renaissance palace for a public toilet.

While the episode was quickly flushed into the deeper debate about cultural degradation, immigration, and the failures of education, very few realized what support Ercole had chosen for his revelation. He probably had no idea himself. What he had seen was a white stone in

xi

a dark brick wall, conveniently placed at chest height: in other words, a blank placard asking to be filled (plate 1). It wasn't even blank; it was incised with the lengths and shapes of the brick and roof tile that the city of Bologna had used as official measurement standards since the thirteenth century. This book is about such standards, their public display, and the oblivion into which they have been plunged.

The day after someone noticed Ercole's scribble, journalists deplored the defacement of a medieval monument. The meaning of those stones, however, had eroded for decades. The process of forgetting that enveloped them was collective, all the more effective because it was undramatic, like a gradual temperature drop in a house that no one notices until the cold fills every corner. Ercole's obliviousness was not special. On the contrary, it was all too common. Those standards in stone exist in almost every Italian city, not to speak of their other European counterparts, and yet it does not matter if you live right next to them or are the occasional tourist; it is unlikely that you have noticed them. Unlabeled, unframed, unwritten about: those stones that were once the focal points of merchants and makers have become indistinctive. And even those who have bumped into their occasional mentions in history books are unlikely to realize how they functioned, since in the ten years I spent researching the topic, I found not a single passage explaining it. But, as my teacher Marvin Trachtenberg once said to me: When you don't find the book you are looking for, it is time to write it. This book is dedicated to him. I hope he will enjoy my attempt to sandblast the decades of silence that have vandalized those standards. And if that goal sounds rather grandiose, I hope that with this book I will at least be able to restore some of the attention those stones received by the late twelfth century, when medieval governments first recorded their installation of measurements in the open.

This book argues that the public display of medieval standards triggered a series of revolutionary practices that not only redefined the cultural landscape of medieval and early modern Italy but also laid the foundations of today's ideas about precision, reproducibility, and truth. This book is thus both a sociopolitical history of a crucial component of medieval material culture, and a quest for the foundations of objectivity. By looking at the emergence of measurement standards—the ways in which they were made, enforced, and blindly believed in—this book reflects on how power shaped and merged with the real.

Although this book reconstructs the point of view of the powerful—wealthy merchants, economic leaders, and in particular the communes, the cities' independent governments ruled by the most important families—I am also careful to bring into view the ways in which measurements were endorsed from below. By publicly exposing their standards, communes made citizens responsible for the dimensional uniformity of anything exchanged within the city. They set up hefty fines for those who did not respect communal prescriptions, giving half of the proceeds to the accusers. Citizens came to police each other's adherence to the newly standardized measurements. They mentally and practically gauged the height of any new tower, the heaps of grain protruding from bushel baskets, and the size of freshly baked bread. Dimensional changes, even in the length of sleeves or dress trains, became worthy of a chronicle, and preachers shouted that mismeasuring was a sin! Friars also invited pilgrims at the end of long journeys to measure the limbs of saints with ribbons, which they would then carry on their own persons as protection against evil. By studying the practices of measuring, the materiality of public measurements, and their relations to the religious beliefs of the time, this book opens a new critical lens through which to understand the social history of Italian culture between the twelfth century and the start of the Renaissance.

On one level, this book offers a traditional history that draws on archival research and material evidence. It retrieves the origins and reconstructs the evolution of premodern measurements, standards in use in Italy until their replacement with Napoleon's metric system. Copiously drawing from statutory and judicial sources, it maps the times and venues for measuring and the layers of authority that certified the correctness of the results.

The book is, however, unorthodox in the way it opacifies measurements. It treats them not as the response to an inherent social need that humanity has felt from the outset, but as historical events that became the focus of intense reflection between the twelfth and thirteenth centuries. Rather than explaining the appearance of standards as a byproduct of a shift toward mercantile capitalism, as scholarship favors, this book sees it as a measure of governmentality. Metric discipline emerged as a way to reach consensus in an age of factions and profound cultural differences; it served as the plane of communication between antagonistic and often wildly different parties who spoke foreign languages and thought in codes. Measuring was the way by which the canons of a cathedral, who worried about costs, and the masons of the same cathedral, who valued geometrical proportions, found an agreement about the cathedral's architecture. Measuring was how two neighbors came to agree upon a mark between two fields that no division actually separated. It was also the way by which the abstract and the concrete connected.

So this book argues for a need to study measurements in relation to the medieval anxieties that brought them into being and affected their forms. Far from resolving those anxieties, standards kept them alive for centuries and transposed them to the modern standards that substituted for them, which we still observe today. Like medieval standards, our measurements—both the customary units of the United States and the metric system employed in almost every other country in the world—are the products of manual labor, and yet are constructed and encountered as if they were not works of human intention. This book thus fuses a history of material culture with a history of ideas to explain the peculiar hybridity of measurements, both ideal ratios and brushed metal bars. Because of their inherent ambiguity, measurements occupy a distinct place not just in the cultural landscape of the Middle Ages but also in cultural history as a whole. And it is in an attempt to define such a place that this book uncovers their role in supporting specific forms of authority while pretending to exist beyond them.

This book, as I mentioned, foregrounds the interdependence of measurements and power. Such a coupling is uncommon. Because measurements stand at the very foundation of objectivity and truth, they are considered to be neutral. In opposition to such a stance, this book explores them as a superstructure that consistently works its way across discussions about material culture, orchestrating those discourses while rarely becoming evident within them. While looking into such a blind spot of history, this book interrogates the historiography for what has routinely been left out of the picture. It thus extends beyond medieval culture to see the silence about measurements as a political gesture. Why is it that measurements have not found a place in the discourses of art, architecture, and space, given their investment in questions of shape and scale? Why have historical accounts been so resistant to measurements? What is being protected? Reflection on the constituents of premodern standards opens out to further thoughts on the formal and material bearings of politics and its contribution to the construction of what we consider accurate, truthful, and real.

Indeed, this book argues that the lack of attention given to measurements historically has not been merely an effect of the methodological challenges they pose to scholars—that is, an effect of dealing with tools that by definition draw from fields as disparate as law, mathematics, and trade. Rather, such negligence has been the paradoxical evolution of the very discourses that promoted the public displaying of measurements. Throughout the Middle Ages, measurement standards were often contested. Aiming at easing their acceptance, governments calibrated their discourses and actions so as to induce people to perceive them less as impositions from above than as innate articulations of their bodies and the natural world. Measurements eventually disappeared from the walls of Italian cities once people conceptualized them as ideal ratios extracted from the surface of the earth. The geodetic systems of measurement of the eighteenth century are a product of such a shift. With them, medieval standards came to an end. This book is thus a reflection on the propagation and popularization of old standards, as well as a cross-fade history of their disappearance from cultural consciousness.

The short chapters of this book are arranged into four parts, each titled and ordered to produce a sense of spatial expansion. We move from the *safes* that protected measurement standards to the *squares* in which they were publicly exhibited. We then proceed to explore how measuring practices transformed the fabric of *cities* and urban life to the ways in which agricultural *fields* served as models for spatial order. The increasing radius of the analysis, from securely guarded strongboxes to territories that expand far beyond what the eye can see, provides a narrative. Yet such an enlargement is also a necessity when dealing with tools that are both transnational in scope and intimate in practice.

The four scalar shifts of this book are also arranged in a backward temporal structure. Starting in the late eighteenth century, this book recedes to Roman authors writing around the time Christ was born. Its chronological core, nevertheless, is the period between the 1150s and the 1350s, on which the second and third parts focus. Even when the book leaves these two centuries, it quickly returns to them.

This way of proceeding is homogeneous with the critique of measuring as the making of straightforwardness and the pursuit of clean-cut extremes. A measurer is not simply quantifying a dimension that is out there, but is constructing the outlines of figures, deciding where each starts and ends. This book presents measuring not as an operation that merely determines the quantity of something out there, but as a formal activity that creates rather than certifies. And it is to relativize this sense of beginning that the book has multiple starts. It may be too late to tell you, but you could have skipped this preface and started reading from part 1, about the eighteenth century. Or you could skip that part too, and begin instead with part 2, on the Middle Ages. I have organized this book as a concatenation of themes and scaffolded information, so that readers can shape their own opinions without relying on any other text. (Diverging from traditional history books, I have folded in a review of the literature along the way, engaging with the authors who are essential to my argument in the notes, where I have also tucked subsidiary information.) Yet, in keeping with the critique of measurements as the provisions of beginnings, I encourage you to read each chapter separately, skipping through the pages and finding new interpretations. Historical accounts, after all, are hardly ever straightforward. What looks like a beginning often turns out to be an arbitrary entry point into the action—one that has been chosen for reasons of convenience, and which fades away with the stacking up of evidence. This is particularly true for the wordless history of measurements.

To say that measurements have a history is almost like calling Ercole's scribble a graffito. You will see. This book is built less on historical records than on evidence of slippage: a loss of steadiness in the action, a crack in a stone, or a blip in a notary's entry. The history this book is trying to recreate is built on very few words, and even fewer of them explain anything. Someone at some point made clear what measurements would be used and how they would be used, and such a declaration was then quickly forgotten, like the ancient origins of a well from which people drew up water for centuries, believing it to be not just water but the liquid core of life itself. But words are missing from records about measurements because, generally speaking, silence is all there is with which to deal with the obvious, and this book is an exploration of the obvious.

Curious word, "obvious." Its meaning is far from straightforward. Even without researching its deep origins, I can tell you it is a compound word, from the Latin *ob* and *viam*, often translated as "in the way." Obvious is something you encounter so frequently that you don't need to search for it—like ATMs in California, where I live, or sunglasses in Italy, where I am from. The obvious springs forth without any trouble and brings with it a sense of comfort, just as some political measures must be obvious if they respond to blinding needs.[2] But there is little obviousness in how the meanings of "obvious" split up. Grammarians call the word a "contronym," as it means two opposite things. Already in the fourteenth century, the stress on the prefix *ob* became heavier and the encounter with that word less nonchalant. What was "in the way" went from being a relaxing sight to being an unavoidable obstacle. The verb "ovviare" meant to remedy an unfortunate situation, to set it right—less like the erection of a wall than like the ironing of a creased linen shirt.[3] Words transform to such an extent that to retrieve their etymological origin is a blunder, an arduous effort to get hold of the wrong end of a stick. And the same is true of measurements, usually taken to be the tools of banality. With this book, however, I intend to tilt the light in which measurements are usually seen, so that they emerge not as obvious instruments, but as the artificially hot irons that smooth out the real—eliminating the creases that nonetheless reappear quickly after leaving the board across which the real has been momentarily stretched.[4]

PART I

Safes

I

Thinking through History

This book may seem to start at the end. But then, the story of Italy's medieval measurements does not begin in the twelfth century. Rather, it kicks off in the rabble-rousing speeches that inflamed the last days of the eighteenth century, when many of the modern states that made up the Italian peninsula ousted the standards employed locally for centuries and substituted them for gleaming new ones. The promoter of those reforms supported Maria Theresa of Austria, who in 1756 had the measurements of her empire rectified.[1] Her son, Peter Leopold, Grand Duke of Tuscany, imposed a new system based on Florence's main linear standard, the *braccio*, on all the cities of his domains.[2] And in 1783 the state of Milan, then also under Austrian control, forced the Milanese yardstick, redefined for the occasion, on the rest of the territory.[3] Other countries followed.[4]

Those reforms required the synching of numerous institutions and took decades to complete. Their complexity was such that sometimes they never came to an end. (This is, for instance, the case of Milan, which after fifteen years spent collecting and replacing all linear standards, never undertook the reform of its weights.) Yet the administrative challenges were relatively straightforward in comparison to the cultural rewiring precipitated by new measurements. It was not just a matter of producing hundreds of iron rods and dispatching them to provincial councils. It was not even solely about training functionaries with new manuals and convincing recalcitrant landowners to have their estates surveyed from scratch. Rather, it was about making people accept new ways to approach space, yearn for such a change, and delight in its possibilities. It was about making people thrive on renewal and long for changing themselves in return. Far from representing a mere technical change, the new measurements rejigged thinking at such a fundamental level that concepts such as accuracy, identity, and even truth were radically transformed in the process.

Before introducing new standards, administrators had to search for the tools they wanted to replace. Such a goal was far from undemanding. Reformers went through areas that had multiple "official" yardsticks as well as metrological deserts, where it was hard to find even one well preserved standard. They undertook painstaking searches through city archives for long-forgotten tools, only for their findings to raise more questions than they expected. As the standards never quite matched—some rods longer than others, some so weathered as to frustrate any attempt even to guess their original dimensions—the researchers were forced to take some decisions.[5] In the following chapters of this first section, I will explore what motivated their choices. Nevertheless, I say straight away that their reports, which recount the mishaps in retrieving standards while propagandistically proclaiming the necessity for reform, are the accounts on which today's historical narratives still depend.

The history of premodern measurements was thus written at the moment of their substitution. Old standards came to be identified with the dictatorial gestures the modern ones rejected. Their material degradation was decried as proof of the barbarisms that an enlightened government set to uproot on its quest for truth. Premodern standards were denigrated according to a spectrum ranging from nightmare to nuisance, labels that have been repeated ever since.[6] The detailed studies undertaken by Marxist economists between the nineteenth and twentieth centuries did little to restore the extraordinary charge with which premodern measurements were invested.[7] After examining even the best of these studies, readers rarely see little more than a revival of the rationalist ideology of the eighteenth century, according to which standards either match or do not. And this is why this book starts from the modern period, as a way to offer a critique of its black-and-white mentality and move past the positivistic impasse it generated.

Amidst the metrological reforms taking place in eighteenth-century Italy, the most successful was the metric system. We still call it a "system" because of the interdependence of all its standards: one cubic decimeter contains one liter of water, which weighs a kilogram. Moreover, it is rational: it defines its components in relation to principles of geophysics and mathematics. Indeed, the meter was established as one hundred-thousandth of a quarter of the earth meridian. And it was divided decimally, rather than by the twelfths as was customary in premetric times, so to facilitate the application of calculus to matter. As both a geodetic standard and a tool to mathematize the real, the meter turned the world into a playground for engineers.

It took the whole of the 1790s for the meter to become a reality.[8] Its proposal was discussed at the outset of the decade by the members of the Académie des sciences in Paris, one of the most important research centers in the world. The academicians had many ties to the government, and pressed for the meter's introduction in 1795 (a hastily organized attempt in 1793 ended up in a fiasco). Yet it is not that year that has taken on an inaugural character. Rather, scholarship sees the launch of the metric system in a string of meetings that spanned several months between 1799 and 1800. Usually labeled as "the first international conference" ever organized, those scientific presentations served to validate the data that backed the construction of the standards, after which they were introduced again through the population with a heightened sense of confidence.[9]

Because the meter was set to become a universal, everlasting standard, it was of paramount importance to both calculate it with the utmost precision and win the approval of the international scientific community. So the French invited several foreign scientists to the last phases of the debate, those who would review and approve the conclusive results. Once the correspondents arrived in Paris, however, they found nothing to see. The French scientists were late in submitting the data, and, pressed by time, they tweaked them so as to confirm the dimensions of the meter as launched in 1795.[10] Even if the adjustment was slight, an inconspicuous thousandth of a millimeter, it was enough to turn the meter into a corrected ideal. Rather than searching for truth, scientists submitted to political demands.

After all, even before the international conference started, the meter had already been made available not only in France but also in some of the territories conquered by Napoleon. By the time scientists approved the results of the conference, many administrations of northern Italy were already referring to the metric system. And in less than fifteen years, the whole peninsula was familiarized with it, abandoning not only the local measurements of medieval times but also the modern standards that had replaced them only a few years before.[11]

As one of the triggers of modernity, the institutional story of the meter has been described in detail. Yet its impact on practice, even if felt far and wide, has been more difficult to track. In a way, the meter was as revolutionary as plastic. As plastic later became the go-to material for anything from acrylic boats to shellac records, desensitizing people to material properties, so the meter quantified rolls of silk in the same way as it measured land and the distance between stars. The technical precision of metric machinery, moreover, made all previous standards appear rudimentary, classifying old data as approximate, and unleashing a new sense of scientific complacency about the past.[12]

For all its exhilarating novelties, the meter was initially short-lived. As an innovation and a symbol of the Napoleonic empire, it was banned at the Congress of Vienna in 1814–15. It was, however, reintroduced in 1840s France, fueling the ambitions of economic liberalism.[13] And it is from this second energetic relaunch that it has became the global phenomenon we observe today. The history books on the meter present the ensuing century as a time in which it was embraced by an ever-growing list of countries, from Mexico to Japan.[14] Italy was a early adopter, as it introduced the meter with the nation's unification in 1861, soon after which the metric system became a compulsory subject in primary schools. Massaged into the brains of children from an early age, the meter has since shaped every action on Italian soil, as well as in the vast majority of the world. Today only Americans, Liberians, air pilots, and cricket players perceive it as a foreign construct. British drivers still compute distances in miles. For most people in the world, however, it has become almost invisible.

The global triumph of the metric system is often invoked to explain the general lack of interest in measurements, both historically and theoretically. In contrast, the next few chapters argue that such an oversight was not an effect of success, but its prerequisite. It is indeed possible to track down a set of coordinated historical actions that tried to make measurements transparent. Since the seventeenth century, scientists extracted new standards out of physical laws and calibrated their discourses so as to induce people—an indistinct, faceless mass whose definition changed according to whoever invoked it—to accept them either as products of nature or as innate articulations of their bodies. While being extracted from the earth, the meter was buried into the ground as its rhetoric played on its geodetic qualities, fertilizing culture from an invisible underneath.[15] This first section, "Safes," recounts this laborious process of interment as a way to reveal the pathology underlying the disappearance of measurements from cultural consciousness.

So, in a way, we could say that this book opens with a twofold critique of the meter.

On the one hand, it does not accept metrological knowledge at face value, but studies it as a truth process that relies on authorities' capacity to persuade communities to accept its results and the cultural processes they endorse.[16] The meter thus emerges not as a neutral tool, but rather as an epistemological filter that adapts seeing and thinking to the political paradigms on which it depends. By offering a study of an element that constantly works its way across the discussion of bodies and space, orchestrating much of those discourses while rarely becoming evident within them, this section questions the seemingly impregnable convictions that underlie many descriptions of the real. This is why this section is titled after the safes in which measuring rods were kept in the eighteenth century, and which also provide a metaphor for that hidden place in our psyche in which we store our metric knowledge.

On the other hand, I also consider the meter in relation to the other measurement reforms of the eighteenth century. It is an unusual move, since the revolutionary content of the meter is so hyped that it is often taken as a one-of-a-kind break. Yet the commonalities between the meter and its predecessors are greater than their divergences. In a way, the meter spread so rapidly because people had already accepted many of the changes it supposedly introduced and the politics it supported.

When we hear that measurements do not matter, that they are easy or technical, what we hear is the echo of a deliberate political voice, which grew in volume by some kind of accord between disparate communities and fields. As measurements are, now as then, employed to quantify every aspect of the real, they cut through boundaries. Like this book as a whole, this section deliberately keeps a wide angle. Only towards the end does it narrow its focus on questions about architecture, sensorial experiences, and history. The reason for this reduction is that no field more than architecture has been invested in questions of limits, proportions, and matter, and yet in no other field was the disappearance of measurements as precipitous. After a century in which even occasional tourists were encouraged to measure monuments as a way to appreciate them, size was excised from nineteenth-century discourses on buildings. This section finds the reasons behind such suppression in Napoleon's contested politics of measurements, which, far from being a confined episode, had a bearing on the ways in which architecture was approached for more than two centuries.

2

Measurements, Epistemological Filters

Imagine that the world shrank to the size of a tennis ball and that all humans—except you, the ball holder—also shrank, so that the world appeared unchanged to them. Those humans would keep living as they did: looking out for wild beasts and contemplating the sea as an infinite expanse. But for you, their fears and dreams would appear insignificant: their monsters would seem microbes, and the sea as if made of just a few dewdrops. Or think of the opposite: that the earth and the sky and everything in between grew enormously, so that each star would have the diameter of the sun as you had once known it, and the gigantic animals would not even notice the pebble on which you sat, and which you would believe to be a mountain.

These examples are not mine, but those of the French philosopher Nicolas Malebranche (1638–1715), who argues that seeing is deceiving. As giants and Lilliputians perceive things differently, so they value them differently. Malebranche quickly disposes of the idea that we should assess our perceptions in relation to our bodies (there is no "our body," since bodies are all different) to deduce that the assessment of magnitude plays an essential role in the construction of knowledge.[1] In other words, scale works as an epistemological filter, making people separate what they ought to know from what they can ignore. Objects that loom large make their presence feel pressing and urgent. Malebranche, who defines humans in relation to their survival instinct (hence his preoccupation with wild beasts), considers scale in relation to danger. As scale shifts, however, it tricks. When beasts appear smaller and less terrifying, people cannot be sure that it is because they are far away. Perception can lead to erroneous judgments and should not be trusted. The only way to achieve truth, Malebranche concludes, is to measure. But to measure precisely is difficult, as it requires "an exact standard: a simple and perfectly intelligible idea, a universal measure that can be employed for all sort of subjects."[2] The question is metaphysical: Malebranche asks whether permanence and moral justice can ever be achieved through human tools and skills. In 1675, when the last installment of his *De la recherche de la vérité* came out, the question also had a practical ring, as time-defying standards then were but a scientist's dream.

In Malebranche's time, every scientist worked according to national standards. Such variety frustrated international collaborations, threatened reproducibility, and hindered empirical authority. How to demonstrate experiments, if they could never be reproduced accurately? The damage had been contained as, over time, some standards had become more common than others. For instance, the *toise*, a long cane employed for land surveying, and to which Malebranche also refers, became popular outside of France. The Brescian engineer Girolamo Francesco Cristiani confirmed it when he wrote that in Italy "no mathematician uses any standard more than the *toise*."[3] Scientists also became skilled at conversion. When in doubt, they asked correspondents to send over their standards so that they could verify them directly.[4] The mobility of scientists across European capitals provided occasions for facilitating such checks. Yet at every passage—for every parcel sent, for every scientist that embarked on a journey, for every conversion from a familiar to an unfamiliar set of standards—errors crept in. And such inaccuracies constituted only one of the problems caused by the variety of yardsticks. Standards were not the "perfectly intelligible idea" that Malebranche yearned for, as they were immanent to specific physical objects. This meant that each deformation of the standards could compromise the recoverability of past experiments. The accuracy of the sciences was threatened by a scratch, a blow, or even an imperceptible drop in temperature.

To overcome such a menace, many scholars of the seventeenth century—from the Dutch astronomer Christiaan Huygens and his correspondents at London's Royal Society to the Warsaw-based Italian Jesuit Tito Livio Burattini—put forward proposals for a universal standard.[5] This new tool would derive not from an object, but from an incorruptible law of physics. A popular idea was to take the length of the pendulum that swung in a second. Yet, as Jean Richer and Edmond Halley pointed out at that time, a pendulum oscillated more slowly at the equator than at the poles, making it unfit as a universal standard.[6] The abbot Gabriel Mouton then suggested taking a portion of the earth meridian and dividing it decimally (figure 2.1)—an idea that gained consensus, especially in France.[7]

Scientific academies vehemently discussed measurements, arrogating to themselves the task of determining their dimensions and their making. Rulers consented to such claims, as they understood that power was inseparable from questions of quantification, making scientists crucial for the efficiency and security of the modern state.[8] Scholars have observed that such an appropriation, which made measurements pass from common trading tools to descriptors of the universe, coincided with an increase in scientists' political engagement. Indeed, the metric system became a reality, and in such a short period of time, also because France's revolutionary government was composed, to a large extent, of scientists.[9] And it is through scientific journals that the meter spread through Europe and arrived in Italy, much before it was part of the equipment strapped to the backs of Napoleon's soldiers.[10]

Yet, little attention has been paid to the standardization of perception triggered by measurements. Still, in 1850 the director of Milan's technical school, Pietro Baraldi, exhorted his students to "get used to seeing" millimeters since they represented the degree of precision of modern machinery.[11] As labeling systems, metric divisions instituted the distinctions they appear to merely demarcate, constructing notions of uniformity and values that did not exist before. In 1853 the archaeologist Luigi Canina lamented that Trajan's monumental column in Rome lost its sense of grandiose perfection the moment its height of 100 Roman feet was rendered as 29.635 meters.[12]

Measurements exert pressure on the real while carrying its stamp. As both physical and epistemological modules, standards shape perceptions as much as they describe them. Measurements are not just

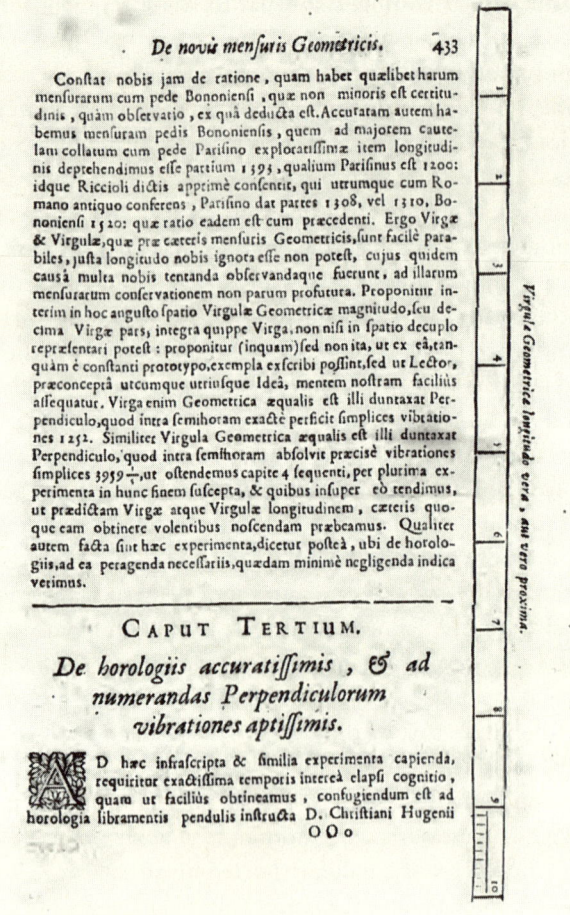

2.1 | Gabriel Mouton's proposal for a universal standard of length. From his *Observationes diametrorum solis et lunae apparentium* (1670).

descriptors, and here lies the shortcoming in Malebranche's reasoning. Measuring may reveal the errors of intuitive perceptions, but it produces deformities nevertheless. If it does not seem that way, it is because measurements are supported by institutions and discourses that certify their findings. In other words, the difference between size and scale, or between objective dimensions and relative ones, rests on their discrete relationships to politics. While scale claims to depend on relative, often subjective impressions, size immediately connects to the political and cultural discourses that validate measurement standards in the first place. Measuring has thus a privileged relationship to politics and, as we will see in the next sections, it is a way, as effective as it is subtle, by which power exerts control over the physical world.

3

Metrological Blurs

In 1768 Paolo Antonio Paoli published *Avanzi delle antichità esistenti a Pozzuoli, Cuma e Baja*, an impressive volume, half a meter tall, reproducing the classical ruins of the Procida promontory, some twenty kilometers west of Naples. The theatrical placidity of the illustrations—stone fragments clothed in shrubs, contemplated by tourists recumbent like river gods—was a fiction. The whole region was being cut open: walls smashed, stones rolled out of place, dirt thrown up in the air. People entered the earth to exhume the secrets of the ancients.[1] The journey to the past was an expedition to the center of the earth.

Paoli himself opened his book with a view of the entrance to the Crypta Neapolitana, a seven-hundred-meter-long tunnel that he presented as a depthless wound, a crack into an abyss. Drilled in the tuff hill of Posillipo in Augustan times, it was still in use in the eighteenth century, when it linked Naples to the Phlegraean Fields—the gate of hell, according to Virgil. Virgil, who was said to have created the tunnel overnight with his magical arts, was buried near the entrance, his ashes kept in an urn in the middle of a squarish vault (figure 3.1).[2] When Paoli visited the chamber, as many tourists did, one of the side walls had been knocked down, and the precious relics were gone.[3] The theft makes his survey appear like the report of a crime scene. He records the height of the wall up to the start of the vault (19 Neapolitan palms) and that of the vault (6 palms).[4] If we take the size of the eighteenth-

METROLOGICAL BLURS

3.1 | Virgil's sepulcher. From Paolo Antonio Paoli's *Avanzi delle antichità esistenti a Pozzuoli, Cuma e Baia* (1768), plate 9.

century Neapolitan palm as roughly 27 centimeters (I am rounding it slightly up), it makes an overall height of 6.75 meters. The scale of the illustration is then skewed. To be proportionate to the figures, the vault should have been at least a third lower. But then readers knew vistas were hardly accurate, and that they had to examine the accompanying floor plan (figure 3.2) if they were interested in scalar precision.[5]

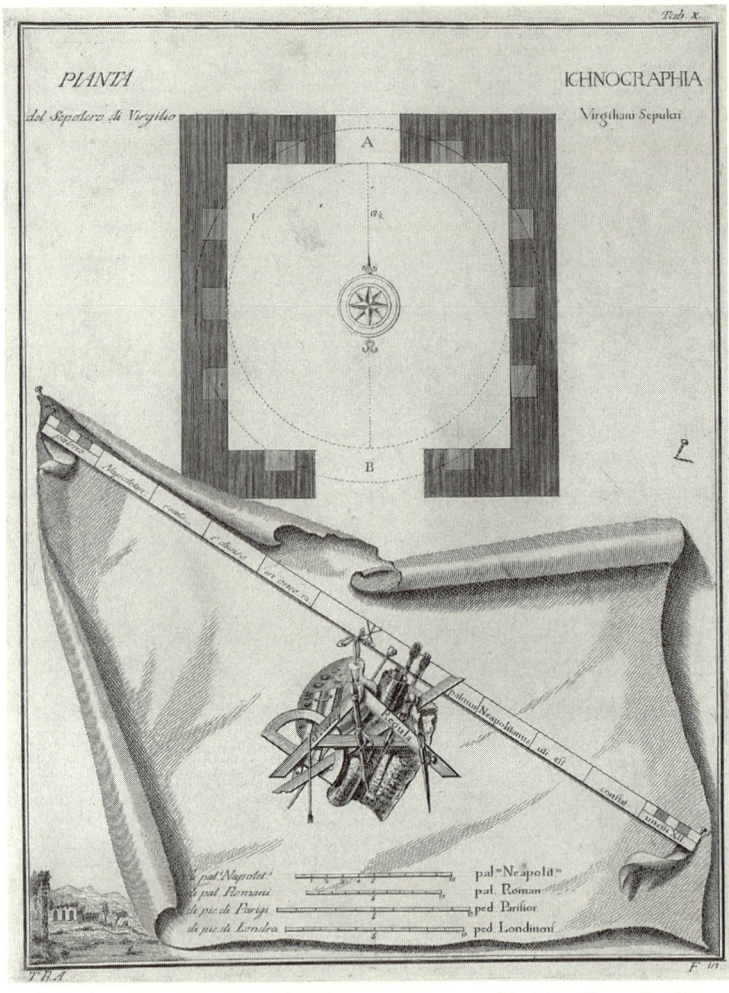

3.2 | The plan of Virgil's tomb and the life-size reproduction of the Neapolitan palm. From Paolo Antonio Paoli's *Avanzi delle antichità esistenti a Pozzuoli, Cuma e Baia* (1768), plate 10. The palm is nailed to a ripped piece of parchment, which at the bottom also offers a comparative chart with Roman, French, and English standards.

Scholars insist that the eighteenth-century custom of illustrating a site twice, as a view and as a blueprint, addressed two types of publics: traveling aesthetes and architects.[6] The ruins offered a double education in taste and technique, in line with Malebranche's dichotomy of seeing and measuring or, if you prefer, the architect Robert Adam's differentiation of "fancy" from "truth."[7] Johann Wolfgang von Goethe, however, disagreed. He considered the separation superficial, concluding that both types of "reproductions give a false impression: architectural blueprints make them [the monuments] look more elegant while perspectival sketches make them more ponderous than they are in reality." Goethe jotted down his thoughts in his diary while promenading through the Doric temples of Paestum, south of Naples, concluding that "it is only by walking through them that one can attune one's life to theirs and experience the emotional effect which the architect intended."[8] Goethe took all forms of representation as fallacious. They shape our perception to reiterate the disciplinary biases they pretend to simply describe. It was only by rejecting both vistas and layouts that tourists could appreciate the starker beauty of the Paestum sanctuaries, unjustly denigrated as anticlassical.

Goethe wrote in response to publications such as Paoli's. Under his plan of Virgil's tomb, Paoli includes a ruler, with its characteristic star-shaped mark at its center, pretending it is nailed to a piece of parchment (figure 3.2). The two inscriptions along the edges reveal it to be a reproduction of the Neapolitan palm of twelve *once*. Which means that, even if you do not know the dimension of that standard—even if you never set foot in Italy—you could reconstruct Virgil's burial site by referring to this 26.9-centimeter ruler (a *regula*, as Paoli calls it in Latin). As a result of this life-size representation, Paoli produced a metrologically self-sufficient volume, of great appeal to an international audience. We can infer that his readership's geographical distribution was broad not only from the different scales he included at the bottom of the page, but also from contemporary correspondence. When the Austrian statesman Anton Wenzel asked his Italian agent, Ferrante Sbarra Franciotti, to get him a copy of Paoli's work, he was told it would be arduous, as it had been written for the English.[9]

Paoli's inclusion of a life-size standard reflects the internationalization of eighteenth-century Italian publishing. Before him, Francesco Muttoni had included half a Vicenza foot (*mezzo piede vicentino*) in his bilingual abridgment of Palladio's treatise (figure 3.3).[10] And many travelogues included life-size standards. The 1739 edition of Maximilien Misson's much-read and translated *A New Voyage to Italy* reproduces the dimensions of the most common Italian standards (figure 3.4).[11] As the plate concluding the book still promises, it does so

3.3 | Giorgio Fossati, half a Vicenza foot and other measuring tools. From Francesco Muttoni's *Architettura di Andrea Palladio* (1741), plate 2.

to turn tourists into locals. It thus appealed to a desire of contact and authenticity, even if the inclusion of the standards ultimately accentuated a sense of remoteness as it reduced trading operations to lines printed on paper, indifferent to the many other cultural norms on which such complex activities depended.[12]

The promise of authenticity, however, was delusional also in a technical sense. Indeed, Misson's Neapolitan palm measures 23 centi-

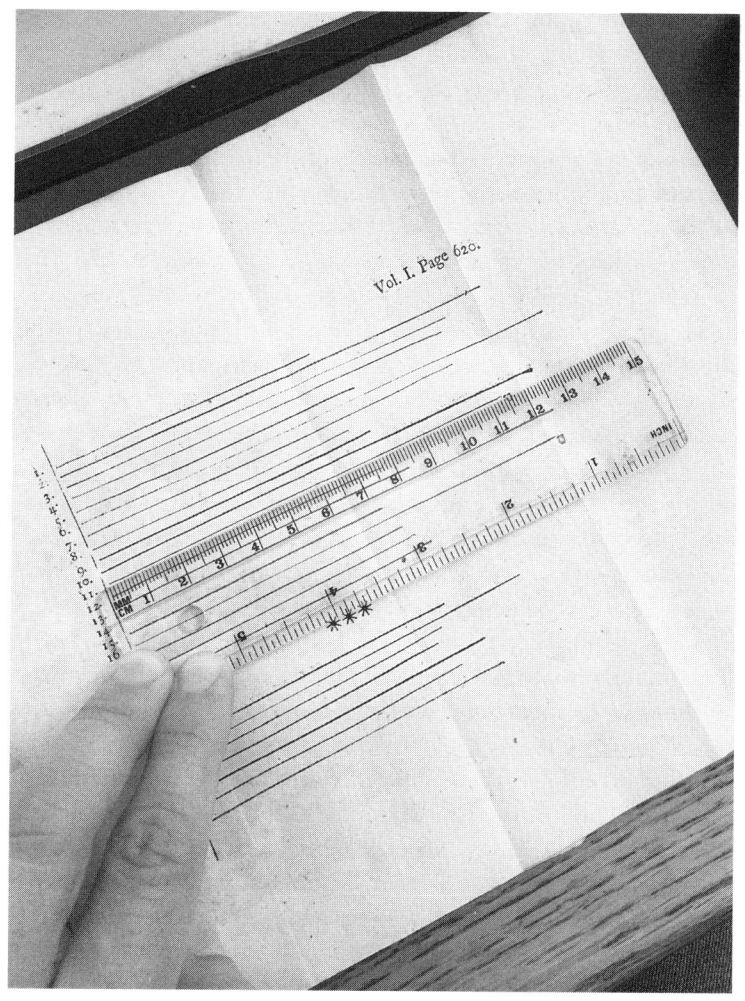

3.4 | At the British Library, measuring the Neapolitan palm (segment no. 11) as represented in the fifth edition of Maximilian Misson's *A New Voyage to Italy* (1739).

meters, that is almost four centimeters shorter than Paoli's.[13] Misson's son, who expanded his father's edition after his death in 1722, confessed that he never saw the Neapolitan standard. The tiny, reversed omega (Ω) at the end of the line (figure 3.4) is a symbol that stands for second-hand knowledge.[14] Yet, the deviation between Misson's and Paoli's standards does not simply reflect the discrepancy between touristic and local knowledge. Rather, it is an instance of the endless dimensional variations in eighteenth-century Naples. Antonio Genovesi, the esteemed professor of economics in that city, admitted that he did not know the extension of its kingdom because he could not figure out the real dimension of the Neapolitan standard. "I cannot but give some vague calculations until the royal arm (*il braccio sovrano*) gives us more precise measurements."[15] (Notice the pun here: "the royal arm" is both a reference to the royal standard of the *braccio* and an exhortation to the king to flex his metaphoric muscles to undertake a much-needed measurement reform.)

We have to wait until 1838 to find a systematic report on the measurements of the Kingdom of Naples. And the conclusions are not good: Naples had been a metrological bedlam for over a century. Ferdinando Visconti, the compiler of the report, could name four different official standards of the palm.[16] One was the governmental standard, an oxidized bar kept in the Zecca dei pesi e delle misure in Castel Capuano, Naples's court of justice. Visconti claimed that it measured 263.5 millimeters. Mapmakers and land surveyors, however, employed a different palm, of 264.55 millimeters, which they called "geodetic," as it corresponded to one seven-thousandth of an astronomical mile. The seven-palm *canna* that was walled in a column of the cathedral (figure 3.5), and which most Neapolitans took as the valid standard, gave instead a palm of 260 millimeters. And finally, two graphic representations on paper preserved in the abbeys of Monte Cassino and Santa Trinita at Cava de' Tirreni—and which constituted the oldest available evidence—showed that the palm was 268 millimeters.[17] All other standards were so weathered that Visconti could not determine their original dimensions. Which implies that there were even more standards.

Despite Visconti's display of precision—the dimensions that he gave went up to one-twentieth of a millimeter—his evidence was patchy, to say the least. Visconti never saw the bar of the Zecca, which had disappeared by the time he undertook research; he could only count on a description published in 1812.[18] He may have measured the *canna* of the cathedral when it was extracted from the pier for restoration, but when it was reinserted, it got plastered so extensively that no one could ascertain its dimension with the precision he flaunted.

3.5 | The seven-palm *canna* plastered in a pier of Naples Cathedral. Photograph by the author.

Today, it still looks like a badly cicatrised wound. And the geodetic palm was never an actual standard, but a ploy so that cartographers could draw their maps according to a dimension that had a simple relation to the meter, also a geodetic standard. At the time, rumor had it that such a palm had been invented by Ferdinand I of Aragon, the king who standardized Neapolitan measurements in 1480. Yet, as far as

I can tell, the only mention of Ferdinand's reform dates to 1787, when the idea that standards ought to be related to the earth meridian had become commonplace.[19]

So Visconti's claims were volatile, his argument hallucinatory. His report disguised a desire that the Neapolitan palm must have existed in a fixed form, somewhere. When we move out of official circles, however, doubts loom large. Never mind if Misson's son had been conned about the real size of the Neapolitan palm. What to say of the value by Cristiani, "the most rigorous Italian intellectual after Genovesi," as a contemporary French historian claimed?[20] In his much-praised metrological treatise, Cristiani wrote that the Neapolitan palm was 262.5 millimeters.[21] And Alexis-Jean-Pierre Paucton, compiler of an even more esteemed metrological treatise, gave it as 262.79.[22] Giovanni Carafa's monumental map of Naples, which historians celebrated as a landmark in the history of cartography—included a replica of numerous international standards, including the Neapolitan palm of 269.3 millimeters (figure 3.6).[23] When the architect Nicola Carletti, who completed the decades-long project, drew the Parisian foot, he must have used another version of the Neapolitan palm, though, as the conversion gives 264 millimeters, which is almost half a centimeter shorter.[24] Such differences may seem slight, but they raise numerous questions when one is reading a map that is scaled 1:3,808, and in which one centimeter corresponds to the size of an urban block.

And the errors—of transcription, of conversion, of calculation—crept in at every stage. In 1811 the Roman palm was considered to be 249 millimeters, whereas Carletti rendered it as 226.[25] Or at least that is what I measured one day in July 2016 after unfolding the copy of the Carafa map owned by the British Library in London. Under my transparent ruler, the Neapolitan palm looked like 268 millimeters. Was my cheap plastic tool gauged imprecisely? Had the paper shrunk over the years? Of course, ascertaining sameness through reproductions is a failed project from the start. Visconti and I measured different impressions of the Carafa maps, using standards produced in different centuries which are only nominally the same. (Labeling various tools by the same word is the first gesture in creating an illusion of uniformity, which ultimately repeats errors.) The Neapolitan palm thus did not exist as a standard, but as a range: it existed across different

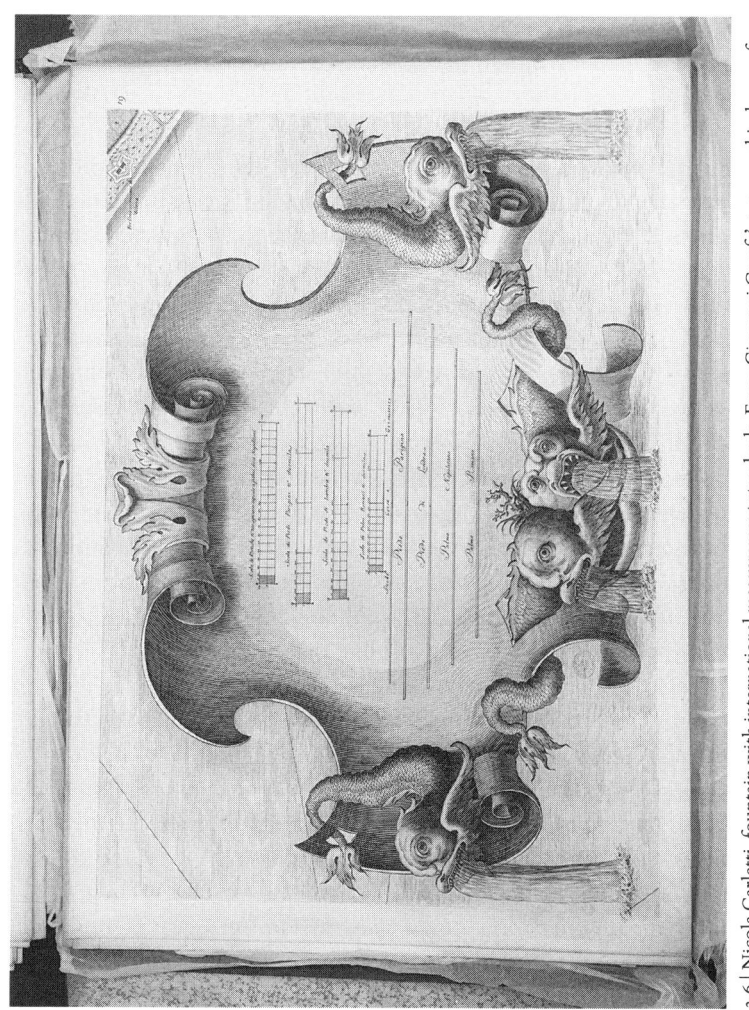

3.6 | Nicola Carletti, fountain with international measurement standards. From Giovanni Carafa's topographical map of Naples and its surroundings (1775). Photograph by the author.

objects, which were never perceived as a multiplicity because the use of one automatically excluded all others.[26] In a paradoxical twist, the very tools thought to provide the foundations of accuracy were all different.

And it was not only Naples that lived in a metrical blur. In the 1750s the astronomer of the Grand Duchy of Tuscany, Leonardo Ximenes, found out that the measurements attached to the wall near one of the doors of the Bargello, then the prison and police station of Florence, did not match the metal bar preserved in the city's tribunal.[27] And in 1770s Milan, Cesare Beccaria, the magistrate in charge of the standardization of measurements, confessed that he could not find two identical standards of the *trabucco*, the city's unit of length. He guessed its dimension by referring not to the official standard but to a rod owned by the director of the cadastre completed twenty years earlier.[28]

Metrological imprecision was thus endemic. As standards were defined in relation to one another, their dimensional range increased with their multiplication. Like Visconti, Ximenes and Beccaria aimed at putting an end to the local variety by undertaking a systematic comparison of all standards. Their operations will be described in the next chapter. Still, I say straight away that they did not succeed, and their reports conclude by wondering whether total standardization could ever be achieved materially, or if it only ever existed as an idea.[29]

4

The Silent Maneuvering of the *Tavole di Ragguaglio*

In the late eighteenth century, Florence, Milan, and Naples all resorted to the so-called *tavole di ragguaglio* to popularize their new standards across their states. Meaning "charts of equalization," from *re* plus *ad* plus *eguagliare* (that is, "to reduce to parity again"), the *tavole* allowed for the rapid conversions between the newly reformed standards and the measurements they replaced (figure 4.1).[1] If a customer wanted to buy, let's say, two Milanese *braccia* of cloth, the sellers would pull out their copy of the *tavole* and go to page 58 to discover that the required amount corresponded to 1.19 meter. As the conversions were not immediate—each standard could be divided in sixths, twelfths, or twentieths, without following any seeming logic—the *tavole* also included an introduction that explained the four basic operations, and so doubled as math textbooks. Not only did they aim at homogenizing marketplace practices across a state; they also checked and standardized the mathematical education of their users.

Because of their utility, Napoleon introduced the *tavole* after each conquest. They were issued together with the 1801 metric legislation of the Repubblica Italiana, and then were republished in an expanded version with the establishment of the much larger Regno d'Italia in 1803.[2] At every territorial enlargement, at every state that became a new *département* of the French empire, a new set of *tavole* came out.[3] Taken as a bulk, they are boring, ideologically charged administrative reports. However, their technical nature, their specialized

MILANO
Ragguaglio fra il Metro ed il Braccio di Milano.

Braccia di Milano	Fanno in misura nuova (Metri, Palmi, Diti, Atomi)	Parti del Braccio (Once Punti Atomi)	Fanno in misura nuova (Metri, Palmi, Diti, Atomi)
1	0, 5 9 5	3	0, 0 0 1
2	1, 1 9 0	6	0, 0 0 2
3	1, 7 8 5	9	0, 0 0 3
4	2, 3 8 0	1	0, 0 0 4
5	2, 9 7 5	2	0, 0 0 8
6	3, 5 7 0	3	0, 0 1 2
7	4, 1 6 5	4	0, 0 1 7
8	4, 7 5 9	5	0, 0 2 1
9	5, 3 5 4	6	0, 0 2 5
10	5, 9 4 9	7	0, 0 2 9
20	11, 8 9 9	8	0, 0 3 3
30	17, 8 4 8	9	0, 0 3 7
40	23, 7 9 7	10	0, 0 4 1
50	29, 7 4 7	11	0, 0 4 5
60	35, 6 9 6	1	0, 0 5 0
70	41, 6 4 6	2	0, 0 9 9
80	47, 5 9 5	3	0, 1 4 9
90	53, 5 4 4	4	0, 1 9 8
100	59, 4 9 4	5	0, 2 4 8
200	118, 9 8 7	6	0, 2 9 7
300	178, 4 8 1	7	0, 3 4 7
400	237, 9 7 5	8	0, 3 9 7
500	297, 4 6 8	9	0, 4 4 6
600	356, 9 6 2	10	0, 4 9 6
700	416, 4 5 6	11	0, 5 4 5
800	475, 9 4 9	12	0, 5 9 5
900	535, 4 4 3		
1000	594, 9 3 6		

Metri	Fanno in misura di Milano (Braccia, Once, Punti, Atomi)	Parti del Metro (Palmi, Diti, Atomi)	Fanno in misura di Milano (Braccia, Once, Punti, Atomi)
1	1 8 2 1	1	3
2	3 4 4 1	2	6
3	5 0 6 2	3	9
4	6 8 8 2	4	1 0
5	8 4 10 3	5	1 3
6	10 1 0 3	6	1 5
7	11 9 2 4	7	1 8
8	13 5 4 4	8	1 11
9	15 1 6 5	9	2 2
10	16 9 8 5	1	2 5
20	33 7 4 10	2	4 10
30	50 5 1 3	3	7 3
40	67 2 9 8	4	9 8
50	84 0 6 2	5	1 0 1
60	100 10 2 7	6	1 2 6
70	117 7 11 0	7	1 4 11
80	134 5 7 5	8	1 7 4
90	151 3 3 10	9	1 9 9
100	168 1 0 3	1	2 0 2
200	336 2 0 6	2	4 0 5
300	504 3 0 10	3	6 0 7
400	672 4 1 1	4	8 0 10
500	840 5 1 4	5	10 1 0
600	1008 6 1 7	6	1 0 1 3
700	1176 7 1 10	7	1 2 1 5
800	1344 8 2 2	8	1 4 1 8
900	1512 9 2 5	9	1 6 1 10
1000	1680 10 2 8	10	1 8 2 1

PER LA REPUBBL. ITALIANA

10 Atomi fanno 1 Dito.
10 Diti fanno 1 Palmo.
10 Palmi fanno 1 Metro.

PER MILANO

12 Atomi fanno 1 Punto.
12 Punti fanno 1 Oncia.
12 Once fanno 1 Braccio.

4.1 | A conversion chart, or *tavola di ragguaglio*, for calculation between the meter and the Milanese unit of length, the *braccio*. From the Napoleonic *Tavole di ragguaglio* (1803).

nomenclature, and the rate by which they were published turn them into revelatory documents of the blind spots of metric history.

The *tavole* always start with a eulogy, as bombastic as as it is repetitive, of the French astronomers who succeeded in measuring the earth meridian. This is followed by a report of the *ragguaglio* itself: the laborious process of obtaining local standards from provincial authorities and comparing them to the meter. The operation often took place in a laboratory, but sometimes, when the measurement standards could not be moved, the scientist—because it was always one man who was in charge of the whole *ragguaglio*—needed to travel.

In 1811, Feliciano Scarpellini went to the Palazzo dei Conservatori on Rome's Capitoline Hill, in whose courtyard stood a stone slab inscribed with the *canna romana*. Having researched metrological manuals, Scarpellini expected the incision's dimension to oscillate between 98.87 and 99.33 *lignes* of the French *pied*. Yet he did not know how to verify such a dimension, as the incision was "jagged" (*slabiato*) and "crudely carved" (*grossolonamente inciso*), with "uncertain" (*incerte*) extremities and irregular internal divisions.[4] Scarpellini did not lose his spirit. He traced a thin line parallel to the canna and translated its extremities by using hairs as plumb lines. How he did that is unclear. He justified his silence by writing that "it would be useless to describe the patient operations that I undertook to determine the extremities of the *canna* standard."[5] Instead, he conveyed scientific professionalism by listing the tools he employed: a massive ruler and numerous pairs of iron dividers, whose temperature he carefully checked with a thermometer every other hour. Scarpellini sweated under Rome's meridian sun. He countered any subjective impression by translating dimensions—both as a whole *canna* and as a succession of palms—multiple times, and by calculating the average for each finding. And yet, when he got to the point of explaining where he had placed those hairs, those compass points, or the baseline of the set square (not an easy task, as the rough extremities of the incision must have looked like craters under his magnifying glass), his scruples ran out.

Then how did Scarpellini end up confirming the impossibly precise dimension (2,234.218219 millimeters, or 99.33 French *lignes*) of the *canna* as calculated by Roger Joseph Boscovich? Boscovich was the most eminent scientist of eighteenth-century Rome.[6] He measured the degree of meridian that crossed the city, which served for mapping the State of the Church and determining the size of the earth.[7] His book, *De litteraria expeditione per Pontificiam Ditionem*, had been translated into French by Jérôme Lalande, the astronomer who became one of the most vocal supporters of the metric system.[8] The measurement of 99.33 *lignes* was not a midpoint between the extremes Scarpellini had found in the scientific literature. Rather, it was borderline. It was a problematic value too. To confirm it, Scarpellini had to overlook Lalande's observation that the French *toise* employed by Boscovich was shorter than the official one in use in Rome.[9]

The problem we encountered in Naples then reoccurs in Rome. The standards were many and they never matched; their material discrepancy was ultimately unsolvable. So Scarpellini maneuvered his ways for constructing consensus. He literally worked by hand, overcoming the impasse by shapeshifting an uneven incision into a straight line, which he then translated into numbers. But it is not by means of physical operations that Scarpellini found a solution. Instead, it is through his omissions that he fudged the results and confirmed the dimension of the *canna* on which the survey of the Papal States depended. The things he left out of his report are more interesting than those he put in, as they reveal measuring to be not a passive recording but a transformative act. Measuring is predicated on a chain not of physical gestures but, above all, of judgments. And judgments do not operate independently; they adapt to the context by which they are shaped.

It is to avoid disagreement that the stone slab with the *canna* that Scarpellini gauged is no longer in the Palazzo dei Conservatori. He must have destroyed it. I cannot know this for certain, and no document says so, but this is the fate that had been reserved for most pre-metric measurements. The Neapolitan *palmo* in the Castel Capuano and the Florentine *braccio* of the Bargello are long gone. Even Beccaria's gorgeous silver standard of the Milanese *braccio* was destroyed, probably around the time the Brera astronomer Barnaba Oriani completed the Napoleonic *ragguaglio* in 1801, when the meter was introduced in Milan.[10] The physical destruction of standards was necessary to avoid any controversy, and it is no surprise that Italy's old medieval standards have been preserved only in provincial centers, where they did not cause much trouble. It is also no surprise that in the 1877 *tavole di ragguaglio* produced after the unification of Italy, all the sections on local measurements start with Napoleon, who made premetric standards worthless.[11]

Historians consider the 1877 *tavole* as the most precise ever compiled, and indeed the scientists who worked at them collected information very scrupulously. Yet their renderings of Milan's *braccio* as 594.94 millimeters, of the *canna romana* as 2234.218219, and of the Neapolitan *palmo* as 264.55 concealed the twisted history of those values. Those dimensions were uncritically lifted from the Napoleonic charts without anyone reflecting on the political circumstances of their production or, more generally, on the interpretative nature of any *ragguaglio*. Sure: the 1877 scientists could not find much information about what had occurred before Napoleon. Still, no one spent a word to articulate such a shortage.

But there is more, as the impossibly precise quantities of the *tavole* not only enact a reduction from multiplicity to oneness, but also negate any materiality. Ximenes, the scientist who collected all Florentine standards, expressed discomfort when comparing metal bars to dimensions identified by the distance between two flaps sticking out of a wall (for a comparative example, see figure 10.3) as different typologies of standard behave differently.[12] Only if transposed to the ideal world of ciphers could standards be equalized. And this is what happened.[13] The numerical rendering of standards facilitated error spotting, as two numbers either match or do not. This possibility allowed governments to fine numerous publishers for printing *ragguagli* with mistakes, even if those were not "errors" but deliberate rounding-off to simplify metric conversions. In 1808 the government of Parma decided that only the Imperial Typographic Office could print *tavole di ragguaglio*.[14]

Measurements, however, continued to be thought of as immanent to physical standards. In 1875, all the countries that adopted the metric system agreed to recast the meter as a set of identical crossbars that would be distributed to all states. It took fourteen years for the director of the metric school in Rome, Napoleone Reggiani, to receive the exemplar destined for Italy. It was 1889, and Reggiani compared it to the then incumbent national standard, a bar originally donated by France to the Kingdom of Sardinia when the latter switched to the metric system in 1845. To Reggiani's horror, he discovered that the old standard was longer. For thirty years the meters of Italy had been one meter plus one-third of a millimeter in length.[15]

Such errors never go away. They return at the end of the nineteenth-century triangulations of the Pyrenees and the Alps, when surveyors from either side of the mountain chains compared their data and discovered that they did not match.[16] When Visconti went to examine the standard by which, thirty years earlier, he had measured the base for the map of the Kingdom of the Two Sicilies, he found that it was a fraction of a millimeter shorter than what he remembered. The error meant that the base of his triangulation had to be corrected by three meters, and the size of the state shrunk by many square kilometers.[17] He realized that measurement standards breathe. They expand and contract, and so do our pupils, pulsating like stars. As the champion of standardization he was, Visconti could only admit defeat with a heavy heart. He never published anything about his error. And, like many colleagues who had discovered similar contractions in their tools, he expressed defeat only in private letters.[18]

Most scientists, surveyors, and mapmakers sooner or later recognized that measurements have a fragile core, even if in their reports they presented them as the foundation of certitude. Those reports are histories of permanence and numbers, as established by the eighteenth-century *ragguagli*, which shun any discrepancy. Beccaria remarked that the slight divergence between the different standards of the Milanese *braccio* might have been problematic for precision machinery, but not for fabric sellers for whom the standards were roughly the same. That is to say, for whom imprecision had to be accepted as part of life.

"Being roughly the same" produces a relation between the members of a set that projects them towards abstraction, since it is only through ideals that those members can overcome their differences.[19] Measurement standards are thus quasi-objects, ambivalent entities that exist between physical objects and ideal ratios. Yet nowhere but in the eighteenth century was such a relationship more skewed toward the abstract.

The eighteenth century is the period during which all national standards were redefined in relation to the virtual size of the earth, and those of the past converted into impossibly precise numbers. It is because of this extensive groundwork that economic historians of the nineteenth and twentieth centuries have often taken premetric standards as stable, such as the Florentine standard for grains, the *staio*, often thought of as virtually unchanged for hundreds of years.[20] But the *staio* was never fixed. The examination of late medieval mercantile records reveals that it existed in multiple formats and that can only be approximated with a 10 percent error.[21] And comparative checks prove that such a discrepancy varied from city to city. If it was contained for Milan and Bergamo, it was considerable for Cremona and Brescia.[22] Such variances may have narrowed down over the centuries. Medieval standards fluctuated more than the Roman palms in Carletti's and Scarpellini's hands. It was a mere 4 percent for the Neapolitan palm, if we exclude Misson's value. Yet at what point do the blurry edges of measurements become sharp? Who or what decides when a discrepancy is a mistake, or when a standard is true enough? "True enough" is a vague predicate whose veridicality rests on judgments. It does not matter how microscopic their discrepancies are, or that the members of a scientific community agree to ignore them, because even the judgment of experts relies on strategies and discourses belonging to an external order of knowledge.[23] Discussion of measurements is never just about measurements. It is never technical, but it opens to questions about frameworks of reference, theories of knowledge, and the stuff that shapes our beliefs.

5

Naturalizing Measurements

To speak of measurements in the eighteenth century is to speak of nature. Their inalterability is that of the earth, whose surface united the civilizations of the past and the future generations. In his aspirational speech to the French Parliament about the need for a new standard, Charles-Maurice de Talleyrand asked his audience to "interrogate nature."[1] And it is through nature that the 1793 report from the Commission des poids et mesures justified the creation of the new set of measurement units. The meter, it reads, "has been taken from nature, it has deduced from the dimensions of the earth itself so that it will always be possible to find it again and reestablish it in case it went lost or was altered."[2]

In the rhetoric of the meter, nature was a concept as loose as it was pervasive, metamorphosing at every turn in unexpected ways. Nature, after all, had two faces. It was seen as both a bare set of unchanging laws and as an eternal enigma—two facets that supported the duality of measurements, both ideal ratios and chthonian standards.[3] Scientists defined their decimal division as natural, since men do not have twelve fingers.[4] (Such reasoning had a strong appeal after Étienne de Condillac argued in his popular *Traité des sensations* (1754) that it was through touching and handling that men truthfully engaged in the real.)[5] And the metric law of April 7, 1795—much copied, not only in administrative bulletins—explained that the earthly origin of the meter justified everyone in saying, "This measurement is mine" (*Cette mesure m'appartient*).[6]

This egalitarian effect was particularly compelling in the new French Republic, which was eulogized as the political expression of the imperturbable laws of nature against the arbitrary and corruptible legal codes of the past.[7] Repeated in mercantile manuals and gazettes, sanctioned by a class of lawmakers that saw measurements springing from the very same source as that of their political forms, the idea of the meter as natural became second nature.

The product of this convergence was the metric bar of 1799 (plate 2), the "crown jewel of metrology," as well as the standard that, after the international conference, embodied the new system across France.[8] Despite its being regularly celebrated for its unprecedented precision, no one seems to have studied this extraordinary standard as an artifact. And yet its manufacture reveals much of the ways in which nature has coalesced around measurements, and of the cultural disappearance of measurements.

Kept in the iron safe (*l'armoire de fer*) of the Paris Archives with other relics of the French Revolution, the rod is hardly ever displayed or photographed. Frankly, there is very little to see. It is a flat platinum rod with no graduated divisions, no marks, and no inscriptions, not even the name of its maker, Étienne Lenoir.[9] Its minimal aesthetic is in keeping with the Jacobin rhetoric of moral purity and material deprivation, which favored natural simplicity in dress and manners.[10] Yet the removal of any decoration is also functional to the reductionist ideology of the meter, which aims at becoming a self-sufficient point of departure.

See how different the platinum bar is from one of the provisional meters that Lenoir made out of brass in 1794 (figure 5.1) and which carries his name along with a decimal scale and three stamps: a revolutionary Phrygian cap, a diminutive allegory of France, and a quarter-circle diagram (a reference, again, to the fact that the meter was calculated in relation to a fourth of a meridian degree).[11] This is a standard that was validated by external authorities.

Or look at how different the platinum bar also is from the reformed *klafter* that Maria Theresa had distributed across her empire forty years

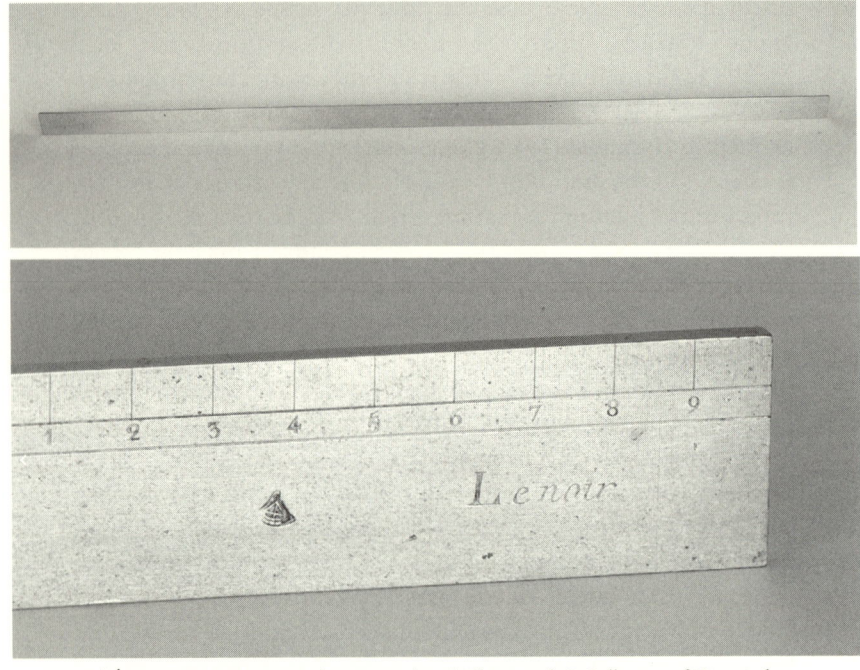

5.1 | Étienne Lenoir, prototype brass meter (1794). Photograph © Collection of Historical Scientific Instruments, Harvard University (inv. No. DW0535).

earlier, in 1756 (figure 5.2). Its flared rococo cartouches celebrate her and her husband Francis as protectors of the Austrian people, and the restorers of measurements to their much-needed precision.[12] Shell and floral festoons adorn their names, presenting the couple as lords over the seas and the earth. Fleuron glyphs punctuate the divisions on the standard, turning it into the metal equivalent of an opulent typographic work. This is less a standard than a trophy.

Specialists call this typology of standards "line-standards," as the physical objects serve as supports to the standards inscribed on them. The French metric law of April 1795 prescribed the new unit to conform to such a model, and commissioned an object "in platinum with the meter traced upon it."[13] Yet the 1799 standard ended up being an "end-standard," for the dimension of the unit coincided with the physical edges of the bar. Since then, the meter has been indissoluble from its material, platinum.

5.2 | The *klafter*, the length standard with the large cartouche splayed out at the center, and Maria Theresa of Austria's other measurements (1756). Photograph by Tomaž Lauko, © National Museum of Slovenia, Ljubljana.

Discovered only a few decades earlier, platinum quickly became the purest metal known to men. Platinum's purity, however, was not natural: it was labored. In the form in which it was quarried, it was contaminated by granules that seemed impossible to remove, as platinum resisted the highest heat and was immune to acids. According to Jan Ingenhousz, the scientist who studied the metal more than anyone else at the court of Vienna of the 1770s and 1780s, the moment platinum could be refined, it would make gold "pass from the first to the second place among the metals."[14] Indeed, platinum had one enticing advantage. Differently from gold, which had to be alloyed with non-noble metals like copper in order to attain the necessary rigidity, platinum held shape by itself.[15] Platinum was immediately recognized as being heavier than gold; and, as scientists took higher levels of specific gravity, or density, as indicative of purity, platinum soon surpassed gold in purity.[16]

It was because of its purity—that is, its density—that the French metric commission supported platinum for the new standard.[17] The official report of 1793 recommended choosing the metal that was "the least susceptible to be altered, the least expandable in the heat and the least contractable in the cold."[18] Scholars have repeated this point—that platinum was chosen because of its inalterability—over and again.[19] Platinum was, however, not only hard and pure but also invaluable. It was both priceless and without commercial value, as the Spanish held the monopoly of its extraction in Ecuador, a region which French explorers then identified as Peru.[20] And, once it was made malleable, platinum became the physical demonstration of the human capacity to penetrate nature. Its manipulation was as formidable as the measurment of the earth meridian. Indeed, platinum was the material equivalent of the meter. The connection was not only symbolic but also historical, as the platinum quarries were revealed to Europeans thanks to the very mission, across the Peruvian Andes, that established the length of the earth meridian that had originated the meter.[21]

Launched in 1735 and running for ten years, the expedition had involved a team of four members of the Parisian Académie and two Spanish naval officers. These officers guided the scientists through what was a foreign land and, at the same time, kept them under surveillance. Besides triangulating three degrees of the earth meridian, the Peruvian exploration confirmed Isaac Newton's theory that the earth was not perfectly spherical, but was squashed at the poles like a tangerine.[22] And the French savants found that gravity decreased when one drew closer to the equator, thus ruling out the possibility that a pendulum could serve as the universal standard.[23]

Amongst the French scientists sent to Peru was Charles-Marie de La Condamine, who, after returning to Paris, spent two decades lobbying for "a fixed and invariable measure, unaffected by the unfolding of centuries and the distance between places."[24] He argued that such a standard was the *toise du Perou*, the rod that served to produce the most precise geodetical mensuration ever undertaken.[25] (The *toise* had to be geographically identified because, at the same time of the Peruvian mission, the Académie had set a similar, parallel mission to Lapland, which, however, ended up being regarded as less rigorous than the operations in South America.)[26] La Condamine succeeded, and in 1766 the *toise du Perou* substituted the *toise du Châtelet*, the worn-out standard that had been displayed on the facade of Paris's police headquarters for almost a century. Eighty replicas of the Peruvian *toise* were sent to the French provincial parliaments.[27] Twenty-four extra duplicates were shipped abroad, including to the scientific cabinet of Florence, where Ximenes saw it.[28] Boskovich, who received a rod in Rome, regarded the *toise du Perou* as the most authoritative standard of his times, and in Milan, Oriani took it as the reference standard when he compiled the Napoleonic *tavole di ragguaglio* in 1801.[29] If the meter became such a success, it was also because it profited from the process of standardization across European academies set in motion by the *toise du Perou*.[30]

While La Condamine composed his report, the Spanish naval officers who escorted him, Jorge Juan and Antonio de Ulloa, wrote theirs. Their *Relación histórica del viaje a la América meridional* was a tremendous editorial success and was translated into French shortly after its 1748 publication. As the first detailed account of the Spanish colonies, it revealed platinum as a most tenacious material, unscathed "when struck on an anvil of steel" and "extracted only with infinite labor and charge."[31]

Platinum may have entered Europe's cultural consciousness as a mysterious element, yet everyone knew about its exotic provenance.[32] Its Peruvian origin was spelled out from the very title of William Lewis's first study of its chemical properties, and was repeated constantly, making it impossible for one to handle a platinum object without thinking about geography.[33] The *History of the Two Indies*, a collection of essays on foreign trade edited by the abbot Guillaume Thomas Raynal in 1770 (and reprinted often), described platinum as a material that came "from the new world to the old," thus adding another explanation for choosing it for the newest, most perfect standard.[34] Platinum was not just the purest material; it was streaked with an exciting sense of novelty.

The stories of platinum and the meter are closely intertwined. Father of the metric system, Lalande was the translator of Lewis's publications on platinum.[35] His colleague Joseph Dombey, who was chosen to bring the meter and kilogram standards to the United States, joined a Spanish botanical expedition to Peru in 1778 that, besides recording new shrubs and collecting Inca pots, brought back thirty-eight pounds of platinum.[36] Some of the metal served Louis XVI's goldsmith, Marc-Étienne Janety, in perfecting his experiments. By adding arsenic, Janety found a way to not only melt platinum but also remove its iron and copper impurities.[37] He demonstrated his method to the Académie des sciences in 1790, the very same year in which the French parliament approved the metric system. He then worked with Lenoir to fabricate the two platinum *toises* (*du Perou*, of course) for measuring the distance of one degree of the Parisian meridian, with which the 1799 international conference confirmed the dimension of the meter.[38]

But while platinum made the meter emerge from the chthonic depths of the Andes, it also hid it behind its mirroring surface. Most of the platinum Dombey brought from Peru was used by scientists for the construction of a sixty-foot telescope that, it was hoped, would compete with that of the London observatory.[39] That project was discussed at the international conference of 1800, along with the meter. Because of its inalterability to both temperature and chemical change, platinum was ideal for making mirrors. Newton had said in his treatise *Opticks* that dense white metals with compacted molecules reflect white of the first order, and the instrument maker Pierre-Bernard

Mégnié had proven him right by constructing much-celebrated precision mirrors in Madrid.[40]

It is thus not surprising that, if you look at the metric bar of 1799, you can see yourself. The mirroring quality of the meter solicits viewers' investment in the object, producing an identification between meter and whoever holds it. Symbols of *honnêté* and of truth, mirrors offered a moralizing guide both in their aristocratic ideal and as social tools. Their reflecting surfaces could not lie, as Jean-Pierre Claris de Florian dramatized in his fable *L'Enfant et le Miroir*, in which a child destroys a mirror for really showing what he is like.[41] By reflecting their viewers, mirrors provided the basis for the social contract of equality and fraternity.[42] The floral celebration of Maria Theresa on the Austrian *klafter* (figure 5.2) was replaced by everyone's anonymous face. Remember the explanation in the 1795 decree that was issued together with the metric bar: The new measurement would make everyone exclaim, "This measurement is mine."

Platinum was thus not just impossibly hard and pure. It did not simply remind scientific élites of the Peruvian mission that was so important for the definition of the meter. Rather, platinum made the metric bar vanish in both a literal and a metaphoric sense. Claiming to be both timeless and universal, the platinum bar employed, as it still does, materiality to paradoxically withdraw from it. Mirroring a mirror, which the famous French architect Jacques-François Blondel called "the representation of emptiness," the metric standard precipitated its physical disappearance.[43] It is a paradoxical object whose materiality appeared from the outset as excessive. But then it was designed to achieve such an effect. It did not refer to an immaterial ratio by means of a thin line, as the line-standard it was supposed to be, but rather hid the dimension of the meter within its material edges. Erased of any human sign, it gave the illusion that the meter was not manufactured. Made from a metal that was extracted from the earth and which was thought impossible to manipulate, it appeared as an object made by nature herself.

By contributing to the dematerialisation of measurements, the platinum meter is thus the physical equivalent of the *tavole di ragguaglio*. It is a quasi-object that cannot be approached without being translated into an abstract form. (What is there to see, after all? The bar is just one meter long.) Or, if you prefer, it is a conceptual object whose actual appearance is secondary to the dimension it conveys. And because its physical appearance was worked so carefully as to trigger such a tension towards the abstract, the metric bar disappeared inside a safe, paving the way for its substitutions across the territory by other, lesser replacement bars. The metric bar is kept in a safe not merely for protection, but to fulfill its potential as a quasi-object that exists between matter and ideal. Its withdrawal from physical life, precipitated by its very materiality, is fundamental for maintaining the idea that all measurements are identical. It is the ultimate demonstration that, in order to perform effectively, measurements have to disappear from cultural consciousness. And at the same time, it does not demonstrate anything at all, as there is nothing to see.

6

Metrologies

In a way, measurements were never invented. They have always been contained by the earthly crust and employed for trading for as long as the world could remember. While the French academicians debated whether to elevate the *toise du Perou* to the rank of official global standard, Alexis-Jean Paucton, in Paris, and Girolamo Francesco Cristiani, in Brescia, each compiled histories of measurements that argued for their universality. Their volumes are quite different, almost opposed in a sense. Paucton's *Métrologie* (1780) is monumental. It lists the numerous currencies of antiquity and transcribes many laws of the medieval period. It is a work of history and includes the type of reflections on history that only the thinking of a lifetime can produce. On the other hand, Cristiani's book *Delle misure d'ogni genere antiche, e moderne* (1760) is his first, and is quite concise. Cristiani wrote it with one specific goal: to demonstrate that he possessed the technical knowledge to become an engineer of the Republic of Venice. Yet, despite their differences, the two works can be studied together, as each tries to demonstrate that measurements are natural events.

Paucton insisted that the ancients had derived their measurements from the earth long before the French.[1] "The Ancients had a natural standard of measurements," he writes, "which had been taken from the distance of a degree of the meridian. . . . Inscribed in nature and of the length of one four-hundred-thousandths of a meridian degree, this standard was universal and common to Asia, Africa, and Europe."[2]

The earth takes a double meaning here: it is the ecumenical base for all measurements and for the collectivity of people searching for a universal standard.

The idea had been heard before. Nicolas Fréret, a scholar of mythology and one of Paucton's main sources, discussed it in an essay published by the prestigious Académie royale des inscriptions et belles-lettres.[3] And Voltaire, in his *Essai sur le moeurs et l'esprit des nations*, recounted how the Abbasid caliph Al-Ma'mun commissioned a team of scientists to measure the size of the earth, from which he extracted a new standard.[4] In pointing out that Al-Ma'mum achieved the feat nine hundred years before the French, Voltaire argued that great cultures think alike.

The notion that all geodetic (Cristiani calls them "trigonometric") standards are reenactments of the same reasoning is one of the tenets of both proposals for new standards and books on metrological history.[5] The process was, after all, reversible: if ancient standards were partitions of the size of the globe, by knowing the divisor, modern savants could retrieve the measurements of the past. It was inebriating to think that the earth could serve as the plane of contact between past and present.

But if measuring the earth was a monumental feat, which only the very best among scientific academies could achieve, antiquarians could retrieve the dimensions of old standards on their own by measuring ancient monuments. Cristiani suggested that the Roman foot could be recuperated by measuring the distance between Augustus's surviving milestones. And, as classical authors such as Strabo, Hyginus, and Pomponius Mela wrote and modern scholars such as Charles Arbuthnot repeated, a side of the Great Pyramid of Giza corresponded to one *stadion*.[6]

If the pyramids, the Roman milestones, and the earth were the embodiments of the measurements of the past, it was Paucton and Cristiani's tabulations that turned their recuperation into reality (figure 6.1). Their impossibly precise charts presented the standards of the past in the same way, regardless of materials or states of preservation.[7] Indeed, Paucton and Cristiani's tabulations recall in every way the *tavole di ragguaglio*. While reading them, we are unsure whether we are facing an antiquarian text or a scientific chart. But then, the

Delle Misure Lineari Antiche, e Moderne 23

	Parti	Piedi	Pollici	Linee	Decimali
Il Piede di Parigi Reale	1440	1	0	0	0
Dordrecht	1042	0	8	8	2
Egiziaco	1920	1	4	0	0
Ebraico	1636	1	1	7	6
Altro ch'è Palmo minore detto Topach	406	0	3	4	6
Altro ch'è Palmo maggiore detto Zereth	1218	0	10	1	8
*Ferrara	1779	1	2	9	9
Filaterio Regio	1571	1	1	1	1
Firenze, che è Braccio	2580	1	9	6	0
Franca Contea	1583	1	1	2	3
Francfort ful Meno	1260	0	10	6	0
*Infpruch	1488	1	0	4	8
Genevra	2592	1	9	7	2
Genova, che è Palmo	1098	0	9	4	8
Granoble nel Delfinato	1512	1	0	7	2
Grecia	1363	0	11	4	3
Halla in Saffonia	1320	0	11	0	0
Harlem	1267	0	10	6	7
Heidelberga in Germania	1220	0	10	2	0
Leiden	1382	0	11	6	2
Liegi	1276	0	10	7	6
Lione	1512	1	0	7	2
Lipfia	1397	0	11	7	7
Lisbona	1287	0	10	8	7
Lodi, che è Braccio	2023	1	4	10	2
Londra	1351	0	11	3	1
Lorena	1292	0	10	9	2
Lubech	1260	0	10	6	0
Lucca	2615	1	9	9	5
Macedonia	1567	1	1	0	7
Maçon in Borgogna	1483	1	0	4	3
Magonza	1335	0	11	1	5
Malines	1017	0	8	5	7
Mantova, che è Braccio	2055	1	5	1	5
Maftricht	1238	0	10	3	8
*Meftrina	1810	1	3	1	0
*Milano (da Fabbrica / che è Braccio)	2633	1	9	9	3
*Altro detto Decimale	1155	0	9	7	5
*Altro detto Aliprando	1926	1	4	0	6
Mildeburgo in Zelanda	1330	0	11	1	0
Modena	2812	1	11	5	2
Monaco	1042	0	8	8	2
Montpellier detto Pan	1050	0	8	9	0
*Napoli, che è Palmo	1164	0	9	6	4
Naturale	1100	0	9	5	0
Norimberga (di Città, detta Civile / di Camp. detto Operario)	1346	0	11	2	6
	1226	0	10	2	6

Il Pie-

6.1 | In his *Delle misure d'ogni genere antiche, e moderne* (1760), Francesco Cristiani argued for the possibility of correlating all standards, regardless of time. He constructed his argument also by means of homogenizing tables, such as the one pictured here.

two operations were leaning on each other. Paucton took data from La Condamine's astronomical treatise, and La Condamine commenced his disquisition on the size of the earth with a history of measurements, like Paucton's. "It has been a desire of all times," La Condamine generalized, "to have a fixed and invariable measure, to which neither the succession of the centuries nor the distance between places could bring any alteration."[8] It is almost as if scientists and antiquarians were answering each other's questions.

They certainly provided each other with the tools they needed. Antiquarian comparisons of ancient measurements had been made possible, as Cristiani explained, by the *toise*.[9] Not only did the French standard provide a reliable and widely available yardstick, but its preeminence among scientists persuaded antiquarians that their conversations were exact. No one seems to have remembered Malebranche's warning that standards affect results as questions shape answers. Rather, convinced of the accuracy of the *toise*, Cristiani and Paucton set to compare all the dimensions of the world's standards, ordering them into genealogies with the same candor with which geologists and botanists classified species.[10] Paucton, for instance, felt he could confidently map out the movements of the Drusian foot—a standard of thirteen and one-half inches of the Roman foot. He retrieved standards of that dimension in Jerusalem, Brabant, Rhineland, and some areas in northern Italy, presenting them as residues of the journey of the Drusian foot, which he assumed had remained unchanged over time.[11]

But sometimes it was not the scientists who provided the associations, but the linguists, and Paucton also proceeded by means of etymological chains. He reflected along the line of *mansio-manso-mansa*, and assumed that the *pygme* was the standard employed by pygmies.[12] He switched between words and dimensions, since, as Cristiani explained, both were "products of Reason, which is the only one that can provide us with the means to conciliate the antique and the modern measurements."[13]

The repartition of measurement standards into categories that spread transnationally is one of the undying myths of the metrological literature. As tools of commerce, standards are taken to move as widely as the goods they accompany. In the seventeenth century,

Richard Cumberland argued that the English cubit of eighteen inches was descended from the Babylonian standard of the same name, which had been adopted by the Greeks and the Arabs before arriving in London and other European countries.[14] And the president of Turin's Academy of Sciences, Prospero Balbo, argued that the nineteenth-century Piedmontese still employed the foot of the eighth-century Lombard king Liutprand, adding that that standard was the original Egyptian cubit, which had moved to the Etruscans and was then adopted by the Lombards after they conquered Tuscany.[15] The connections were delusional, the historic leaps implausible. To justify his reasoning, Balbo could point to only a few dimensional similarities. But he did not even need them, as the underpinning assumptions to his reasoning—that modern Western states were the continuation of the glorious Eastern kingdoms, that the Mediterranean was a connecting sea, that standards went through centuries unscathed—were mainstream.[16] So his reasoning caught on. Writing after the Congress of Vienna, at a time when European states reintroduced premetric standards, Balbo set to return the Piedmontese foot to the astronomical perfection of the Ptolemaic original. He proposed to correct its length so as to make it correspond not only to what he thought was the Egyptian standard, but also to 1/19,440,000 of the meridian degree. (The retouch simplified its relation to the metric system, which Balbo knew well, since he had participated in the international conference in Paris fifteen years earlier.) It was because of the simplicity in converting the new/old standard of Piedmont to modern geodetic standards that Balbo's proposal was approved in 1826, giving rise to the oddity that Italy's largest nineteenth-century economy ran on ancient Egyptian measurements.

So standards moved through the Mediterranean sea as they moved through time. Water was not just a means of transport; it was the very medium through which measurements were thought. As tools for conversion, measurements liquidized the real, producing homogeneities between faraway lands in space and in time. The water fountain that framed the international scales of Giovanni Carafa's map (see figure 3.6) was not just decorative; it visually argued for the seamless fluidity of standards. Measurements are, after all, watery from the start—or at least so claimed Cristiani, who described them as streams flowing out of one spring. "The primordial measurements of the most antique, disciplined governments," he wrote, "are the source from which all those in use today are derived. It is impossible to call by name their antediluvian author given the thick obscurity of those primitive times. We, however, know from the Bible that they were known to Noah when he built the Ark according to the dimensions that God gave to him."[17] So measurements date back to the Biblical flood. And, as René Descartes and Thomas Burnet considered the deluge as the originator of the world, to find the origin of measurements in Noah's flood was to elaborate on the idea that all standards are geodetic. Indeed, the so-called theory of the Earth interpreted the deluge as a break in the earthly crust which had produced, after the waters subsided, the mountain chains and irregular oceanic expanses of contemporary maps.[18]

Still, standards could not be simply taken as articulations of the earth, since classical writers had emphasized their origins in human will. Pliny claimed that measurements had been invented by Palamedes, the man who discovered Odysseus's tricks and had codified writing.[19] (Writing is akin to measuring, as Euripides explained, since it translates information across distance without corruption.)[20] And Strabo nominated Pheidon, lord of Argos. Both Palamedes and Pheidon, two Greek kings, felt too recent for Paucton. "It is certain that this invention [of measurements] is much more ancient," he wrote.[21] "It must be as old as the earth," he concluded.[22]

While denying that measurements had been invented, Paucton and Cristiani did not completely dismiss this classical tradition. Their treatises, after all, tried to bridge the various narratives about measurements. They did not create self-sufficient, systematic accounts by critically editing information. The value of their works, instead, rested in finding points of connection. And the most important link they found—the one that, like the earth, firmly joined past to present and culture to culture—was the human body.

The linear measurements, wrote Paucton, are "standardized by nature and registered on the proportions of the human body."[23] Measurements emerged not only out of the earthly crust, but also from people's skin, turning their discoveries into acts of self-discovery. The names of standards—*dita* (fingers), *palmi* (palms), *piedi* (feet), *braccia* (forearms)—revealed the interdependence of measurements and body. Many ancient writers stressed this convergence. From Pythagoras and Hero of Alexandria to the much-read architect Vitruvius, there was hardly any classical source that did not describe standards as limbs.[24] And the myth returned in numerous medieval and early modern sources. When researching measurements today, it is hard not to run into a text that repeats what has become a platitude.

The belief is powerful, as it reconciles the two natural origins of measurements; it produces the illusion of a body was commensurate to the earth. "The geodetic foot has the idiosyncratic peculiarity to also correspond to the natural foot of a man of average size," Paucton wrote.[25] The myth of the Vitruvian man thus expanded, taking up a timeless, global dimension despite the fragility of its foundations. In modern times, one critique came from osteological studies that argued that human skeletons had shrunk over the centuries, some physicians going as far as accepting the existence of giants in antiquity.[26] But then, when Paucton and Cristiani reflected on human size, they were thinking of one specific, idealized body: the body of the king.

The king's body served as a way to reconcile the many, inconsistent findings of metrology. It did so because it was, essentially, both a paradox and a fugitive thing. Even if real, it was mostly approached through imagination, and thus aggrandized or shrunk at one's will. The king's body was exalted as both average and extraordinary. And when measurements were approached as its articulations, they emerged as naturally authoritative.[27] They revealed themselves as gestures of power: produced and preserved for the well-being of states. "In all the states of the world, regardless of whether we cover the antique chronicles or we glimpse the current governments, we will notice everywhere and at all times the same vigilance in the preservation of the measurement standards."[28] Never mind Paucton's generalizations; the logical shift that is worth stressing here is that governments were presented as caring as nature.

Governments were nature's allies and continuators. Only states had the capacity to survey the earth and make standards withstand time. The convergence of nature and governance obliquely exposed the desire of the state to earn the right to shape society and prescribe the limits of its citizens' activity.[29] It also implied that to criticize the state was to criticize nature, as well as the scientific and philosophical doctrines that were at the basis of the rhetoric of reason that enabled the retrieval of measurements. Measurements did not lie outside of criticism because they were as old as the earth or part of the human body. Instead, they were presented as a priori elements because the logic they supported blended with the power structures of the time.

7

Measureless Art

Not everyone supported the project of naturalizing measurements in late eighteenth-century Europe. The enforcement of those measurements had been too harsh and sudden to be condoned. The association of the metric system with Napoleon—after whom the mathematician Pierre-Simon Laplace wanted to name the new standards—was as strong as the bond between the *pied du roi* and the king.[1] Revolts broke out. The Duchy of Parma, conquered by the French in 1797, considered the meter an imposition and never got used to it, if we trust official accounts of a decade later.[2] In 1812 Napoleon himself agreed to reintroduce in France the old standards (*mesures usuelles*), as a way to assuage the population plagued by famine and wars. And with the Congress of Vienna, many ex-Napoleonic states repealed all French laws, and initiated a return to the previous measurements.

Many people were relieved. Several intellectuals had found the propaganda of the meter, with its mathematical minutiae and absurd aspirations, nauseating. The politician François-René de Chateaubriand complained, "If you meet a man who speaks to you in meters, decimeters, and hectares, rather than *toises*, feet, and acres, you are dealing with a prefect."[3] His depiction of measurements as a concern for petty bureaucrats returned throughout the nineteenth century. In *Hard Times*, Charles Dickens mocked the schoolmaster Thomas

Gradgrind, "ready to weigh and measure any parcel of human nature, and tell you exactly what it comes to."[4] And Adam Mickiewicz wrote that measurements only fitted corpses.[5]

While those attacks expressed frustrations of different kinds, they all seemed to share a belief that life was what lay outside of quantification. In his essay "Von Deutscher Baukunst," Goethe stated that those who measured monuments rather than approaching them through feeling occluded the source of art, which could only flow freely out of emotions, away from any attempt to quantification.[6] Penned in 1772, Goethe's attack was directed towards antiquarians like Paoli, who recommended that tourists should explore monuments with canes and measuring lines.

In his travel manual, Misson also suggested that tourists should carry with them a "cane divided into several measures."[7] He also suggested bringing "a piece of packthread, well twined and waxed, fifty fathom long, and exactly divided into feet by knots, by the help of which I could in a moment measure long distances, the height of some towers, the bigness of pillars, and what else I thought fit."[8] Another influential traveler, Thomas Nugent, advised to make use of "maps, measures, prospective glasses, a mariner's compass, and quadrant, and to be able to take the dimensions of things."[9]

In *Avanzi delle antichità*, Paoli showed such well-equipped visitors in action at the Piscina Mirabilis, a freshwater cistern built by Augustus for the imperial fleet, which was harbored nearby in Roman times (figure 7.1). No description of the monument went without mentioning its impressive dimensions, not even the paragraph-long account of its rediscovery published in London's *Gentlemen's Magazine* fifteen years earlier.[10]

Paoli's illustration represents measuring as a hierarchical two-step process. A servant has climbed to the top rung of a ladder. He holds a cane whose opposite end touches the top of the vault. The string falls vertically, and another servant firmly presses its ending to the ground. The cord is then passed to the two gentlemen in the foreground, who determine its length by measuring it against a yardstick. It is unclear whether the two scenes are following one after the other—so that the gentlemen would realize that the Piscina is thirty-nine Neapolitan palms high (roughly ten meters and a half)—or whether the two pairs

MEASURELESS ART 51

Tab. LXI.

*Veduta interiore d'una conserva d'acqua | Aquarum receptaculi quod vulgo dicitur
Detta volgarmente Piscina ammirabile. | Piscina admirabilis interior Prospectus*

7.1 | The measuring of the *Piscina Mirabilis*, a Roman cistern at the western end of the Gulf of Naples. From Paolo Antonio Paoli's *Avanzi delle antichità esistenti a Pozzuoli, Cuma e Baia* (1768), plate 61.

are measuring different strings. But such distance is crucial, as it is precisely what the engraving represents.

There is no interaction between the servants and the gentlemen, because the confidence of the latter can only derive from not taking part in the operations of the former. As Scarpellini revealed, measuring is a messy business: it requires one to determine where a dimension starts and where it stops, even if such decisions remain questionable. The certitude of measuring belongs to those who have not themselves measured.

Goethe knew this. In opposition to the uncritical measuring of tourists, he presented measuring as a deforming process that has little to do with truth. And in opposition to a perfectly quantified world, he wrote of a world where scalar opposites meet: where observation of ants crawling on a tomb can lead to explosive images of mountains heaped into the clouds.[11] "Von Deutscher Baukunst," the essay from which I took this vision, is a text of self-enlargement and self-derangement that, with its constant leaps, shatters the logic of the ruler.

Goethe's readiness to turn the large into the small had numerous imitators, especially in Italy.[12] Pietro Selvatico referred to Goethe's essay—today we would say he plagiarized it—in his 1856 opening lecture at Venice's Academy of Fine Arts, which he then directed. Selvatico, one of the most influential art historians of his time, attacked antiquarians for measuring the ruins of the past in soulless, mechanical ways.[13] By doing so, he made claims for the expulsion of measurements from the budding discipline of art history.

Selvatico was not alone. His predecessor at the helm of the Venetian Academy, Leopoldo Cicognara, did not report one single measurement in a lecture on Palladio, even though he agreed that Palladio's buildings "proceed from measurements, mathematics, and proportions."[14] In his thirteen-volume history of architecture, Amico Ricci mentioned the dimensions of edifices on only three occasions, copying them from other sources and not even converting them to the same standard. He justified his disinterest by defining metrology as pedantic.[15] And Goethe's contemporary Gian Francesco Galeani Napione explained that "the judgment of a trained and experienced eye is both prouder and more delicate than any pair of compasses or measurement."[16] The romantic sensitivity of this sentence, in which the celebration of sight comes with a depreciation of measurements, could hardly be more obvious. Napione presented measuring and seeing as incompatible, in direct opposition to Leon Battista Alberti, the Renaissance thinker who, three hundred years earlier, had famously written that "the eye measures these dimensions by visual rays as if it were a pair of compasses."[17] Napione reversed the syntactic order of Alberti's sentence to detach measurements from questions of perception.

While carefully removing any trace of quantification, early nineteenth-century Italian art critics presented buildings as visions or, better, "fictions." The term was employed by the philosophers Arthur Schopenhauer and Friedrich Schlegel, who interpreted the Gothic revival not as a taste, or as a new aesthetic, but as the direct effect of an unprecedented flow of images.[18] Their analysis was still valid decades later, when Camillo Boito, Italy's leading nineteenth-century intellectual, called for a "pictorial conservation of medieval buildings," by which he meant that the goal of restoration was to preserve the aesthetic effect of monuments, not their dimensions.[19]

The publication in the early nineteenth century of the first art-historical surveys contributed to the neglect of measurements. The *Histoire de l'art* by the archaeologist Jean Baptiste Seroux d'Agincourt (1823) and the *Storia della scultura* by Cicognara (whose second edition was published also in 1823) reproduced artworks and buildings without any indication of scale.[20] Their large plates included many objects whose size was inconsistent with one another. In a particular sense, such plates were the visual translations of the unsystematic discourses of their authors, who quickly moved from planar typologies to decorative details, describing buildings and artworks as fragmented entities in which dimensional and spatial relationships did not matter.

The writing of art history in Italy thus coalesced at a time of, and in response to, a depreciation of measurements as authoritarian, mechanical, or both. Describing the dimensions of an ancient monument in meters was then a way to disengage with its aesthetics, to reduce it to the plane of industrial reports and the *tavole di ragguaglio*. Modern measurements trivialized, and such an idea continued for much of the century and beyond. In a way, the lack of art historians' engagement with questions of size and scale even today can be traced back to that opposition.

Measurements thus disappeared twice. Scientists, politicians, and antiquarians folded them into the surface of the earth, while art and architecture critics refused to engage with them, presenting measuring as a sterile activity that was to be excised in order to recuperate a truthful experience. The goals of the latter were in opposition to those of the former, and yet both groups contributed to expel measurements from cultural consciousness. By either assuming that measurements were integral elements of the natural world or dismissing them as technicalities, both teams remained blind to the political role of measurements, which validate the power structures from which they are thought to derive. But to counter such a stance—to start forming a critique of measures as truth frameworks rather than scientific tools—it is useful to move to the twelfth century, when measurements were not locked out of sight in iron safes, but were displayed in the open.

PART II

Squares

8

The Pratissolo Deal

Let us start with the notary record, stripped of its formulaic Latin. On December 6, 1263, six people gathered in Pratissolo. The meeting-place may have had a name, but it was really little more than a thoroughfare cutting across a chessboard of fields in the Po valley, north of Reggio Emilia. Two of the six people were there to do business. On one side, Ubertino del Poggio, owner of some fields; on the other, Rolandino Muti, agent of the commune of Reggio, interested in purchasing nine of them. Three members of the group were locals, summoned to carry out the identification of the estates. They, too, had reason to be present. Since some of their fields adjoined some of Ubertino's, they wanted to make sure none of their land was snaffled in the process.[1] All properties concerned were named and their tenants interviewed. Gerardo Spadinasaco, the sixth member of the party, directed the technical survey. He took the distances with a long rod called *pertica* and, before registering the data, he double-checked that the sides of each land parcel were consistent.[2] Disputes occasionally flared up but were resolved quickly. The whole process took two days.

On the third day, the three countrymen (Gerardo Magno, Giberto Guarzoni, Maso de Gislini) and the surveyor Spadinasaco were summoned to the palace of the commune in Reggio. There they rejoined Rolandino, and in a large hall on the upper floor, they swore an oath in front of the *podestà* Anselmo da Rivola, the commune's chief magistrate. They pledged to abide by the precepts of the commune, and promised to assist in the transaction impartially and without deceit (*bona fide, sine fraude*): neither the commune nor Ubertino were to be favored.³ Business then moved to the loggia downstairs (*sub palatio*) for negotiating the price. A promise of neutrality was not enough this time, and the four men were put under threat of paying one hundred *soldi* in case of a disproportionate estimate. The group then moved—again—a few steps away to the church of San Giovanni Battista, where the four men reconfirmed Ubertino's ownership of each of the fields, repeated their value, and gave their measurements in *biolche*, Reggio's unit of surface measurement.⁴ The party was then dismissed, only to be summoned again two weeks later. On December 23, Ubertino and the four men from Pratissolo appeared in front of Rolandino, some notaries, a delegate of the *podestà*, and four members of the Elders. In front of this solemn crowd, the four men confirmed the information that had been recorded in the church of San Giovanni as a preamble to the final act of sale. As payment, Ubertino received a fourth of the mills located east of the San Lazzaro gate. The total settlement was said to be worth three hundred imperial *lire*.⁵

The documentation concerning the Pratissolo sale is contained in the so-called *Liber grossus*. This massive volume, assembled from around 1270, gathers many documents about the communal administration from the twelfth century up to 1352, the year of its last record. Sales of fields were routine, but what I term the "Pratissolo deal" stands out because of its level of insight about governmental logistics.⁶ It shows, for instance, that the commune valued its members highly: they obtained information about a place or an event from locals, considered their representatives personally responsible for errors, and regarded consensus as proof of correct judgment. The documents also reveal that a business transaction was a journey through architecture. The Pratissolo party toured in and out of the communal palace, moving through as many as three sites in one day. Given the accuracy with which the notary recorded each venue, place mattered. Take the Baptistery of San Giovanni (plate 3). This is where the size of the fields was recorded, a fitting procedure to take place there. Far from being an exclusively religious building, the baptistery acted as one of the city's economic fulcrums.⁷ The rite of baptism attached people to a place, cementing their tithing responsibilities. And, as a 1217 document informs us, the building overlooked a small square (*platea parva*) where the Saturday market took place.[8] Like other churches in medieval Italy, San Giovanni sometimes worked as an extension of the market, where transactions took place under watchful saints.[9] And this is why the left pillar of the San Giovanni facade was incised with Reggio's two main linear standards: the *pertica* and the *braccio* (plate 3).[10] Those who needed to check the correctness of their measuring tools stopped by San Giovanni and inserted them into those incisions. If they matched, they knew their measurements were correct. The standards are perhaps the reason why the Pratissolo group moved to the baptistery to record the dimensions of the fields so that the notary could double-check Spandinasaco's rod, the *pertica*. The *biolca*, the surface unit in which the Pratissolo fields were registered, corresponded to seventy-two square *pertiche*.[11]

We do not know when the standards appeared there, but the communal statutes of 1265, drawn two years after the Pratissolo deal, report that another standard, the *passus*, was incised "near the door of the cathedral" (*in reza ecclesia maioris*).[12] (Cathedral and baptistery are today enveloped by the towering bishop's palace, but they were two disconnected buildings in the thirteenth century.)[13] Reggio's civic museum holds a medieval standard of capacity (the *mina*) carved in a stone, which may also have found its place around the market area.[14] It is, however, the baptistery that developed a privileged relationship with measurements. This relationship was so close that the Reggiani invoked John the Baptist against commercial fraud and falsity.[15] The physical presence of the standards altered the local perception of the titular saint, who is not associated with measurement checks in the rest of Italy. It also transformed the perception of the church. In the 1470s, when the bishop took over San Giovanni to enlarge his palace, emptying its architectural body and translating the baptismal functions to the cathedral, he preserved its facade, as it served as the wall of measurement standards.[16]

9

Opacifying the Invisible

Once you know standards are there, you find them everywhere. On the south tower of Fidenza's cathedral runs, parallel to the ground, the local *pertica*, the standard for ground distances. *Braccia* and brick molds are carved in a marble block sticking out from the socle of Bologna's Palazzo d'Accursio, the seat of the medieval government (plate 1). Standards appear after turning around a pillar in the courtyard of Piacenza's communal palace and grow vertically on the posts of the canopy in Verona's Piazza delle Erbe, like the grooves of a Corinthian column (see figure 13.1). They fan out on the apse of Modena Cathedral (figure 9.1), and dot the door and window frames of Todi's communal palace.

Despite their pervasiveness, those incisions have become invisible. Made obsolescent by the idea that modern standards are inscribed in nature rather than by unassuming scratches on a wall, they have been filled in or removed. In Tuscany, in Lazio, and in other Italian regions, they were substituted at the dawn of the unification by slabs reproducing pages of the *tavole di ragguaglio* (figure 9.2), as if measurements were numbers, all homogeneous even if varied according to place and trade.[1] Those hanging gravestones bury the notion of the city as an economically independent entity while celebrating the centralized metric state. Such an idea reverberates even when the substitution does not take place. In Ferrara, old iron standards greet visitors to its moated fourteenth-century castle, where the tools were moved

9.1 | Measurement standards on the apse of Modena Cathedral. Photograph by the author.

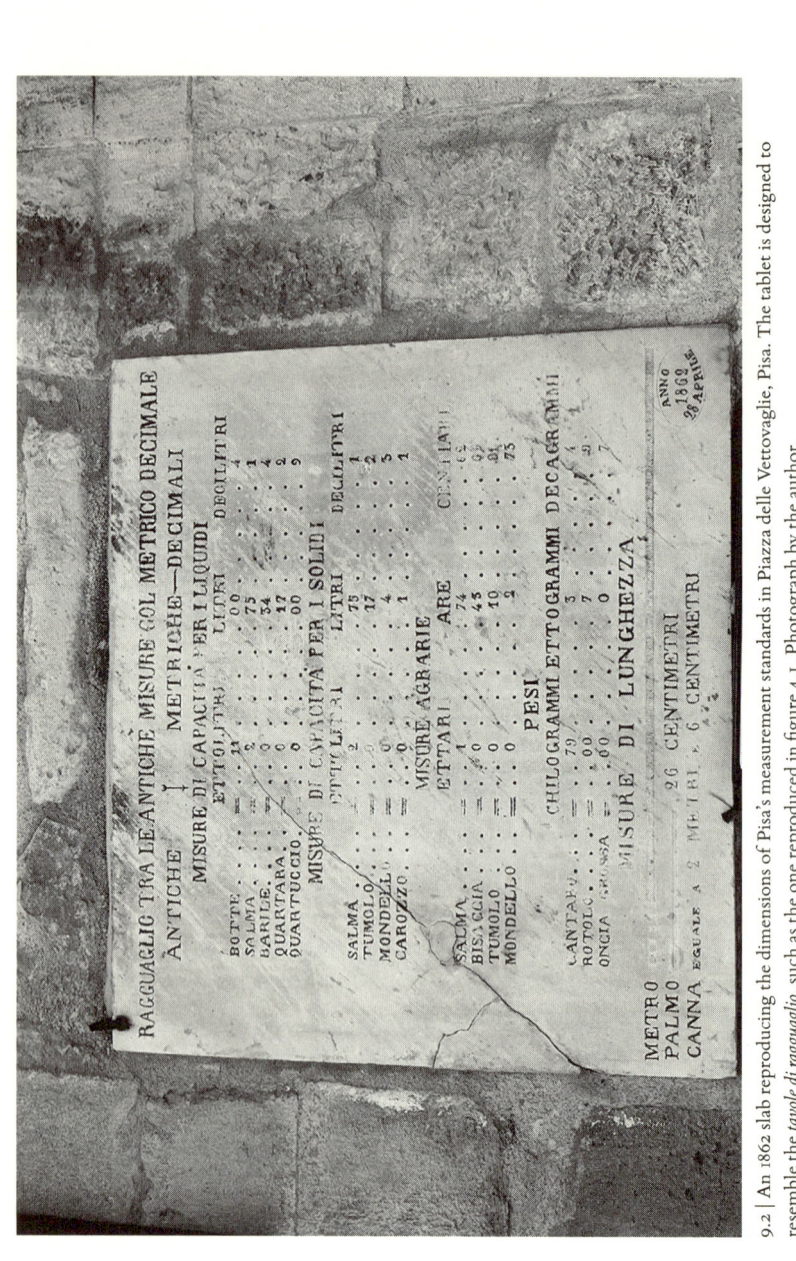

9.2 | An 1862 slab reproducing the dimensions of Pisa's measurement standards in Piazza delle Vettovaglie, Pisa. The tablet is designed to resemble the *tavole di ragguaglio*, such as the one reproduced in figure 4.1. Photograph by the author.

in the nineteenth century, forging a fictional parallel between local measurements and medieval rulership. The disregard for measurements has carried on in writing, and even the few historians who do mention them regard them as indications that a market was nearby.[2] Measurements are thus only gestured to as signs, their presence taken to refer that an authority was present, which is what the communal statutes of the Middle Ages would have us believe.[3]

In writing the chapters that make up this part of the book, I also looked at statutes—many of them—but without taking them as innocent sources. Dodging their hammering dos and don'ts, I read their discussions of measurements critically to argue that standards were not just administrative tools but objects of labor. The Pratissolo deal described in the previous chapter offers a glimpse of the energies necessary for measuring in medieval Italy: rods folded and unfolded, in and out of buildings, carried through fields and into towns, laid on the ground, or pressed against walls. A wearing process, measuring was labored through the territory, changing it in return. But the very possibility of measuring had also been laborious: the governments of Italy's medieval cities—the communes—fought for the right to measure. And after securing it, they worked the standards through the population. Their usage was not straightforward; it was questioned, and their policing shifted pace and hands. Sometimes standards stopped being used altogether, and for long periods of time, their displays were seen as mere scratches on the wall, until they returned to sharp focus.

The incisions of standards on walls show the labor of measures. They are the results of handiwork, and yet are constructed and encountered as if they were not products of human industry. Measurements came to be seen as twofold: ratios belonging to the realm of ideas, and the most physical of tools. Such ambiguity makes measurements occupy an unusual place in the cultural landscape of the Middle Ages. In attempting to locate that place, I uncover them not as mere aids to trading operations, but as tools that supported power while pretending to exist beyond it. Like the communal statutes that discussed them in more detail than any other medieval document, measurement standards were the very tools by which ruling elites shaped reality, materially and ideologically. Showing such a convergence is a way to expose that the commune's success rested not solely on military power or political support, as scholars have demonstrated, but also on material training.[4] After being carved on the wall, measurement standards worked like social and cultural grips, engendering practices of continuity despite the turnover of medieval politics. They fostered a sense of justice even when it was inadequate or missing, becoming pivotal lines for the tracing of a communitarian horizon.

As measurements were both subject to power and powerful in their own right, the questions I will be asking in this second part of the book—which I have titled "Squares"—are about process: Who conferred judicial value to standards, who had the authority to do so, and how did such an authority translate at both the material and social levels? To answer these queries, I will read the regulations from numerous Italian cities against the physical standards that have survived. I realize that keeping a wide-angle view on history is problematic. Modern historians of the Middle Ages hardly ever take up such an approach, preferring to focus instead on one city and on a short span of time.[5] They keep the work manageable. Given the historical complexity of late medieval communes—with their seasonal shifts from consular governments to popular associations, incessant familial feuds and internal battles amid an ever-changing international scenario—to restrict the view would seem the only option.[6] And yet a narrow approach erases measurements from the radar of history. As tools for the construction of evenness, measurements reveal their character only when moving in and out of the areas in which they are taken for granted. So this part of the book is meant to deliberately sprawl spatially, chronologically, and disciplinarily in order to catch a glimpse of what is taken for granted. Widening the lens through which medieval history is usually focused creates problems at the level of texture and narration. As an effect, some details of the narrative will be blurry. I have tried to keep things crisp while maintaining a broad perspective. Yet, as a way of centering, my attention returns again and again to the elusive medieval standard displays and questions about the politics of elusiveness. If in part 1, "Safes," I explored how the modern state benefitted from the naturalization of standards, I now look at how medieval powers labored measurements through the population, setting up the conditions for their dematerialization and reaping benefits that governors may not even have considered.

10

The *Pietre di Paragone*

One of the earliest surviving records of public installments of size standards comes from Faenza. The city's chronicler, a canon of the cathedral who lived in the first quarter of the thirteenth century, mentions that in 1195 the foot used for measuring land was installed near the doors of his church.[1] That foot is gone today, as is the medieval cathedral, destroyed in the 1470s and replaced by a grand new building by architect Giuliano da Maiano.[2] Similar records, however, multiply from the beginning of the thirteenth century. In 1204 the statutes of Bergamo's local weavers reminded its members that all *parietes*—the local standards for fabrics—"must be shaped according to the antique standard that is above the ground, near the main door of the church of Santa Maria Maggiore."[3] The standards can still be seen in the same spot today (figure 10.1), even if it is hard to say whether their current Mondrianesque configuration was medieval. (We do know, however, that the current iron bars postdate Napoleon's campaign to Italy, as they are engraved with their dimensions in meters.)

After Bergamo, public measurements are mentioned in the statutes of Ravenna, Brescia, and Mantua.[4] The communal chancelleries speak of "mensurae," "tabellae," "moduli," or "forme," without reference to a standard vocabulary.[5] There is often no linguistic consistency even within the legislative output of a single city. And while it may be possible to see some regional similarities—*scede*, for instance, is a term that returns in central Italian cities such as Siena and Perugia—the

10.1 | The length standards on the north side of Bergamo's Santa Maria Maggiore. Photograph by the author.

one common element, over almost two centuries, is the fact that they were incised in stone.[6] The statutes of the town of Amelia, in Umbria, speak of "mensure sculte."[7] Those from the region's current capital, Perugia, mention the "campione e mesure de pietra."[8] In Mantua the record specifies that the stone had "to be left natural" (*in lapide vivo*), without plastering or painting.[9] And still in the mid-fifteenth century, the famed Florentine architect Filarete recommended that readers of his treatise insert some blocks of stone into the facade of a church so that measurements could be "sculpted and carved out of them" (*scolpite nelle pietre e cavate*).[10] I could cite other examples.[11]

Stone was the material that withstood time, that guaranteed the firmness (*firmitas*) of a legal act.[12] In 1116, when Pope Paschal II feared that Emperor Henry V would conquer the Canossa hamlet and thus alienate one of his strategic strongholds, he had a stone carved with the record of his ownership.[13] And in 1173, during the war against Emperor Frederick Barbarossa, the citizens of Ferrara chiseled their public rights in the stone walls of their cathedral for claiming them back in case of defeat.[14] Stone did preserve them: Ferrara's are the earliest communal statutes handed down to us.

Besides solidifying local customs, stone petrified traders. The inscription that wraps the east gable of San Giacomo di Rialto, allegedly Venice's oldest church and the fulcrum of its marketplace, forbade merchants to tamper with tools and disregard contracts.[15] In IIII, a stone inscription on the facade of Lucca's cathedral reminded money changers and spice dealers of their pledge to "commit no theft nor trick nor falsification within the courtyard."[16] And in 1234, the commune of Perugia announced a new form of taxation, based on the value of possessions rather than on pre-fixed amounts, by carving it in stone.[17] The slab, called "the stone of justice" (*petra iustitiae*), was mounted on the belfry of the cathedral to fortify the bond of the commune with its citizens.[18] Near the *petra* were incised, also in stone, the local measuring standards, the very tools through which to fulfill the governmental pledge (figure 10.2).[19]

Because stone is the support of all those medieval standards, I call them *pietre di paragone*. The expression is a conceptual condensation, as it does not embrace all the standards produced throughout medieval Italy. It is also a critical concept, rather than a merely descriptive one, as it is not found in the sources. It is derived from them but it also imposes critical demands upon them, as well as upon the objects they refer to. Calling attention to solidity is thus a way to both recognize the aspirations of medieval governments and place them under scrutiny, especially in the moments the *pietre* moved in space, thus triggering on the part of the reader a movement of thinking between words and objects that structures the critique this part of the book aims to offer.

Such movement, from object to concept, was mirrored in the medieval checking of standards. Communal officials, merchants, and surveyors inserted their wooden tools into the incisions in the *pietre* as if they were placing them back in their original molds. If they matched, the wooden rods were stamped with the symbol of the commune, thanks to which they were recognized as genuine. To facilitate the operation, Bassano's statutes of 1259 also prescribed that officials affix thin metal plaques at the extremities of the carvings so as to more clearly individuate their "ends."[20] Other communes did it too (plate 3c).

10.2 | The length standards of Perugia on the belfry of San Lorenzo, near the *petra institiae*, on the right. Photograph by Stefano Dottori.

While medieval statutes described the physical appearance of the standards (both the *pietre* and the wooden rods for everyday use), the process of checking received very few words. In 1216 Milan, the representatives of the city's traders decreed that the two length standards, the *passus* and the *corda*, had to match the incisions at the fish market opposite to which the consuls resided. In cases where the sellers' tools did not match the incisions, the consuls confiscated and discarded them and fined the crooks. "Passus falsus, sive corda falsa intelligantur, quae non inveniuntur juxta mensuram petrae de pescharia."[21] Please pay attention to the logical construction of the sentence: the standards were false if they were not found to match when placed next to (*juxta*) the measures incised in stone at the fish market—that is, in the *pietra di paragone*. Justice was an effect of spatial contiguity.

The statutes never speak of degrees of error. They only consider the result—the question of whether the standards matched—thus implying that the process was constructed to yield such an outcome. The thirteenth-century jurist Alberto Gandino, in his treatise on wrongdoing, articulated such a black-and-white approach when stating that "a false measure is not a measure as a false weight is not a weight."[22]

There may be not much evidence about the ways in which officials checked standards. Still, you may recognize this process, as it was memorably dramatized in the ending of the fairy tale *Cinderella*, where the protagonist is asked to insert her foot into a slipper. Many would know Charles Perrault and Walt Disney's versions of the story, but the idea of using a shoe as a mold to check Cinderella's identity comes from Giovan Battista Basile's seventeenth-century version, *La gatta Cenerentola*.[23] When Basile wrote his rags-to-riches tale in Naples, merchants were checking the sizes of products, including shoes, by placing them on the respective *pietre di paragone*. Basile turned a market test of dimensional correctness into the climactic revelation of a mysterious woman's identity. His finale does not simply gesture to a commercial check: it ticks all the statutory requirements. It is a precise reenactment of how measuring was undertaken in premodern Europe, with Cinderella's foot placed in the mold by an official inspector and the whole operation carried out in public to improve citizens' moral standards, as offenders were barbarically punished.[24]

Incising dimensions in stone frustrated criminals. Faced with an empty groove, a vandal's only recourse was to chip away a piece of stone from the frame, increasing its size and thus achieving the opposite of what any forger would wish, which was to fabricate shorter, not longer, standards. While an object can be manipulated, a void is incorruptible, because, as Aristotle wrote, "the void is thought to be place with nothing in it."[25] Communal officials inserted iron plaques at the extremities of the *pietre* to further thwart vandals, but a stronger sense of safety came from rendering the standards as negative impressions. At a time when cities could be robbed of their treasured tools—the precious bells that marked the time, the iron chains that kept harbors safe at night, the miracle-working banners that protected soldiers on the battlefield—this proved a brilliant solution.

To render measurements in the negative was a way to make them appear as something abstract. Euclid's *Elements*—the foundational geometry treatise that reached new peaks of popularity after its mid-twelfth-century translation by Abelard of Bath—defined, in its first page, measurements as immaterial ratios. And Gerbert of Aurillac, the author of the first geometrical work that was not a mere commentary on Euclid, also defined measurements as two-dimensional. "The foot is a line," he wrote, "and with it we measure lines or a certain length, without taking into any consideration of its height or its width."[26] Measurements, he continued, "should be understood only mentally."[27] They could not be experienced in reality, since everything in the world necessarily has three dimensions. Measurements were thus theorized about as being deprived of matter, and it was to trigger an analogous process of abstraction that the *pietre* were carved in the walls and constructed as negatives. Many of the *pietre* did not even preserve the bodies of the standards, but only their extremes—the distance between point A and point B—demarcating an expanse of air by means of metal plaques sticking out of a wall (figure 10.3). They looked like the numbers in medieval manuscripts, framed by two dots.

Abstract matrixes and, at the same time, physical molds constructed through the removal of stone, the *pietre di paragone* were ambiguous objects from the start. Their centuries-old stories oscillate between physicality and abstraction, the two sometimes reaching an equilibrium though often one state took over the other. After all, only a

10.3 | Metal bits indicating the dimensions of the *braccio* and *piè* in the courtyard of Ancona's Palazzo del Governo. Compare them to the one in plate 3c. Photograph by the author.

few knew the subtleties of Euclidean geometry. The friar Bonvesin de la Riva (1240–1313/15) appears not to have been one of them. In his chronicle of Milan, he identifies a *cubitus* with the metal rod that represents it, assigning to it not only a particular length but also a specific width.[28] But then some governments relied not only on the *pietre* but also on physical standards. A document in Viterbo says that the *pietre* on Santa Maria Nuova could be restored to their original dimensions thanks to the length of wood in the possession of a certain Angelo Burgundione.[29] And the Parma statutes of 1265 reveal that the commune kept some tangible standards in a chamber of its palace and made them available in case of disputes.[30]

While physical standards served as guarantees, however, communal governments dedicated most of their energies to the *pietre*. The stone well-curb that served to measure grains in Pistoia was checked by its *podestà* every January.[31] And while the *podestà* of Reggio Emilia had to duplicate the local *pietre* within two months of his election, he did not need to inspect the bricks and tiles that, it is documented, were preserved by the abbot of San Prospero.[32] Within three months of his election, Bologna's *podestà* had to survey the incisions "under the vault of the communal palace."[33] It is unlikely that such norms were to be taken literally. The *podestà* was more of a debater than a street detective, and the inspections were likely to have been carried out by officials. Still, the fact that the *podestà* was held responsible points to the primacy of the *pietre* over any other standard. This last requirement appears in Bologna's statutes of 1250, which also record the size of the local standard brick as nine by four by two *once*, stressing that if the standard were damaged, its shape could be retrieved from the twelve-*oncia* foot incised amongst the other *pietre*.[34] The possibility of inferring the size of the brick from another incision reveals an interest in finding ways to preserve standards that avoided relying on corruptible objects. Modena did something analogous when in 1327 it represented measurements in drawings, turning to illustrations as a way to record its standards.[35]

The hiding of the Modena standards in a book and the locking of the Reggio bricks in an abbey are indicative of the anxieties arising from the public displaying of measurements. Such worries grew, especially toward the second half of the fifteenth century when the *signori*, the lords who replaced most communal councils in governing the Italian cities, destroyed many *pietre* and replaced them with physical artifacts.

In 1468 the *pietre* that had stood in the middle of Modena's market square for more than two centuries were dismantled and sent to the city's *signore*, Borso d'Este, for use as building blocks in his new garden. The statue of justice that topped the structure was fixed to the corner of the communal palace, where it still is today.[36] It is unclear when the standards reappeared on the apse of Modena cathedral (see figure 9.1). If it happened in 1547, the year of their first attestation there, the citizens did not have access to official standards for almost eighty years. Other tools must have circulated. Bosto d'Este commissioned new standards, of which six weights survive in Modena's civic museum.[37] But, symptomatic of a change of approach to measuring, he had them made as tangible objects to be kept in a safe and exhibited when necessary. Without the *pietre*, measuring existed in a sort of blur. Tommasino dei Bianchi, the officer who oversaw the market in the year 1530, complained that traders did not know if they had to weigh grains or measure them by volume.[38] With the dismantling of the *pietre*, gone also was the office in charge of their policing: it was reintroduced only in 1547, together with the reappearance of the *pietre* on the cathedral apse.[39]

Modena does not seem to have been an isolated case. Other cities became enveloped in a metrological mist, thus upending the simplistic labeling of the Middle Ages and the Renaissance as, respectively, an age of chaos and one of order. In 1493 Parma, it was impossible to find "one single baker or wine seller or butcher or miller or fisherman or shop keeper who had correctly gauged standards and scales and who sold his goods to customers fairly."[40] And in 1554 Mantua, two edicts describe the standards as being so corrupted that the sellers could deceive even themselves.[41] That alarm prompted the ruling Gonzaga family to commission twenty-two new standards from the sculptor Bernardino Arrigoni. Arrigoni was a virtuoso and the casts he produced, which have all survived and are today kept in Palazzo Te, should be ranked among the pinnacles of his craft (figure 10.4). Hanging garlands and large laurel leaves framing the arms of the Gonzaga, the bishop of Mantua (Ercole Gonzaga), and Margherita Paleologa (Duchess of Mantua by marriage to Federico II Gonzaga) celebrate the standards as the gifts of power. While defying falsifica-

10.4 | Bernardino Arrigoni, weight standard (fifty *libre*) in the shape of an amphora (mid-sixteenth century). Collezione Gonzaghesca, Pesi e Misure, Palazzo Te, Mantua. Photograph © Museo Civico di Palazzo Te.

tion, the profusion of ornamentation also closed off any possibility of the standards being replicated; they alone embodied the correct measurements of Mantua. But measurements, as the local jurist Francesco Negri Ciriaco pointed out, ought to be considered valid only if they exist across two or more exemplars, not just one.[42]

II

Measurements, Made and Remade

If I call the incisions *pietre di paragone* in modern Italian rather than drawing from the myriad of Latin terms found in the documents, it is also to foreground the complex relationship of those incisions with temporality. The expression is deliberately anachronistic, like the standards themselves, perpetually remade and impossible to date. When looking at those weathered slabs, we are not looking at the medieval originals that modern guidebooks pretend they are, even if we may find them in the same spot as described in thirteenth-century sources. Rather, we are looking at the last link of an unknowable chain of substitutions. After all, the very documents that help us locate the *pietre* also tell us that they were replaced, over and over again. The statutes drawn in Bologna in 1250 and in Reggio in 1268 specify that the officials in charge of their restoration had to replace them with new ones as frequently as every few months, each new copy acquiring the status of their prototypes.[1]

As both geometrical abstractions and regularly replaced objects, the *pietre* were manufactured and approached as defying the present. Authorless and yet authoritative, they projected any act of measuring over a timeless horizon, in keeping with the idea that justice could only exist across time. And yet, like the chipped blocks of stone that they are, the *pietre* were immanent in history and often celebrated as such. In Assisi they carry the name of Angelo di Latero, the leader of the commoners' faction (the *capitano del popolo*), who prided himself for having restored them to precision in 1348; the date was carved next to his name.[2] Dates are sometimes present on measurement displays. Still, when they appear, they should not be taken to identify a prototype. Dates, after all, can be reproduced as easily as the incisions. Rather, they hint at a stiffening in the process of substitution, a moment where the accuracy of measurements was in danger and needed to be reaffirmed through a specific event or person of authority.

While the temporalities of the *pietre* as physical displays are complex and fugitive, their existence as a collective phenomenon finds a point of origin in the loosening of the royal grip on standards. The king of Italy plays a fundamental role in the history of measurements, as it was he who originally held the right to preserve and provide the standards, a quintessential component of public justice. Before coming to power in 1155, Frederick Barbarossa had his chancellery issue diplomas in which he insisted that measurements were part of the king's privileges, the so-called *regalia*. In a diploma issued in Marengo, standards are listed as his prerogative, together with transit tolls and other commercial taxes over public roads, navigable rivers, bridges, and markets.[3] Still, measurements were not insisted upon with the same frequency of other privileges because, while he aimed at reviving royal authority in Italy, Barbarossa conceded the right of measuring to vassals and faithful parties.[4] In 1155 he allowed the Genoese to keep trading with their own weights and lengths, a liberty that was again requested and again granted in 1162.[5] In 1159 he conceded to the commune of Asti the power over *mensure*.[6] And five years later, the consuls of Ferrara obtained all rights concerning commercial justice, including—and they spelled it out—authority over the *pietre* and their derivatives.[7] Also in 1164, Frederick gave the right of measurements also to Obizzo Malaspina, lord of a territory that stretched between Piacenza and the Tyrrhenian Sea.[8]

Frederick granted such prerogatives to cities and lords either in exchange for support or because he lacked the power to enforce them. He came to rule in a world where his predecessors had taken advantage of martial clans and clerical enclaves rather than resolutely blocking their growth.[9] In the very years when he gave measurement rights away, he was occupied in suppressing many riots against military associations in Italy, among which the most powerful was the Lombard

League, a mutual-support consortium of numerous northern Italian cities.[10] The struggles intensified to new levels in the 1170s, when the emperor himself crossed the Alps and waged war against the league. The conflicts were terminated with the Peace of Venice in 1177, but the final agreement was struck only in 1183 in Constance. That meeting recognized the judicial autonomy (*plena iurisdictio*) and the passing of *regalia* to the league members in perpetuity.[11] The resolution document was so important that each of the victorious Italian cities had it transcribed, word for word, as the opening page of its *libri iurium*, the collection of its new rights.[12]

While standards had been scratched in the walls of cities since at least the eleventh century, it was the shifting of *regalia* to communes that made the *pietre di paragone* take center stage. All the documents about the public installation of the *pietre* that I have mentioned (Faenza, Bergamo, Mantua) and those that I footnoted in the previous chapter (Brescia and Milan) refer to League members and postdate the Peace of Constance. And even in Genoa, which was not in the League, the record of the palm on the cathedral facade dates from 1184, thus hinting at the new political climate created by the peace.[13]

The fact that all those documents refer to the incisions of *pietre* on the exterior of cathedrals also ensues from the *regalia*. To back his privileges, Barbarossa turned to jurists from Bologna, a city then famous for its *studium* dedicated to the research into the *Corpus iuris civilis*, the compilation of Roman laws that served as the backbone of late medieval law.[14] The *Corpus* had been systematized at the time of Emperor Justinian, a ruler who spent much energy in clarifying the legal usages and displays of measurements.[15] One passage, for instance, states that "lengths and weights were to be preserved by only one church for each city" (*uniuscuiusque civitatis ecclesia*).[16] The passage served as the model for an entry of the statutes of Pistoia drawn in 1176, which declared that a standard of volume was to be kept by "each parish church in the district of Pistoia" (*unamquamque cappellam civitatis Pistorie*).[17] And another norm of the *Corpus* adds that standards should be exhibited publicly so that any tributary could know how much to pay.[18] By associating standards with churches and pushing them in the open, the Italian communes updated the prescriptions that they found in the *Corpus*.

The history of the measurements from Pistoia is particularly indicative of such a shift. Back in 1159, when Barbarossa reconfirmed the bishop as the sole beneficiary of the local market, Pistoia's volume standard, the *humina*, was kept in the sacristy of the church of San Zeno.[19] The statutes of the *podestà*, compiled no later than 1180, indicate, however, that the standard had then been moved to the center of the marketplace. By then its jurisdiction had also passed from the bishop to the *podestà*, who was deemed responsible to "locate and maintain the linear and volume standard for grains."[20] The latter was called *pilla* and no longer *humina*. It was also described as "de civitate Pistoria," an indication that the tool had to be regarded as a civic rather than a diocesan tool.[21]

Such changes—of name, location, and person in charge—must have been common, as Pilio of Medicina, a renowned jurist trained in Bologna, reflected upon them in his law manual, titled *Quaestiones*. One of the cases that make up this famous treatise asks the following question: Could someone who cultivated somebody else's farm for one Bolognese bushel of grain still be paid in that quantity after a change in the official dimension of the bushel? And, if this were the case, would he incur the penalty set by the commune for not measuring in the official standard?[22] The case, indicative that measurements were worthy of judicial attention, was far from a mere academic exercise. Rather, it registered the split between old and new measurements opened by the Peace of Constance around the time when Pilio was composing his text.[23] Pilio's final response—that payments ought to be in the standard valid at the time of the contract—relativized measurements while discouraging their being changed at will. He also made sure that they would not be taken for granted. And indeed they were not: his approach remained the usual one for at least two centuries. In 1350, when the lord of Milan, Galeazzo II Visconti, decreed that all his territories had to observe the standards in use in Milan, he added a sentence that excluded all existing contracts drawn beforehand that referred to other standards, in line with Pilio's prescription.[24]

So communes (and, later, lords) complied. Standards mattered. Statutes spelled out their locations and the timings of their checks. Governments incised them in stone to preserve them. The *Corpus*, whose significance notaries and lawyers debated, engendered the installment of standards on the external facades of cathedrals, while the *podestà* became responsible for their conservation.

It may be important to recall that the *podestà* was not elected from among the citizens. He was instead a professional administrator foreign to the city who, after his mandate was over, moved with his entourage to another. His travels exposed him to numerous ways of doing things, and contributed to the homogenization of measuring practices among cities. As trading tools, measurements benefited from standardization. And yet, as the next chapter will explain, it would be wrong to see their making only as an effect of external forces.

12

In the Open

The *pietre* emerged not only through the engagement of cities with imperial rights, Roman law, and foreign professionals, but also in the smoothing out of their internal conflicts. The Italian communes were, after all, fragmented entities. Starting as associations of landowners and military aristocrats who entered into complex relationships with royal appointees and bishops, the communes changed with the rise of merchants and artisans. By the beginning of the thirteenth century, these categories of people organized themselves into patrolling associations and especially *societates artium*, artisanal guilds bound together by oaths and common property. Sometimes intertwining to the point of indecipherability, those associations enlarged political participation, fighting noble families and producing a constellation of commoners' parties and institutions that modern historians simplify under the label *popolo*.[1]

As major productive and commercial forces, the *societates artium* were deeply invested in measurement standards. Indeed, it is not surprising that Bergamo's first record of the *pietre*, in 1204, appears in the statutes of its weavers' guild. With the consolidation of the *societates artium* and other popular associations, the commune absorbed them, making their needs its own, appointing their members as its officials, and shaping its statutes around theirs. Their assimilation was far from straightforward. While triggering change, the *popolo* was often on the brink of collapse, especially in the early thirteenth century when it was threatened by the takeover of powerful families and the rapid shifts of international politics. Measurement standards were often caught up in the way. As vital judicial tools and sources of duties, they were the very instruments through which factions expanded their influence. They did not end up in the hands of those with power. Instead, the tools themselves conferred power on those who wielded them.

In 1164, when Pisa was in the midst of a building frenzy (the monumental baptistery was underway and the city walls had just been completed), the local stonecutters and tile makers organized themselves into a guild. The self-legislating association could have raised wages if the commune did not intervene to deny its legitimacy, accusing it of "going against public honor." (*Contra communem honorem factas* is a stupendously ambiguous expression that feigned public interest while designating the interests of the aristocratic elites who made up the commune.) In a preventive move, the commune of Pisa appropriated the verification of the shapes of bricks and roof tiles.[2] As bricks were sold by quantity rather than weight, the commune made sure that the guild did not reduce their size and thus elevate the costs of building and maintenance. The initiative was not a first. Ten years earlier, after putting down an insurrection spurred by the Visconti family, the commune appropriated the Visconti's abiding rights, which included the weighing of iron from the island of Elba, and checks on bakeries, wine sellers, and all the arts.[3] Beyond the commune's reach were only the standards for grains, which the canons of the cathedral had securely held since 1147 and managed to have reconfirmed in 1192 by Emperor Henry VI. In the latter occasion Henry granted Pisa some other *regalia*, thus extending the concessions of the Peace of Constance to cities that were not part of the Lombard League.[4]

Being granted the right to measure did not necessarily imply the capacity to collect duties, and modern historians have documented numerous clashes between right-holding communes and land-ruling lords.[5] By the mid of the thirteenth century, however, some communes strengthened their authority. In 1262, Siena kept the molds for bricks in the Biccherna, the city's top financial institution, which was in charge of keeping track of all municipal revenues and expenses. The Biccherna not only stored the matrixes but also displayed six prototypes. The *camerlengo*, the chief of the Biccherna, was personally responsible for having new bricks fired every April, before all makers gathered in his presence and swore to respect the size and quality of the models. The *camerlengo* also appointed three inspectors to go from kiln to kiln.[6] Their posts were still filled in the 1290s, during which time the commune embarked on a campaign against fakes. The inspectors were also mentioned in the communal directives of 1310, which reconfirmed the *camerlengo* as the sole overseer of the official molds, turning the bricklayers from superintendents into mere artisans to be kept in check.[7]

It is thus as part of the process of appropriation of standards by communes that I interpret their insistence in requiring all measurements to carry the communal seal. In Parma, rods and scales were stamped with a seal depicting the Virgin, and in Mantua one depicting Virgil.[8] (The commune often tried to identify with respected figures of the community's cultural landscape.) Such stamping conferred authority upon the physical tools. When Emperor Frederick II took upon himself the role of reordering measurements throughout the south of Italy, he requested in the Constitutions of Melfi of 1231 that all standards be stamped with the symbol of the royal court.[9] And when commenting on Frederick II's prescriptions, the jurist Andrea d'Isernia (c. 1230–1316) concluded that violators should be punished as offenders of royal rights, thus collapsing the distance between standards and royal authority, and merging cause with effect.[10] Indeed, it is often difficult to separate the authority of the measures from that of their rightful owners or their guards. Measurements worked as power conductors, distributing power through all those involved in maintaining them in existence. They often appeared to be self-sufficient sources of power, and it is as a testament to this effect that the Visconti family of Milan included the *sextarius* for grains on their banner, until they ceded its right to the commune in 1225.[11]

The powerful agency of measures is well exemplified in the standard for grain that the commune of Venice constructed in 1263, the earliest such standard to have survived to our time (plate 4). Emblazoned with the symbol of the lion, it carries the inscription "ANNO D[OMI]NI MCCLXII M[EN]SE FEB[RUARII] + T[EM]P[OR]E D[OMI]NI N[OST]RI RANIERII," which means made "in February of the year of the Lord 1262, at the time of our doge Ranieri [Zeno]."[12] (I must add, though, that in this case "February 1262" means "February 1263" by modern reckoning, as Venice then started the new year in March.) The use of Latin—that is, the language of notaries—is in direct continuity with the statutes, which scheduled measurement checks during winter months and considered their accuracy to be their governors' personal responsibility. The Venetian standard, then, is not just an execution of the statutes but their physical equivalent: it reproduces their logic and at the same time validates them, enacting a circular process on which the authority of the statutes and standards rested. The statutes, after all, circulated among the population in oral form. They were shouted regularly at crossroads by communal criers, and publicly sworn in guild rituals. But as these means were fleeting (only the guild's top administrators and the main notaries of a commune treasured written copies), so the standards reproduced, on their surface, the words that legitimized their existence.

This necessity explains why Rome's *rugitella*, a standard for grains carved out of an antique funerary stone (figure 12.1), included the symbols not only of the commune but also of the *banderesi*, the fourteenth-century patrol that guarded them.[13] This standard, together with another for lime that is now lost, was at the feet of the Capitoline hill.[14] The presence of both standards, as well as that of public scales and weights, turned the area into the city's metrological depot. There, in fact, also stood the Temple of Saturn, which is where the Romans preserved the official steelyard. When a Christian church took over the temple, it did not suppress the area's identity: in the 1140s it was called San Salvator de Statera—that is, "San Salvatore of the Steelyard."[15] In a way, it was difficult to weaken such an association, given that the area served for the public weighing of the goods sold not only in medieval Rome but also in the towns under Rome.[16] In the 1560s, the *rugitella* and the lime measure were under the porch of the Palazzo dei Conservatori, where the Venetian printer Aldo Manuzio (the Younger) recorded them adjacent to the third and seventh columns.[17] They were still there in 1573, when the jurist Luca Peto installed a slab with the linear standards Feliciano Scarpellini would measure with an enormous pair of dividers in 1811, as we saw in chapter 4.[18]

The receding of Rome's standards from a public street to the facade

12.1 | Rome's volume standard for grain, the *rugitella de grano*. Tabularium, Musei Capitolini, Rome. Carved out of an ancient urn, the so-called Agrippina's Sepulcher, it projects authority through numerous coats of arms, including that of the *banderesi*, the two chiefs of the *militia* pictured on both sides of the SPQR sign. © Sovrintendenza Capitolina ai Beni Culturali, Roma. Photograph by the author.

of the communal palace indicates that the commune appropriated measurements not just symbolically and socially, through stamping and naming, but also spatially and architecturally. Many communal palaces built the *pietre* into their walls. One of the earliest examples comes from Bologna, where the standards were documented under the loggia of its communal palace in 1250.[19] Other cities followed suit, even if the documentations register the *pietre* on municipal headquarters at much later dates: the 1330s in Monza, 1348 in Assisi, and 1377 in Todi.[20] There the standards appear around the main door, framing the entrances and exits of the *podestà*, who lived there, as well as those of the five hundred members of the general council, who assembled in the main hall.[21]

While cathedrals had been the most common location for the *pietre* installed around the time of the Peace of Constance, communal palaces became the go-to location for later adopters.[22] Such a shift manifested itself only when communes managed to take full control of the standards, to the detriment of the guilds, whose energies they absorbed.

The takeover of communes was, however, rarely absolute, and in many cases the available documentation registers an uneven situation in which each category of standards followed different trajectories, in line with the factions interested in them. For instance, at the beginning of the fourteenth century the Florentine *canna*, on which the commune relied to measure cloth, was kept by the merchants of the Arte di Calimala. The commune invested the Calimala members as those in charge of the standard, but it is clear that the power relationship was reversed, as the city's statutes reproduce word for word the instructions of the guild.[23] Padua also offers an interesting case. There, the standards appear on the communal palace in 1277, almost sixty years after the commune undertook the construction of the building, the so-called Palazzo della Ragione (figure 12.2).[24] The statutes prescribed the use of communal standards from 1265, yet the situation must have been rather chaotic through the following decade, as cloth and leather soles were still sold in Venetian units.[25] (A similar directive existed in Verona, whose leather soles also matched the Venetian shapes.)[26] In 1276, under the *podestà* Guidone de Robertis, Padua energetically revamped its weaving industry.[27] Artisans were told to fabri-

12.2 | The measurement standards on Padua's Palazzo della Ragione, near the staircase of the birds (*osei*): the roof tile (left), the crossed bread shape (center), and the brick (right). At the bottom right, the metal plaque marks the upper end of a broken length standard, either the *brazzolaro* (the length for cloth) or the *piede di fabbrica*. Photograph by the author.

cate their cloth according to new linear standards and using a specific quantity of threads. (They were also to dye the cloth using precise quantities of woad: an indigo pigment extracted from the leaves of a Mediterranean crucifer; at least twenty *libre* for large cloths and ten for small ones). Such a shift was meant to transform Padua into a commanding production and trading center, a goal incentivized by rejecting the Venetian standards and instituting new, local ones. It is not by chance that in the following year, 1277, the *pietre* were incised on the facade of the communal palace.

When the *pietre* appear on public buildings, they are less markers of rights of possession than tools by which governments constructed a new idea of the city, pivoting it on notions of industrial productivity. While the post-Constance cathedrals seem to have carried only one standard—this is what the documents say—the communal palaces presented whole sets of them, ranging from cloth lengths to brick sizes and bread shapes. Their multitude can be taken as indicative of the popularization of the commune: the fusing of its agenda with that of the artisanal and industrial institutions. Back then, the *pietre* were not merely representing change, but were the very instruments of it.

In 1230, when the commune of Bologna offered foreign weavers privileged conditions to move into town in an attempt to boost its cloth industry, the majority of settlers (at least sixty-three masters) came from Verona. To facilitate trade, the commune allowed them to keep working according to their standards, and forced the same rules also upon settlers from other cities such as Prato and Florence. Not only did the commune remodel looms and reeds around Verona's customs, but it also adopted the Veronese *pertica* for measuring cloth.[28] The mercantile *braccio* that you can still see today on the dado of Bologna's communal palace measures around sixty-four centimetres (plate 1), a mere four millimeters shorter than Verona's (see figure 13.1).[29]

13

Disciplining Standards

It may be difficult to understand the important status that measurements occupied in medieval life, but the process is made a bit easier by looking at the *Allegory of Good Government*, painted by Ambrogio Lorenzetti in Siena's communal palace (plate 5). There, Lorenzetti presents standards as the premise of politics and social life. Painted in the very room where the city's rulers met and deliberated, the fresco constructs government as a holistic process whose various components cannot operate effectively unless they intertwine practices and develop knowledge of one another. Linking elements through ropes and spatial proximity, the image belongs to the tradition of medieval diagrams. And, like diagrams, it has a beginning and an end. If I start reading the image from the top left, as art historians encourage me to do, I encounter the angel of Wisdom, principle and inspiration of any good government. She clutches the book of knowledge and holds a large scale stabilized by Justice, who sits underneath. Notice that Justice exerts no pressure, but gently presses her thumbs against the inner sides of the plates, producing constancy by means of stilling rather than holding. Lorenzetti depicts her as a mountainous woman who is hard to budge, perhaps alluding to the *Corpus iuris civilis* definition of justice as "persistent and perpetual will" (*iustitia est constans et perpetua voluntas*).[1]

On the plates of the scales, left and right, I see the personifications of the two subcategories of Justice.[2] Wearing violent red is

Distributive Justice, who beheads an armed culprit and crowns an honest man (he holds the palm of virtue). In white is Commutative Justice, who hands down not what many art historians have mistaken as a strongbox, a spear, and a torch, but Siena's standards of measurements: the *staio,* the *passetto,* and the longer *canna* of six *braccia* (plate 6).[3] While Distributive Justice gives people back according to their actions, Commutative Justice provides the tools against deception. These definitions are not mine; they come from Domenico Cavalca, a preacher who was very famous in Tuscany by 1339, when Lorenzetti completed the fresco.[4]

The misidentification of the tools in Lorenzetti's *Good Government* is indicative of the general lack of attention to measurements and has led to a number of speculations, such as whether the damaged labels floating above the two justices were added later and should be switched.[5] But the doubts are unfounded, as Lorenzetti painted the standards with utmost precision. He turned the *staio* upside down, so to make the observers realize that the man in green is holding an empty cylinder, like the Venetian standard (plate 4), and he has rendered it in iron in accordance with the statutes, which forbade making it in copper.[6] Lorenzetti also painted the extremities of both the *passetto* and the *canna* in iron realistically, as the statutes prescribed.[7] (This, by the way, is valid also for many of Italy's *pietre,* and today it is still common to find rusty bits at their extremities, or the greenish ionized stains metal leaves on stone.)

Here Lorenzetti not only is depicting communal tools, but is also attentive that the two forms of justice look equivalent in every respect. He places the *staio* in the front so to align it with the round crown on the opposite side. And he keeps the two iron rods in the back, to parallel them with the two swords (the angel's and the culprit's lying on the ground). In other words, he uses formal analogies to present measurements as the visual doubles of weapons, the traditional tools of power. In his chronicle, Giovanni Sercambi pointed out that after Florence and Lucca conquered Pisa in 1276, they demanded that the city should assume their standards.[8] And the constitution of Siena of 1309-10 stated that the *podestà* had to make sure that all the castles and towns defeated by Siena employed its standards.[9] Standards thus were weapons for coercing the population as well as the ultimate symbols of territorial control. But, unlike weapons, they did not hit citizens from above but were endorsed from below, an operative shift that cast measurements as the population's supplements and counterparts, as Lorenzetti also reveals.[10] The two men handling the standards are, after all, only half kneeling. Contrary to those who passively accept the actions of Distributive Justice, they hold and support the standards, taking an active role in the exchange. We are left unsure whether the angel in white is donating the standards or receiving them. The moment is suspended in the act of balance.

Wearing the iridescent robes and the hats typical of wealthy merchants, those two men are members of the Mercanzia, which the Sienese government entrusted with the official standards of measurement since 1310.[11] Having grown dominant in the early fourteenth century thanks to international trading routes and the contributions of smaller arts, the Mercanzia developed its judicial court and prospered as the key economic institution. This is true for Siena as well as elsewhere. In Piacenza in 1312, the local mercantile guild preserved all the size standards. And in Milan in 1330, the weighing of wool and related issues fell to six officials elected by the association of its *mercatores*.[12]

Communes benefited from mercantile consortiums, the only institutions that, aside from championing metric stability, had the necessary economic capital to carry out the checks. Earlier, governments had handled the control of measurements by appointing officials who performed directly under the *podestà*. These went under different names: *iusticeri* in Venice, *proveditores* in Volterra, *scarii* in Bologna.[13] They checked the standards of both manufacturers and shopkeepers. In some cities their duties ranged more widely. The Bolognese *scarii*, for instance, also controlled the hygiene of the streets, the flow of canals, and the right to carry weapons, a variety of tasks that were eventually redistributed among other officials. (Mentioned in the statutes of 1245, the *scarii* disappeared by 1288.)[14] Finally, they fined counterfeiters and arranged for their public derision—a punishment particularly dreaded by merchants, who feared the loss of credibility.[15]

Punishments could be rather dramatic. In the 1220s the Visconti family, who supervised the size of bread in Milan, stripped the bakers of their clothes and whipped them in the streets if they did not

13.1 | The *capitello* in Verona's Piazza delle Erbe. The *pietre di paragone* are carved on the shafts of the columns and on the steps of the canopy. Photographs by the author.

respect the standard. Thirty years later, the citizens stopped what they thought too barbaric a punishment by paying two *grossi imperiali* instead.[16] Whipping seems to have been fairly common, as it was also the punishment prescribed by Frederick II's constitutions for whoever employed wrong standards. The disciplining took place on the spot where the crime had been committed. If caught again, the abuser would have one of his hands cut off; and if found a third time, he would be hanged.[17] In Lorenzetti's Siena, sellers were fined twenty-five *soldi* (that is, a quarter of what they paid per year to rent a market stand) and their tools were burned in front of everyone.[18] In late-fourteenth-century Verona, the culprit was chained to the columns of a stone ciborium in the middle of the vegetable square between the third and ninth hours—that is, during the opening hours of the market.[19] Known as *capitello*, the ciborium stood on *pietre di paragone* (figure 13.1), as the local standards were carved vertically along the posts, under abbreviated titles: the PER[*tica*], the PAS[*sus*], and the BRAS (for *braccio*). On the steps were, and still are, the molds for the brick and the tile. Caged in the *capitello*, the culprit was then physically framed by the *pietre*, which served—quite literally—as the pillars of justice. It was, after all, under the same ciborium that the *podestà* was publicly appointed, his investiture visually casting him as the moral antithesis of the deceiving merchant.

Punishing sellers, inspecting scales, and collecting fines were time-consuming and unpopular activities, and the commune outsourced them. In Reggio, the revenues on the *mina* for grains and all other measures were on sale each year for both the city and the surrounding district.[20] And before Venice created the *iusticierii* in 1173, the fees of its Rialto market (which included those collected by checking and stamping standards) were given to a group of nobles in repayment for a loan.[21] The right over the same standards in the Venetian market of Constantinople, where the Venetians traded with the rest of the eastern Mediterranean, was instead in the hands of the patriarch of Grado. In 1169 he granted it to Romano Mairano in exchange for five hundred Veronese *lire* per year. A merchant who had become rich selling iron to the Templars in the Holy Land, Mairano struck what seemed like a good deal because of the political tension between Venice and the Byzantine Emperor Manuel I Komnenos, which slowed down commerce. Mairano would recoup his money the moment trading resumed and he could pocket the money for thousands of checks. It did not resume. In 1171, Manuel I accused the Venetians of having started the fire of Galata, the Genoese colony separated from Constantinople by the Bosphorus, and expelled them from his empire, seizing all their possessions. It took Mairano twelve years to pay back his debt.[22]

If I have lingered for a moment upon the handling of Venice's measurements, it has been less to reveal the interconnections between local and international politics than to show the variety of measurement holders. Communes insisted on change. Statutes specified that size checkers could not stay in office for longer than one year, but packed in instructions about how to measure so as to smooth the transition from one appointee to the next. In Pistoia, the copper *quartini* for grains, which were brought to market days from the sacristy of the church of San Jacopo, were to be hit twice with one's hands or feet. They could also be leveled with a scraper, a *rasiera*, but only once.[23] In Florence, the grains had to be distributed thoroughly inside the container, so the statutes recommended that the container should be beaten with both hands.[24] (The grains had to be piled up on some occasions, though, thus showing that the normative was far from being straightforward.)[25] The *podestà* swore to have this information shouted throughout the city every month, to ensure that it was known by everyone.[26] There seems to have been no difference between general products and professional knowledge, and the statutes of the *podestà* of 1324 also include instructions on how to measure cloth.[27] Cloth cutters and measurers had to hold the end of the fabric against the extremity of the standard and make it adhere up to a quarter of a *braccio*, after which they loosened the fabric, but only up to the last eighth of the rod, when they stretched the fabric against the standard again. When measuring a cloth in Reggio, the measurer had to place the roll on a large table, making sure that its borders would not fall outside it, and then, without pulling the cloth, put the *passus* on the fabric rather than next to it, rotating the standard by pivoting on his thumb.[28]

This specification was added to the 1262 statutes at a second moment. It is indeed written in the margins, which notaries used to add corrections to the main text and offer clarifications. Norms were canceled out and reformulated, transforming the gestures and timings of practices that were but straightforward acts of quantification. In Bassano in 1259, salt sellers had to swear to each customer that they would measure without deceit. The pledge must have been deemed essential, as seven years later it was also extended to wine sellers.[29] And in Perugia, all contracts were to include a sentence that confirmed that the standards used for measuring carried inscriptions saying that they conformed to the *pietre*.[30]

Those validations were indicative of a society that was testing processes and solutions to elicit trust. I know of no other period in Italian history in which measuring was approached as a question about process—not even the nineteenth century, when millions of mystified Italians spent months adapting their eyes and hands to the divisions of the metric system. The careful wording of statutory norms of the thirteenth century releases a sense that people objected to some ways of measuring, questioned the ownership of standards, and doubted the reliability of officials.

Consider, for instance, the 1305 description of the routine of the twelve measurers of Pisa who, each year, were selected from the members of the Mercanzia.[31] Every morning they met in their hall and divided themselves into four groups; each group took a *canna*. They then went to the entrance of the San Michele church, which served as the workplace of the *camerario*, the communal treasurer. And there they waited in front of the gate, facing the tower of the Buttari family, anticipating being summoned by customers.[32] The specification not only of their location but also of the direction they were to face reveals the commune's degree of attention. As officeholders rotated through these appointments and were approached by foreigners, their physical orientation was critical to their identification. The officials were the space they occupied and the gestures they performed, successful at enforcing the disciplinary process of measurements because they themselves were the very object of it.[33]

The Pisan measurers could not leave their spot or get distracted. They could not play chess, drink with any buyer or seller, or accept any form of payment that did not match the fees set by the commune. (The measurers did not lose their share of earnings in case of absence, which was covered up to twelve days a year.) And when they measured, all three did so together without no disruption, as pauses could lead to errors.[34] The quantity of the measurers and the brevity of their mandates were ways to reproduce plurality and prevent an individual from taking control of the business of measurement.[35] It was not a form of objectivity, but a way to reduce the fluctuations caused by the fundamental arbitrariness of measuring. It was, in fact the opposite of objectivity, as the combination of practices led to a multiplication of possibilities that eventually eroded the very idea that one single principle was always at work when measuring.

1 | The measurement standards of Bologna. Bologna, Palazzo d'Accursio. Photograph by the author.

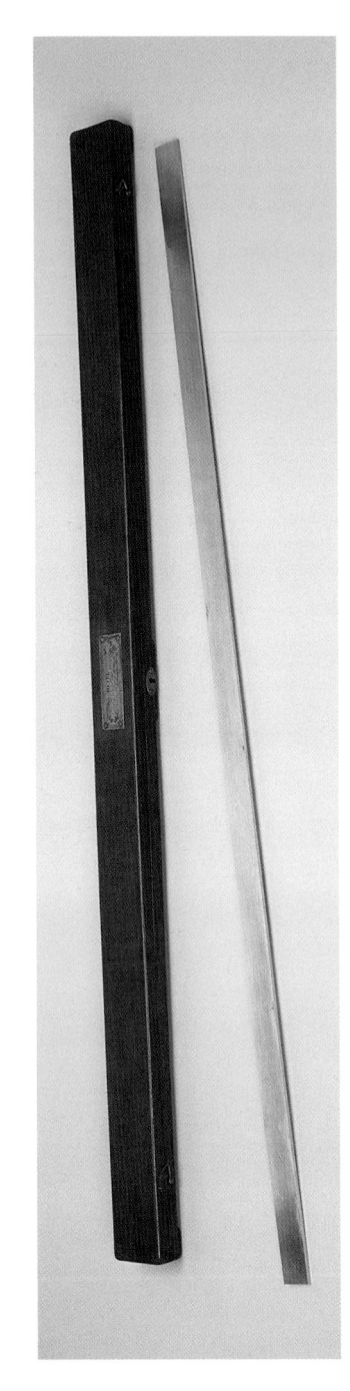

2 | The platinum meter (1799). Photograph © Archives nationales, Paris.

3 | Reggio Emilia's Baptistery of San Giovanni (a), which carries on its left pillar (b) the local standards of length. These are the long *pertica* for measuring land (left) and the short *braccio* for cloth (right), whose top edge is marked by a metal plaque (c). Photographs by the author.

4 | The Venetian *staio* (1263), the earliest standard of volume that has survived in Italy. Archivio di Stato, Venice (catalogue no. 97.63). Photograph courtesy of Ministero dei beni e delle attività culturali e del turismo.

5 | Ambrogio Lorenzetti, *Allegory of Good Government* (1338–39). Palazzo Pubblico, Siena. Photograph © Comune di Siena.

6 | Ambrogio Lorenzetti, *Allegory of Good Government* (1338–39), detail. Commutative Justice gives three measurement standards to two merchants. Photograph © Comune di Siena.

7 | Monk and *vir religiosus* Guido as treasurer (*camerlengo*) on the cover of the financial register of the commune of Siena (1280). Archivio di Stato, Siena. Photograph courtesy of Ministero dei beni e delle attività culturali e del turismo.

8 | The *mensura Christi* oration, in *Supplicationes variae*, a manuscript made in Genoa (ca. 1293). Ms. Plut. 25.3, f. 15v, Biblioteca Medicea Laurenziana, Florence. Photograph courtesy of Ministero dei beni e delle attività culturali e del turismo.

9 | The *mensura Christi* oration (late fourteenth century). Ms 1294, f. 103r, Biblioteca Riccardiana, Florence. Photograph courtesy of Ministero dei beni e delle attività culturali e del turismo.

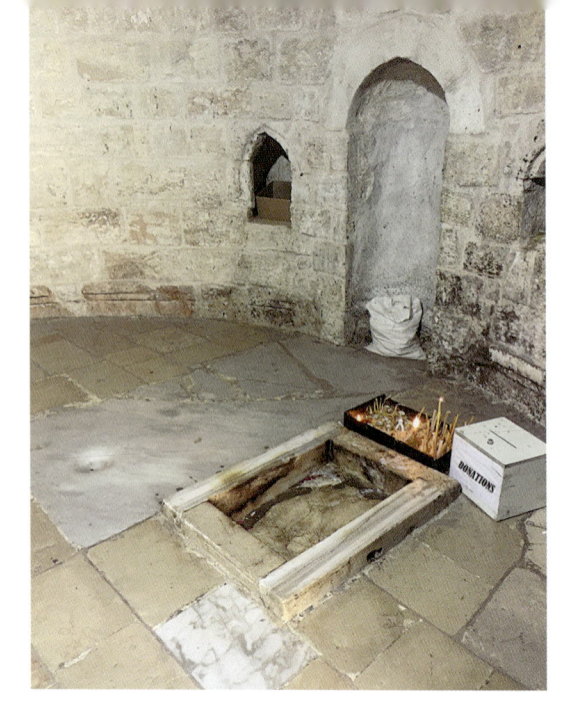

10 | The stone that allegedly bears impressions of Christ's feet before the Ascension. Chapel of the Ascension, Jerusalem. Photographs by the author.

11 | The *mensura Christi* canopy (fifteenth century). Cloister of St. John Lateran, Rome. Photograph by the author.

12 | A ribbon reproducing the height of Saint Rose (eighteenth century). Archivio del Santuario di Santa Rosa, Viterbo. Photograph by the author.

13 | The cluster of churches called Santo Stefano, Bologna. In the center of the picture is the octagonal "replica" of Jerusalem's Holy Sepulcher. Photograph by the author.

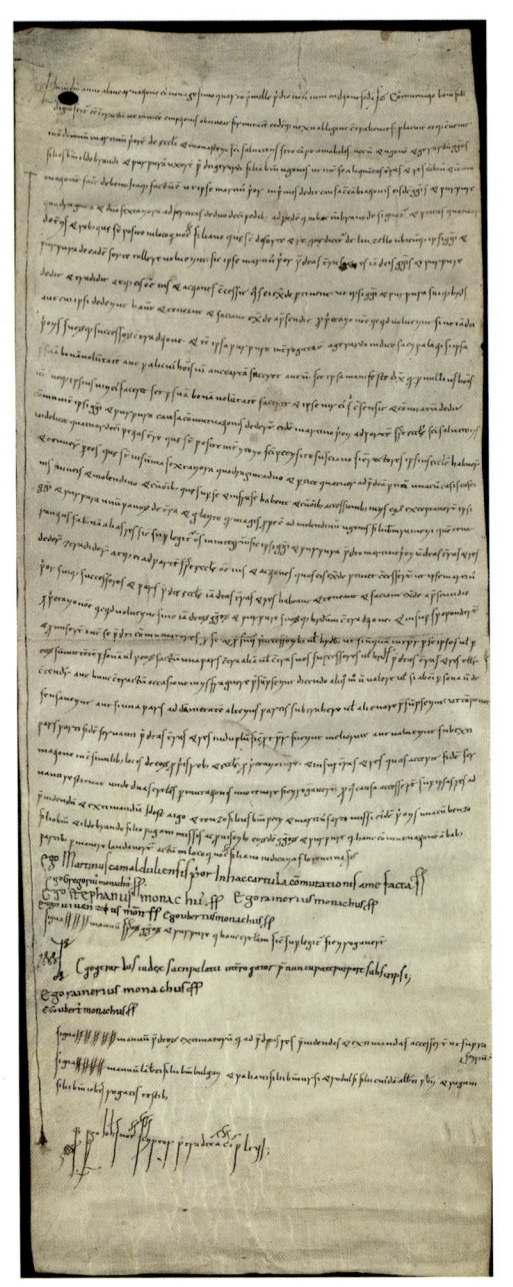

14 | The vertical line running along the left margin of the document is the foot of the San Pancrazio gate in Florence (1094). Diplomatico Normali 83, f. 33r, Archivio di Stato, Florence. Photograph courtesy of Ministero dei beni e delle attività culturali e del turismo.

15 | An irregular territory homogenized by the centuriation reticulum. From the so-called Codex Arcerianus (sixth century). Cod. Guelf. 36.23 Aug. 2°, f. 16v, Herzog-August-Bibliothek, Wolfenbüttel. Photograph © Herzog-August-Bibliothek, Wolfenbüttel.

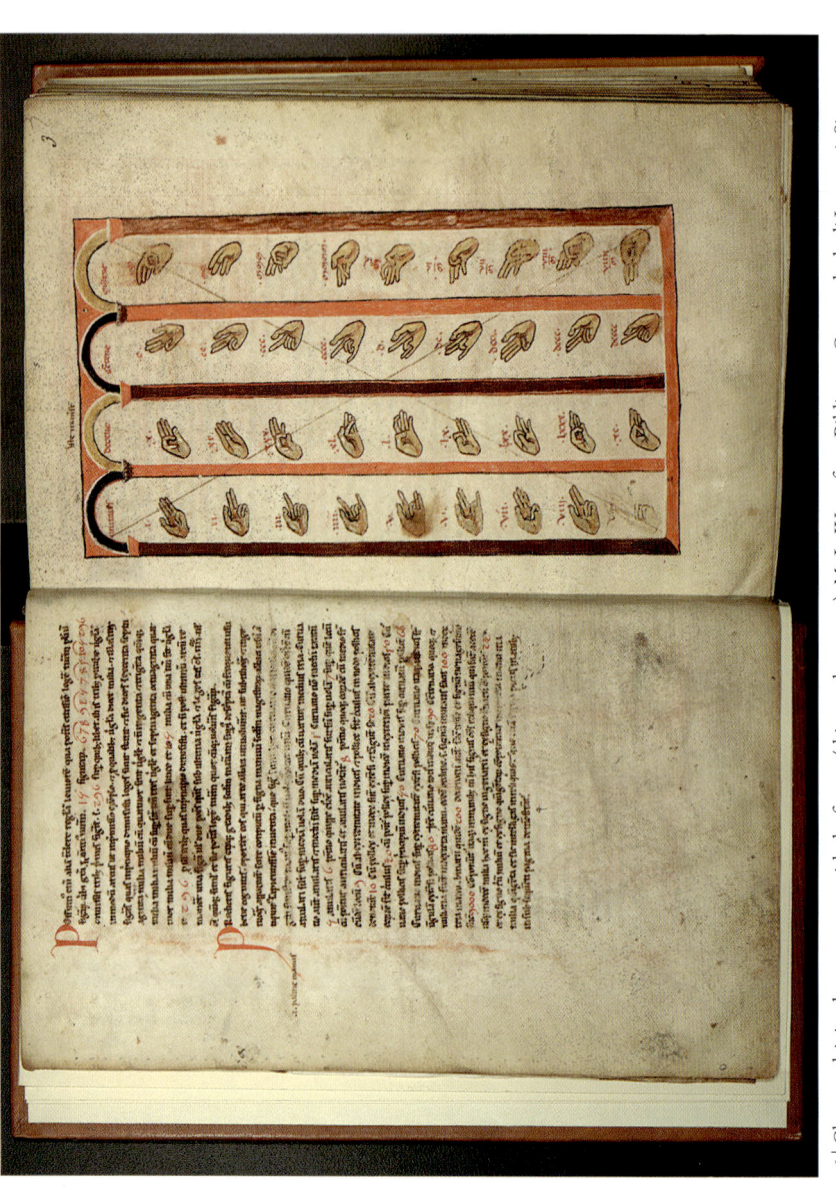

16 | Chart explaining how to count with the fingers (thirteenth century). Ms L. IV 20, f. 3r, Biblioteca Comunale degli Intronati, Siena. Photograph © Biblioteca Comunale degli Intronati di Siena (23.02.2018).

14

The Politics of Measurement

The statutes of the Pisan measurers were bound together with those for the *sensales*, all-around mediators who, identified by the ringlet they wore, could be approached to give opinions on products, match sellers with buyers, and more generally keep an eye on market performance. While their activities were limited to consulting, their fields of operation were those of the measurers, producing a fluidity between the two categories that reveals of the closeness between measuring and judgment.[1] Indeed, in medieval times measuring was a *stima*, an estimation. Size was not gauged objectively in the modern, scientific way, but assessed.[2] As two rods hardly ever matched—their ends were chipped, their surfaces eroded, their wooden bodies bent more than surviving records care to admit—there was always a discrepancy (an error, we would say today, drawing from the vocabulary of positivism), and it was up to the officials to decide, case by case, how to handle each comparison. In Volterra, the *providores* were required to fine bakers five *soldi* for each *uncia* that did not correspond to the standard weight of bread.[3] It is a clause as illuminating as it is rare, since it reveals a standard for precision, about which the statutes usually keep silent.[4] In 1371 the law of Lucca did not even specify the width of the streets, explaining that it was at the discretion of the street officer, the *officialis viarum*.[5] The omission served to accommodate changes over time. Still, the phrasing foregrounds the deep connections between measuring and power, especially since those officials—it is written—had "complete authority" (*omnen auctoritatem*) over them.[6]

Measuring folds politics into the real. Unlike counting, which is a form of discrete quantification, measuring pertains to continuous quantities. Such a dichotomy dates back to the Pythagoreans, but it was popularized through Aristotle's book on categories, one of his few to be available throughout the Middle Ages.[7] Aristotle presented counting and measuring as a pair, and it is as equivalents that they have entered common knowledge.[8] Yet, as Aristotle noticed, counting and measuring operate according to different regimes. Counting identifies a collection of discrete entities, such as the number of marbles in a bag. Their number cannot be questioned, and arithmetic is beyond doubt. But measuring deals with continuous entities that must be broken to be quantified. And this is what measuring provides: a rationale and a set of tools for rupturing something, like a line or a textile, without internal elements that can serve as dividers. "Mensura est id per quod cognoscitur quantitas rei," writes Giles of Rome, a pupil of Thomas Aquinas who rose to become one of the most prominent theologians of the early fourteenth century.[9] That is, "Measure is the thing by which we know the quantity of something"—a sentence that rests on so many indefinite elements (*id, quod, rei*) as to call into question whether measuring is a self-replicating activity. The opposite is in fact the case. Measuring relies on external agents who, so to speak, make the cut. While the objects to be counted are independent from each other before the counting takes place, measuring constructs the very units it pretends to simply record. This is why in his *Metaphysics*, the treatise in which he provides the most extensive description of measuring, Aristotle singles out the identification of the unit, which he calls "the one," as the very principle of measuring. Measuring is, first of all, the identification of the unit on which the practice depends. His words are telling: "For everywhere we seek as the measure something one and indivisible."[10] This point was evident to Uguccione da Pisa, who in his *Derivationes*, one of the go-to dictionaries of the Middle Ages, defines measuring through a mere list of units. Time splits into years, days, and hours; land divides into regions, territories, and fields, and then in *actus*, steps, and feet.[11] To measure is to know which units to search for, and the heterogeneity of Uguccione's list—in which

Roman units for surfaces (such as the *actus*) coexist with body parts (the foot) and vague spatial notions (the territory)—reveals the multiplicity of the categories that apply to measuring.

Uguccione's categories mix the abstract with the concrete. It is so because measuring is thought to move efficiently from one regime to the other even if the physical act of measuring—the translation of standards on cloth, for instance—can only approximate the idealized ratios that standards are supposed to embody. Such an approximation is what differentiates measuring from counting. While in counting the abstract and the concrete converge, the ways in which measuring rods move across the fabric are open to debate: they hide chasms, no matter how sharp the tools or attentive the measurer.

Mediating between the world of mathematics and the reality of the physical world, measuring necessarily involves a leap between objects and numbers, whose equivalence is made possible only by an oversight. The many oaths that the *podestà*, measurers, and sellers swore in front of buyers served to trigger customers' fleeting assurance that the translation from the world of ideas to the world of matter took place accurately.

So it is less measuring or measurement standards than their presiding authorities that constructed sameness. After all, in the Middle Ages the word *auctoritas* referred to someone who knew and reiterated preexistent solutions.[12] An authority was the person who erased the distance between an original and any of its subsequent reiterations, like the *pietre di paragone*, which physically conveyed the feeling that every user always faced the same deceptively simple incisions. Medieval standards carried *auctoritas* because they themselves were constantly remade, thus becoming instrumental in and homogeneous to the authoritative process they triggered.

Measuring is the promise of repetition: the same tool moved over and over again through space and time with a set of gestures carefully shielded not from errors but from unprecedented anomalies. The Pisan measurers returned every day to the same corner of the church of San Michele. They swore to the instructions that had been repeated for decades and swore to use only the official standard, forgetting that the official standard had existed through four official *canne*, and that the statutes did not describe how to measure but relied on the consensus of the measurers to solve any issue that arose. That such decisions defied simple solutions is evident when we move away from the assertive timbre of the statutes to the interrogatory tone of one of Franco Sacchetti's stories.[13] A fourteenth-century diplomat and sought-after *podestà*, Sacchetti had a consummate knowledge of measuring, and at first his tale of commercial trickery may seem merely to dramatize statutory norms. Still, the novella presents darker undertones, casting doubts over Sacchetti's faith in measuring and justifying a second, more attentive reading.

The story, which modern scholars identify as the ninety-second in Sacchetti's collection of three hundred *novelle*—is deceptively straightforward. A certain Soccebonel from Friuli stops by a Florentine *ritagliatore* to buy twelve *braccia* of cloth for a *cappa da barons*, a long mantle that in the fourteenth century was mostly worn by judges and lawyers.[14] It is an odd, extravagant choice that casts Soccebonel as the delusional simpleton connoted by his stupid name, especially as he wants the cloth in a princely sky blue, acting on a suggestion that the Florentine meant as a joke.[15] When the two get to measuring, Soccebonel clumsily tries to get more fabric for the agreed price by lifting it from the measuring rod and quickly pulling it, to which the seller responds by stretching the fabric under the cane to save cloth. Sacchetti casts the scene as a reversal of gestures: the *canna* disappears and reappears under and over the fabric, and the two men act as mirrored figures who pull in opposite directions while hiding from each other's eyes.[16] The cunning Florentine ends up having the upper hand, and, to conceal his wrongdoing, tells Soccebonel to soak the fabric overnight before having it pruned by the *cimatore*, as was customary after any purchase. It is indeed only after returning to the *cimatore* that Soccebonel realizes he has been tricked. Soaking the cloth has further reduced its size, leaving him with what is not even enough for a *saltamindosso*—a mantle too short to cover one's buttocks.[17]

Sacchetti's story follows the traditional conman-is-conned typology of cautionary tales. Its takeaway line, "Credendo ingannare, rimase ingannato," casts Soccebonel as the crook who got a well-deserved punishment. Yet the story is less about deception than about falling short. Soccebonel is repeatedly mocked for his lack of understanding of the Florentine idiom, or for not knowing how to shop.

Deficiency takes center stage especially in the numerous scenes of measuring, an action that never confirms previous results. Not only is Soccebonel's cloth measured in different ways at the moment of the purchase, but it also shrinks after drying, to such an extent that the *cimatore*, who is searching for it among other fabrics by size alone, can no longer find it. At a loss, the characters imagine that the fabric has been nicked at night or while drying. Their speculations are absurd to the reader, who knows the *ritagliatore* is to blame. Still, their dialogue also discloses the discrepancies between facts and belief, casting measuring as a practice that multiplies errors instead of reducing them. The novella thus mocks people's idiotic belief in measuring, while questioning its very capacity for reiteration. The dimensions of the cloth always differ not only from what they are thought to be, but also from what they ought to be.

Sacchetti's critique is particularly biting as it directly engages with the regulations of the clothing industry, which the story contravenes one by one.[18] Tailors were forbidden to sell fabric before soaking it, because of the risk of shrinking.[19] Weavers and dyers even had to record the quantity of thread employed before immersing the fabric in water.[20] And so common were the effects of humidity on measuring that they were recalled even in a generic dictionary such as Giovanni Balbi's *Catholicon*.[21] The contemporary *libri d'abbaco*, the schoolbooks for training merchants' sons, explained how to keep track of the dimensional shifts of cloth throughout its production.[22] Sacchetti's novella is a dramatization of one of their case studies, and yet it also exposes how their procedures fall short.

Falling short is, after all, not only the condition of Soccebonel, of his cloth, and of the industry as a whole, but also an effect of reading. Indeed, Sacchetti switches the standard of reference halfway through the story, presenting the dimensions of Soccebonel's piece in *canne* at the *ritagliatore*'s, and then in *braccia* at the *cimatore*'s, without ever explaining the ratio of one to the other. Reproducing the medieval custom by which each business had its measurement standards, Sacchetti exposes measuring as a labyrinthine path that leads Soccebonel "close to madness" (*fu per impazzarne*). The size of his *cappa*, which is supposed to be as large as the sky (*cappa del cielo*), ends up smaller than the vault of a tiny oven (*tornò che non avrebbe coperto un* cielo *d'un piccolo forno*). The reiteration of the word *cielo* to designate both the immensity of the sky and the vault of a domestic oven translates Soccebonel's deception at the linguistic level (he does get a "*cielo*" after all). It also casts his adventure as a moralistic tale that ironically spans from heaven to hell. Soccebonel, after all, intends to decorate his cape "with the sun, the moon, the stars, and perhaps even a large portion of the Paradise" (*che vi fosse il sole e la luna, e le stelle, e forse gran parte del Paradiso*), but the underhanded tricks of measuring lead him to perdition.

The wordplay on *cielo* opens the possibility that Sacchetti may here be engaging with the sermons of Giordano da Pisa, one of the most influential Dominican preachers of fourteenth-century Tuscany. In a homily he gave in the square of Florence's Santa Maria Novella on March 13, 1306, he attacked people's obsession with money, insouciant as they were of "God's wealth" (*ricchezza di Dio*). To strengthen his point he conjured up the fool (*matto*) who deluded himself that he could buy the sky (*che tu non ne possi comperare il cielo*), a spot-on definition of Soccebonel's character.[23] Sacchetti's novella has a religious ring to it. Indeed, he assimilated preachers' warnings, as well as their commentaries to the Deuteronomic prohibition against operating with two standards.[24] The Franciscan Pierre de Jean Olivi, who lectured in Florence's church of Santa Croce around 1288, elaborated on this specific biblical passage when he called mortal sinners those who employ one standard for selling and one for buying.[25] And in 1304, Giordano da Pisa publicly attacked merchants who "pull the fabrics and squeeze the bowels out of their body."[26] The metaphor for people's financial strangling echoed Emperor Frederick II's Constitutions, which punished merchants who stretched fabrics beyond their natural dimensions.[27] Sacchetti's decision to pivot his tale around an episode of cloth-pulling engaged with those warnings. Yet it also opened up questions about Soccebonel's final position. The traditional reading of the story presents him as guileless.[28] A reading of the novella through Giordano da Pisa's sermons or mercantile statutes would make him a fool who wants to buy a *cielo*. But with the expanding of the interpretative framework, both options seem less and less likely, and Soccebonel emerges as a victim of the market's lawlessness.

15

Measurements and the People

Though it recounts a dispute, Sacchetti's novella features no professional measurer. Medieval statutes would have us believe that all customers turned to officials when in difficulty; yet the tale of Soccebonel implies that the opposite may have been true: measurement checkers were just not that common, especially in retail trade. It is as a way to compensate for their absence that the communal regulations insist that all measuring take place under the sun, whether it happened in a private transaction or at a guild.[1] The statutes of Venice recall that the *fustagnai* gathered in the square of St. Mark's once a month to measure their pieces of cloth (*cavecios*) against the communal length that was brought in by the guild's *gastaldi*, the supervisors of the whole process.[2] And in Florence, the tailors who wanted to cut tunics out of new cloth, rather than recycling previous garments, measured the material "in the presence of two officials as appointed by the rectors and the other members of the art, after which it had to be soaked and measured again."[3]

Magistrates for checking measurements had long been in place. Collegia for corn measurers are known to have existed in Roman times, while Byzantine weights and length standards were inspected by *bullotai*, who helped fiscal inspectors of higher rank, the *episkeptetai* and the *epoptai*.[4] Yet, for everyday transactions, the medieval communes broke with either tradition by relying not only on officials but on a collective system of cross-checking that, instead of radiating from single authorities, allowed any citizen to call any other to order. Parma's statutes repeat continuously, almost to excess, that "any citizen could accuse any counterfeiter" (*quilibet possit accusare contrafacientes*).[5] If proved right, the informer was rewarded with half the fine paid by the offender, an incentive also offered by many other communes. The norm returns in Bologna, Perugia, and Florence, to mention just a few.[6] In Padua the fine of sixty *soldi* was divided between the accuser and the commune. "The plaintiff could be anyone" (*quilibet possit accusare*), read the statutes, even if the expression should not be taken literally, as "anyone" in communal Italy mostly meant "any fully-integrated citizen."[7] The statutes of Bassano even specify that "the accuser had to be a man of good reputation" (*quilibet bone fame possit manifestare*), a designation that was probably implied in other centers.[8] Still, even if the people charged with supervision represented only a portion of the population, the increase in participation opened many eyes.[9] By relying on diffused watchfulness, the communes achieved something close to constant vigilance.

This move seems to have been implemented mostly in the second half of the thirteenth century. While the Paduan statutes of the 1230s relied on the *iusticierii*, by 1269 they included the "quilibet possit accusare" expression.[10] Statutes are notoriously problematic for tracking change. As each version absorbed previous drafts, it is often impossible to date the precise origin of each norm. Nonetheless, the broadening of checkers seems to have occurred together with the opening up of the judicial system to larger sectors of the population.[11] Historians have remarked that in the second half of the thirteenth century, Italian cities witnessed an explosion of trials. Bologna had some three thousand a year.[12] Perugia may have had even more.[13] Modern scholars debate whether such an unmanageable quantity triggered reforms to the judicial system or was an effect of them. The division between civic and criminal courts, the systematic recording of acts, and also the establishment of archives (which helps explain the abundance of documents) were all implemented as part of this revision.[14]

As a way to reduce the number of trials, however, jurists also developed fast-track procedures for such minor crimes as deceitful measuring. It was agreed that accusing an offender required only one informant rather than the two or three witnesses prescribed by Roman law; wrongdoing, after all, could itself be demonstrated by measuring. Fines were thus assigned to those who did not respect the allocated locations for business and what today we would call breaches of quality standards. If the leather of a boot sold in Padua was not perfectly smooth and still had some hair, the accuser could get thirty *soldi*; the same amount for notifying the authorities that its sole did not conform to the correct shape.[15] And to do so—that is, to know the shoe sizes as well as the depth of the local *staio*, the diameter of bread, or the length of rods for linen—the citizens needed *pietre di paragone*. If the display of multiple standards was an effect of the shaping of the communal agenda around the needs of artisanal guilds, the judicial reforms of the second half of the thirteenth century made the *pietre* a necessity.

Please notice that the commune set up hefty fees. With thirty *soldi* a Paduan accuser could buy fifteen *gonnelle* or six mantles bordered with squirrel—that is, the wardrobe of a lord.[16] No surprise, then, that people kept their eyes wide open. As a typical example of such alertness, the chronicler Salimbene de Adam describes Guidolino d'Enzola, a busybody who turns the main square of Parma into his sitting room. "And every day he heard Mass in the cathedral and, when he could, the Divine Offices. He then sat with his neighbors under the community portico near the bishop's palace. He spoke of God, and gladly listened to others speaking of Him. He also used to stop the boys from damaging the bas-reliefs and the frescoes of the baptistery and the cathedral. . . . He acted as though he were an official custodian."[17] While Guidolino was not specifically guarding the *pietre*, their placement in the main square—the pulsating heart of a city's public life and the crowded site of happenings—created the ideal conditions for secure monitoring. Located on the main facades of important buildings or on elevated structures, the *pietre* facilitated citizens' visual patrolling.[18] Whole exposure was a form of control as effective as total concealment; citizens embraced this diffuse public monitoring as an inevitable, almost natural consequence of communal living.

The increasing importance of the population for measurements transformed the perception of standards in turn. At the end of the thirteenth century the friar Olivi, whom we encountered in the previous chapter, reasoned that one merchant's corrupted standard was not on the same level as a corrupted standard employed by every citizen, since only use by people determines the acceptability of standards.[19] It is an extraordinary comment—I know of nobody else who made such a claim—as it recognizes the rationale of measurements not in political authority, but in their social use.

Olivi composed his *Quodlibet*, in which he elaborated on this idea, in the late 1280s or perhaps a few years into the 1290s, when he was visiting lecturer at the Universities of Narbonne and Montpellier.[20] By then he was considered an unorthodox thinker. His sermons, most of which attracted criticism for attacking magnates, had been censored in 1283, and his books burned after his death in 1298. He was partially rehabilitated only with the Council of Vienne, which ran between October 1311 and the summer of 1312. Still, despite the bans, or perhaps precisely because of them, his ideas circulated. Francesco della Marca, the lecturer in theology at the papal court in Avignon, read his work.[21] Bernardino of Siena had Olivi's reflection on measurements and weights copied in a volume—today held in Siena's Biblioteca degli Intronati—on which he added his own comments.[22] Written in the tiniest minuscule, so to pack a lot of text in what was meant to be a traveling library, the hand-sized volume contained key works by Albert the Great, Saint Augustine, and Saint Bonaventure—a list of names that alone witnesses to the esteem Bernardino had for Olivi's *Quodlibet*.

Bernardino learned Olivi's ideas about measurements by heart.[23] While preaching in Siena's Piazza del Campo in 1427, he intertwined Olivi's ideas with those of Giordano da Pisa, urging his audience to remain skeptical of vendors, for "one stretches the fabric over the rod and one uses two standards, one for selling and one for buying."[24] The Piazza del Campo, then also Siena's market square, was the perfect venue for such a speech, as the local linear standard, the *passetto*, was incised along its perimeter three times.[25] Having not one but three *pietre* was anomalous in Quattrocento Italy, but then Siena was hardly a conformist city.[26] It never relied on citizens and diffused monitoring. For checking standards, the commune instead appointed market custodians and secret agents, who—the statutes make clear—could blame anyone, even without proof.[27] Bernardino's urging anyone to keep their eyes open aroused that sentiment of panopticism which the *pietre di paragone* precipitated. As Bernardino spoke in front of the palace where Lorenzetti had painted the allegory of good government (plate 5), to which he publicly referred, his words buttressed those of the commune, providing social and moral direction.[28] But then such an involvement in civic affairs was called for, given that the government had relied on friars as measurement checkers since the middle of the thirteenth century, as the next part of this book will illustrate.

PART III

Cities

16

Divine Measures

If I have refrained from speaking of friars and mendicant orders' role in measuring until now, it is because the relationship of religion and measuring is so particular that it requires a reflection of its own. Religious orders played a crucial role in managing measurements in thirteenth- and fourteenth-century Italy. Through practice, they also developed important ideas that stuck to the objects. Members of the Franciscan and Humiliate orders, in particular, weighed goods and collected tolls at gates while Dominicans preached and wrote against mercantile misconduct. Together, they shaped notions about measurements for many centuries to come.

Much of this activity was on behalf of the commune, which aligned religious and lay officials. After specifying that the measurers of flour and grain should be chosen from among the Penitents, the 1255 statutes of Parma order that they ought to work as "thoroughly as their secular predecessors" (*integraliter sicuti faciebant layci*).[1] Religious and secular measurers alternated, in and out of office. So in a way, their appointment changed little, if at all, from the institutional practices I have reconstructed. Measurers and measurement checkers depended on the *podestà*. Official tools remained possessions of the commune even when kept in churches and monasteries. The story of religious orders and measurements could easily have been included in part 2 of this book.

So to separate friars from the rest of the city is, to a large extent, an arbitrary choice.[2] It is a risky move, too, as it creates the idea that friars had something that the other citizens did not have. Many historians believe in their alterity, and register their appointment as an oddity in need of an explanation. Were friars picked because of their charisma and their neutrality among the rowdy factions of the medieval city? Or were they appointed because of their knowledge and expertise? (Monks were educated, and the religious communities that settled in Italian cities started teaching numeracy and mathematics as early as 1250.)[3] What role did Penitents' vows to poverty play in securing the conditions for fair measuring?[4]

I have asked myself these questions, and I still do; but, as the documents are silent about the motives, the answers can only tend towards the speculative. Further, I cannot help but feel that the historians' preoccupation is anachronistic, as it echoes the separation of scientific thinking and faith wrought by state positivism. Many of our ideas about measurements depend on the modern alliance of politics and science, which, among other things, has also loosened the grip of religion from daily life. The history of science generally, and of measurements in particular, is today expurgated of its devotional streaks. Few recall that the inventors of the meter, Gabriel Mouton and Tito Livio Burattini, were an abbot and a Jesuit.[5]

So in this part of the book I leave the question of motives behind, and pull my argument in two seemingly opposite directions at once. On the one hand, I argue for a time-honored tradition of religious involvement in measuring. It dates from at least 554, when Emperor Justinian gave measurements to both laymen and the clergy through a norm that was eventually included in the *Corpus iuris civilis*.[6] By gathering the historical evidence, I provide a new background to the history of the previous section, which—it may be worth reiterating—started with the movement of measurement standards from the safes of bishops and abbots to cathedral façades.[7] Mendicant orders—the religious orders that settle down in urban centers rather than retreating in isolation—spread throughout Italy more or less simultaneously with the appearance of the *pietre di paragone*. Arriving in a city after its standards were already publicly displayed, they took up those standards, recognizing in them powerful tools with which to relate to other urban communities and take control of life, even in a moral sense. The mendicant orders are the protagonists of this section. And, as a way to bring their involvement in measuring back to normal, mixing their doings with those of the merchants and public officials with whom they were often associated, I have given this part of the book the expansive, profane title "Cities."

While amalgamating religious figures with the rest of urban life, showing how urban proximity contaminated forms of measuring and was transformed by them in return, I also argue that some religious takes on measurements were distinctive. Ecclesiastical institutions did not deal with measurements simply as trading and judicial tools. Rather, they contributed, more than any other medieval institution, to the idea that measurements ought to be trusted. Religion, in other words, turned measuring into an act of faith.

This definition is in direct continuity with the takeaway point of part 2, that measuring offers a promise of sameness. If my emphasis before was on repetition and standardization, in this chapter I zoom in on the concept of promise. How did measurements come to be trusted? It is a difficult question, but a look into religious practices of measuring provides many answers. Issues of measure and self-control, after all, go to the very heart of the Christian faith. Many sermons and theological treatises celebrate the benefits of moderation, warning against excess and stating that justice is ultimately God's will. Yet, despite the plethora of variations and elaborations on these and other themes—the Bible mentions measuring more often than it does charity—measuring belongs to two main currents of reasoning.[8]

First, Christianity recognized measuring as an act of God, and to reflect on its practices and processes meant to access God's geometrical mind. "Who has measured the waters with his fist, and weighed the heavens in his palm?" asked Isaiah.[9] "You ordered all things according to measure, weight, and number," recited the apocryphal Book of Wisdom in a verse that surged above all others as the go-to expression of theologians and preachers when discussing measurements.[10] Saint Augustine commented on it repeatedly, extracting from it multiple lessons, from proof that the world has limitations to the existence of order between all the world's elements.[11] His reflections, many of which were excerpted and included in treatises on geometry,[12] stimulated Peter Lombard, who in his *Sententiae* (1142–58), the most successful theological book of the Middle Ages, defined a body (*corpus*) as something that is measurable in opposition to God, who is incircumscribable.[13] Measuring is both a marker of secularity—the Dominican Thomas Aquinas took the variety of standards between cities and their imprecision as effects of their venality—and proof of the eternal order of the cosmos.[14] The friar Robert Grosseteste believed that to measure the universe through angles and lines was a way to come closer to God.[15] And Bonaventure of Bagnoregio, general minister of the Franciscans, also believed in the convergence of scientific knowledge and divine rectitude.[16] The list could go on and on.

Second, measuring became fundamental for the solution of sin. The reference text, here, is the Gospel of Matthew: "The measure by which you measure will be used to measure you in return."[17] Since the sixth century, priests compiled lists—the so-called *libri poentientialis*—for the quantification of error in which every type of sin corresponded to a number of fasting days (that is, days on a diet of bread and water).[18] Yet, after the turn of the millennium, wrongdoing began to be measured. Alain de Lille, the compiler of a major text of atonement, speaks of "commensurare" to make sure that any error was matched by a penance of equal weight.[19] The gates of heaven could only open the moment Archangel Michael's scales were evenly balanced. The *liber poenitentialis* of Siena Cathedral, dating to the first half of the twelfth century, even gave importance to commercial tools. "Have you falsified or cheated about measurements or weights?" it asked. "Have you sold other Christians your products by using a false linear standard or a false weight? If you did so, or allowed it, you will repent for twenty days."[20] Long before the commune took care of mercantile tools, religious order presided over commercial fairness and to stabilize their role as legitimate adjudicators.

Yet, here is where the story becomes more complex. To separate ideas of cosmic beauty from questions of error is misplaced, as the two had grown intertwined over the centuries. The verse from the Book of Wisdom, for instance, entered legal treatises on wrongdoing. It returns, word for word, in Pisa's *Constitutum de legibus*, among the oldest legal compilations in Italy (it dates to around 1155).[21] And then it appears again in passages about fraud and bequeathing of the *Corpus iuris civilis*.[22] If the Book of Wisdom verse had, and still has, the ring of an inescapable truth, it is because it has shaped numerous channels of life, each buttressing others to the point of indecipherability. It even returns in pre-Justinian texts. The Roman jurist Gaius, one of the main sources for Justinian's *Corpus*, quotes the Book of Wisdom triplet to define what can be sold—that is, everything that consists "of weight, number, or measure."[23] We know that Gaius was influenced by rabbinic material, and the Book of Wisdom, which was composed only a few decades before his legal compilation, seems to have been one of his sources.[24] Tracking the movements even of this one verse shatters the idea that secular and religious legislations are independent processes. In the second half of the twelfth century, when monks started incorporating passages of the *Corpus* into the Decretals, the foundational text for the emerging legal code of the church, they were borrowing from a compilation that had already been shaped by biblical principles.[25] To realize the intensity of such an intertwining is essential to also understand the profound connection between religion and capital. If what can be sold is what can be counted, weighed, or measured, and if these are the exact principles by which God created everything, how long before someone asked if everything that God created could be bought?

One of the main goals of this book so far has been to break away from the idea of measuring as a trouble-free, straightforward act of quantification. Now, my objective has become to problematize the assumption that measuring is simply a commercial activity. Most historians see the plethora of references to measurements in late medieval life as an effect of the hegemony of the mercantile classes.[26] One wrote that their enforcement was a combination of centralized decision making and a "self-regulating mechanism compounded by a balance of terror and a lively sense of mutual advantage felt by all members of the international community."[27] In this guise, measuring is a necessity in a world that has already turned capitalistic. My reading, however, digs around this sense of certitude. If people believed that measurements produced results, it was above all because friars and priests convinced them to perceive measurements in that way. They saw measuring less as producing material benefits than as a condition of God's creation. They relied on measuring to enter in touch with the divine. Measurements became tools not just for trading, but for praying and thinking about God. And even when they were brought to the market, their moral implications did not pass unnoticed. People trusted merchants as a way to practice their faith.

Such acts of faith were not without problems. The coupling of religion and measurements produced anxieties that did not disappear even after measuring entered its modern, secular existence. Quite the opposite: the radical image of scientific research as subversive to religion is largely a fantasy, and measuring offers a privileged way to explore the complex, wayward relationship of science and religion. In particular, science appropriated and profited from one of the achievements of religion's handling of measurements, namely the generative process that measurements engender.

Religion cemented the idea that measurements are triggers to metamorphosis: they translate matter to the realm of ideas, producing concepts that shape culture at large. (They also do the inverse, thus visualizing time, or faith, through space.) In a sermon he gave in the square of Florence's Santa Maria Novella in a late winter day of 1306, the friar Giordano da Pisa told his audience that measuring was the way to compare secular goods to the heavenly rewards. He sustained his sermon with numerous comparisons and acts of translations that bridged between man and God.[28] "Translation" may be too deceptive a term, though. It implies a transparent and legitimate process, giving the idea that measurements inhabit physical and abstract worlds at once. And yet it was this notion of measurements as a bridge that Giordano cared about. He repeated it in another sermon, in which he describes the pain caused by a common, one-*braccio*-long sword and compares it to the pain produced by God's incommensurable knife.[29]

Throughout the Middle Ages, measuring was defined as an operation to find the limits.[30] As theologians repeated, measurability is the condition that separates ideas from bodies, God from humans. And yet, paradoxically, it also offers the operative way to link them. Measuring produced a way to link the unlinkable. As a tool of mediation, it also collapsed the distance between other realities. Preachers insisted that measuring, God's creative reasoning, was to be taken up not only by scholars but also by the ignorant masses living in the city. One of the main consequences of mendicant orders' involvement in measuring was to convince people that measuring brought everyone closer not only to God, but also to everyone else. By confessing and preaching, the friars did not simply spread knowledge of measuring; they also tightened the threads of society.[31]

17

From *Fratres Penitentie* to *Religiosi Viri*

The statutes of Bologna from the early 1250s record that within three months of his appointment, every new *podestà* had to summon three Penitents (*fratres penitentie*) to check whether the size of bricks and tiles conformed to the *pietre di paragone*.[1] The *pietre* were then carved under the vault of the communal palace: their ownership was thus not questioned, which may explain why the commune felt it could assign their control outside the circles of its officials. Something similar happened in the nearby city of Parma, where four *fratres penitentes* were appointed to check standards and oversee sales in grain, flour, and wine.[2] Helped by a group of attendants (*custodes*) of their choice, each of them presided over one gate and at the stations in the corresponding neighborhood to which bakers, millers, and winemakers carried their goods, had them weighed, and got them approved for sale.[3] Such a role was new both to friars and to the commune. In a way, we could call it an experiment. The *podestà* checked the new appointees every two weeks, and the statutory instructions seem written for newbies. The friars, an entry reads, should "work as carefully as their lay predecessors."[4]

The name *fratres penitentie* refers to a specific category of friars: the lowest branch of the order, the so-called Tertiaries. This group was made up of people who followed the example of Saint Francis of Assisi but, unlike members of the first order (who resided in a convent) and the second (who took the habit but lived independently), the

Tertiaries continued to live a married life in their homes.[5] They followed a specific rule, approved by Francis himself, who in establishing Tertiaries continued a tradition dating back to the heyday of monasticism, when some people refused to follow hermits into the wilderness and dedicated themselves to pious initiatives and, especially, penitence.[6] Hence the name "Fratres de penitentia." Tertiaries are recorded in Bologna as early as 1236, fourteen years after Francis preached in the city and eight years after his canonization.[7] In the 1260s, they were fifty-six, a number that remained more or less stable for the rest of the century.[8] So, when the commune of Bologna chose its measurement checkers from within their ranks, it was not relying on an enormous pool of candidates. Each year the ministers of the Tertiaries compiled a list of the members, and presented it to the *podestà* within a month of his appointment. (It is from such lists that we can infer the community's growth and social composition.)[9] We know that the commune reimbursed the expenses and paid the appointees a maximum of three *soldi* per day.[10] This was a modest amount, not even enough for purchasing two small flour *focacce*.[11] The friars were obviously not taking up the role for profit.

Spotting measurements in the hands of penitents was a common sight in the cities of the Po Valley. Yet, north and south of this region, they were more frequently found among the Humiliates, a movement that grew in popularity among wool workers.[12] In 1201 Pope Innocent III recognized the Humiliates as a religious order, moved by their desire to return to the uncontaminated teachings of Christ. Such resolution sometimes clashed with communal normative. The Humiliates did not want to carry weapons or pay for the upkeep of the army.[13] They also refused to take oaths, as Jesus expressly forbade, even when it was just a trivial matter, such as certifying the size of a roll of cloth.[14] Such dissent invalidated core communal procedures, which, as we have seen, extensively relied on witnesses.[15] In Milan, the clashes reached such an intensity that the pope intervened. In a number of letters to the bishop and the Humiliate leaders, he asked them to consider swearing in relation to its context. He asked them to separate between necessary and inessential oaths, but also between secular tribunals and religious courts.[16] We do not know how successful his plea was.

The Humiliates could afford to go against the Milanese commune because, by the time of the papal intervention, they no longer were at the margins of communal life, but were an integral part of its mechanism. In 1234 in Cremona, they worked as *massari*, overseeing the flows of cash in and out of the city's treasury. They took care of the collective finances also in Novara, Alessandria, and Florence (but also in Parma, Modena, and Ferrara).[17] Besides collecting tolls, they ran quality and quantity checks on the goods brought into the city by water. In Cremona they waited at the bridge over the Po River and in Bergamo at stops over its tributaries.[18] Near the water was where the cloth-working Humiliates lived, as wool had to be washed repeatedly before being carded.

In Brescia, the Humiliates resided near the San Matteo gate. They came to be known as *fratres tolonei*, from the name of the tax for weighing the iron entering town (*toloneo*).[19] In Como, a nearby city, two Humiliates collected payments for weighing and inspecting the incoming flour. From the records they left, we know that the commune was afraid that millers would mix the flour with sand. Four other friars went to the city's bakeries to taste their bread. They made sure it was thoroughly cooked and that each loaf matched the shape prescribed by the statutes.[20] One such shape is still preserved in Padua (see figure 12.2).

Such tasks could be enormously laborious. Bonvesin de la Riva, the Humiliate living in Milan whom we encountered in chapter 10, reported that the city had more than nine hundred mills, for a total of about three thousand wheels, each of which gave food to more than four hundred people.[21] Bonvesin also counted the city's tombstones and the chapels, the ovens, and even the cherries consumed in a day.[22] Bonvesin's staggering numbers have raised some scholars' eyebrows. Should we take them as hyperbolic? Maybe. After all, he confessed to recording them to make his fellow citizens marvel at Milan's greatness.[23] Still, the pressing question for me is not whether Bonvesin's data were correct, as to linger on such an issue is to tacitly subscribe to a way of thinking that postulates that quantities can accurately represent reality. Instead, I wonder about Bonvesin's approach. Was he spurred by members of his order to describe Milan through numbers and distances? Was his account—which became popular, by the way—a way to prove that the Humiliates had control of Milan's productivity?[24]

Bonvesin also reported the perimeter of the twelfth-century city walls (1,141 cubits), the depth of the canal (38 cubits), and the size of the main communal square: 130 by 136 cubits, for a surface of 10 *pertiche*.[25] Bonvesin could work out the relationship between sides and area because, like his fellow members, he was experienced at measuring. He knew how to measure, and stated that he himself (*ipsemet*) did measure. When explaining how he determined the depth of a canal in Milan, he even described the rod he employed, which is significant as it not only provides his findings with a realistic texture, but also demonstrates that he had access to standards.[26]

Bonvesin's account reveals the traditional rhetorical baggage of measurers, full of claims to familiarity and precision.[27] His enthusiasm, however, indicates a shift from the attitude the Humiliates had displayed in the past. Forty years earlier, some members had expressed deep frustration with measuring. They also complained that to unmask fraudulent vendors, they sneaked inside shops like spies.[28] They pleaded with Pope Innocent IV, who issued three letters to the commune asking to dispense the Humiliates from secular offices. The second letter of the three, dating to 1251, is particularly interesting as it describes the Humiliates stuck "at the city gate, demanding tolls, weighing grains, and measuring rather than working at their mills as they would like."[29] The letter is a rare certificate of intentions. It is also a record of communal coercion, and I wonder about the nature of the intimidation. Did the commune demand such tasks in exchange for some concessions, or did it take advantage of the Humiliates' profession of penitence, as stated in their rule?[30] The mention was quick, and the pope's reply glided over it, reminding the commune that the Humiliates were a religious order and that their appointment to civic offices should have been discussed with the bishop. Tertiaries, the pope also added, should be dismissed altogether from measuring, a task reserved for members of the first and second orders only.[31] Things, in other words, seem to have gotten out of hand. Facing an unrelenting expansion in the second half of the thirteenth century, and running short of officials, the commune appointed those who should have remained exempt. Milan was not alone in recruiting members of religious orders. In 1265, when Reggio Emilia increased the number of *massari* to two, the commune started choosing them from the Franciscans of Santo Spirito (first order then).[32] The following year, the notaries of Parma updated the city statutes so that measuring could be carried out not just by Penitents, but by any *religiosi viri*. They doubled the number of officers: two had to supervise the market, and two navigated through the city. They also appointed four more inspectors for the rest of the bishopric. In one year this group was required to visit every village, even those in the mountains, and make sure that each parish church kept copies of the standards.[33]

Demographic growth is a way to explain communes' reliance on *religiosi viri* in the second half of the thirteenth century. Their documentary pervasiveness, however, is also an effect of the great range of roles that they came to perform. In Siena, *religiosi viri* assisted the *camerlengo*, the city's top financial officer, in checking the standards of bricks and tiles and in the fabrication of the wooden molds.[34] They went through the city to check that windows would not be taller than one *passetto* and that nothing—signs, bars, stools—protruded from shops for more than a *passetto*.[35] *Religiosi viri* also recorded the size of the fields.[36] The *camerlengo* too, however, was often picked from among them: the designation even appears on the cover of the 1280 register to define the office holder for that year, the monk Guido, busy counting coins (plate 7).[37] The post often went to a monk of San Galgano, the affluent monastery near Siena, but candidates were considered from several other orders. In 1307, for instance, it was friar Magino, a member of the Humiliates.

But here is where relying on the accurate historical term becomes a problem. "*Religiosi viri*" was after all a generalization: an expression that offered the commune enough legal indeterminacy to address a vast range of circumstances and adapt over time. And, as I try to stay close to the documents, I realize that such breadth fails to provide a distinctive pattern. If Parma expanded the involvement of *religiosi viri*, Pistoia reduced it. The statutes that the *podestà* of Pistoia issued in 1296 declared that "no secular cleric should sit in any office" (*ut nulla persona ecclesiastica sit in aliqua balia*), a norm that has been interpreted in response to the Tertiaries' refusal to recognize the authority of the commune.[38] Lay and religious communities could break apart. Some never assimilated. In Pistoia, the supervision of measurements passed from a religious group to a singular judge, the *judex super dannis*, who also checked that the streets were clear of pig shit and the canals free of debris, so to guarantee four feet of navigation.[39] The *judex* received the official standards and gave copies to local artisans. He organized the oath of carpenters and stonemasons. He checked the shapes of bread and the volume of wine every day, sending someone to examine them even in pilgrim shelters.[40] Even then, measurements were not entirely out of religious control. Indeed, the *judex* had to retrieve and return them to the church of San Jacopo, where they were under the control of its Opera, the institution that presided over the financial and architectural well-being of the church. It was members of the Opera board who checked and sealed the standards, independently of the *judex*.[41] We know that in 1313 the standards were treasured in the sacristy, where they stayed up to the eighteenth century: less a medieval practice continuing into modern times than a centuries-old staple in the organization of production, and one that collapses the very distinction between medieval and modern.[42] Soon afterwards—we have jumped to the 1340s—members of religious orders were back in office "so that the post will be done better, more carefully and without flaw."[43] Those words negate half a century of Pistoia's history. They describe the *religiosi viri* as competent, not hostile. And yet, they add little if anything to the stranded situation in which the piecemeal historical evidence has cast us. The history of medieval measuring and religion was tightly intertwined even when on the verge of breaking apart, which seems to be the case the closer we draw to local circumstances.

18

Cutting through Buildings

The issues of measurements, of their administration and their uses, did not simply depend on the shifting balances between the city's factions. There were other, larger forces that contributed to their standardization. Zoom out.

Measuring, for instance, was energetically discussed at the pontifical council held at Lyon between 1272 and 1274, the second to be held in that city. The papal chancellery had been flooded by requests of Franciscans and Dominicans complaining of the mushrooming of urban religious houses. And so in Lyon, members of the ecclesiastical hierarchy discussed their location and the necessary distance between them, constructing, through measurements, hierarchical divisions between the various religious communities. With two bulls, Pope Gregory X forbade any new religious community to settle within three hundred *canne* from the Dominican monastery of Bologna, Saint Dominic's burial place.[1] And he threatened the commune of Assisi with excommunication for having founded a hospital within that same limit from the Franciscan mother house, the burial site of St. Francis. As it was applied to the two "capitals" of the mendicant orders, the norm took up some sort of general validity.

Questions of distance were not taken lightly in the thirteenth century, because space was an essential resource, especially for mendicant and preaching orders, who lived off alms and people.[2] Religious orders needed people: they needed ears to speak to, hearts to convert, and hands to receive from. The sudden explosion of mendicant architecture in the second half of the thirteenth century—churches as large as cathedrals, but erected in a tenth of the time—is a direct consequence of religious orders' ability to capitalize on human contact. Space was most precious at the time because there was so little of it. Medieval cities' dense street grids got denser, limited as they were by the city walls. Their erection, between the end of the twelfth century and the beginning of the thirteenth, coincided somewhat with the spread of the mendicant orders. The walls initially pushed the orders out of the city and, stripped of cash and with few members, the orders searched for suitable locations in the outskirts.[3]

But fifty years later, in the 1270s, the situation was drastically different. Scholars herald the Council of Lyon as a triumph for Franciscans and Dominicans. Pope Gregory X forbade any other religious congregations to increase their membership, thus dooming them to extinction.[4] He had been advised to do so by his four counselors, all of whom belonged to either the Franciscans or the Dominicans.[5] One of them, the general minister of the Franciscans, Bonaventure of Bagnoregio, was even elected bishop of Albano, a first for the Franciscans, who then aspired to climb up the ecclesiastical hierarchy.[6] (Indeed, Bonaventure's successor as Franciscan leader, Girolamo d'Ascoli, became Pope Nicholas IV.)[7] But the hegemony of Franciscans and Dominicans was not just an effect of rethinking membership, cutting some orders off while granting others access to top ecclesiastical ranks. Rather, their increase in power was first and foremost an effect of the three-hundred-*canne* rule, which pushed their competitors out of city centers.[8]

Politics controls the accessibility of bodies to things and places; the working of power is about enforcing hierarchies through the distribution of money or organizing space (as well as time).[9] The three-hundred-*canne* rule did precisely that: it provided a system for the expulsion of most religious orders and the centering of a very few, which then were put in the position of leveraging on their privileges undisturbed, growing in size and power at every opportunity.

Despite what some modern scholars say, I see no direct religious precedent to the three-hundred-*canne* norm. In the 1130s the Cistercians may have set the minimum distance between two abbeys at ten Burgundian leagues.[10] Yet, not only did the norm soon become ineffective (by 1152 the Cistercians stopped building altogether), but as there were around four kilometers in each Burgundian league, the norm constructed space at a completely different scale: one that accounted for cities, not people.[11] The Cistercians approached the territory as medieval kings did, and their sense of distance was probably directly influenced by royal decrees. In June 1081, Emperor Henry IV commanded that no fortification could be built within six miles of Lucca.[12] In 1117 Pistoia co-opted that reasoning and constructed justice according to a set of decreasing radii from the city walls: murders were banned within five miles, churches were protected within four, and the width of streets was guaranteed for two.[13]

The Council of Lyon, however, abandoned these broad monarchical gestures by adopting a much finer spatial texture, one in which even a few *braccia* made a difference. And the inspiration seems to me to have come from the *Corpus iuris civilis*. Justinian's compilation, in fact, included a norm that forbade erecting a building within one hundred feet from another if the former blocked the latter's view of the sea.[14] Space became human again.

Justinian's norm served as the model for the respatialization of Italy's urban centers. Statutes set precise dimensions for the height of towers, the width of windows, and the length of shop signs. Communes pervasively relied on dimensional proscriptions to freeze locations, lock heights, and even regulate the amounts of air and light in rooms. (Indoors luminosity is a consequence of space, as it depends on the size of the windows.)

Canon law adapted to this thick set of civic norms. Gregory X did not as much innovate as search through the forest of regulations promulgated by his predecessors and other ecclesiastical authorities. He identified as the most authoritative ruling a decree issued by Cardinal Simone Paltanieri in 1265, when he was papal legate for a region that included both Bologna and Assisi.[15] The directive, which granted to the Dominicans of Bologna a clear area within a radius of three hundred *canne*, was itself an extension of a privilege by Pope Alexander IV of five years earlier to the Franciscans of Ascoli, a city also under Paltanieri's watch.[16] Alexander IV's ruling, however, had been undermined by the bishop of Bologna, Ottaviano, who in 1268 prohibited the erection of new monasteries within an urban area that he defined by perimeter and not by radius.[17] Ottaviano's decree was advantageous to the Dominicans, as he basically made them the lords of the southeast quarter of Bologna. Aware of their luck, the Dominicans immediately sought approval from Ottaviano's principal, the archbishop of Ravenna. The archbishop also complied because in the meanwhile, Paltanieri's norm had itself been compromised by Pope Clement IV, who, after originally confirming it, eventually reduced the distance to 140 *canne*.[18]

In short, the situation was a mess. It looks chaotic even to me, despite the fact that I have all the directives at hand, spread on my kitchen table. And the Council of Lyon did not help to fix it. Canon courts continued to receive complaints. In 1274 the Augustinians of Saint Eulalia in Barcelona sought the destruction of a new oratory built within three hundred *canne* of their church, even if the pope had explicitly permitted it.[19] In 1286, the Benedictines of Sant'Apollinare in Assisi won a lawsuit against the nuns of Santa Chiara, who requested to have the three-hundred-*canne* rule enforced, but were told that the distance to respect was 140 *canne*.[20] And in 1291, the Franciscans of Pistoia opposed the construction of the local Carmelite church, claiming that it ought to stand more than three hundred *canne* away.[21] A definitive revision was due, and in 1300 Pope Boniface VIII assigned it to the cardinals Gentile da Montefiore and Niccolò Boccasini, a Franciscan and a Dominican respectively. Three years later, when Boccasini became Pope Benedict XI, their decision became law: 140 *canne*.[22]

If I insist on dates, laws, and lawmakers—and I have loaded the narrative to the point that I am reproducing the medieval fuzziness— it is because I want to make the confusion palpable. Chaos is important here, as it blocked historians who reviewed a handful of documents and read them as certificates of inefficiency.[23] But those ecclesiastical rulings were not empty words. The friars did measure.

In the first months of 1327, the Franciscans residing in San Fermo, a grandiose monastery in the center of Verona that originally belonged to the Benedictines, brought the Servites to court.[24] The charge was that the location of their building, Santa Maria della Scala, breached Benedict XI's rule (that is, 140 *canne*). First, they took the case to the local curia, and then, when the Servites claimed that the rule did not apply to them as they were not a mendicant order, they went one step up, to the patriarch of Aquileia. (The patriarch's court was located in Udine, near today's Slovenia; the length of the journey speaks for the seriousness of the matter.) After the patriarch confirmed that the law did apply to the Servites, the Veronese judge started measuring.

It is important to add now that all papal norms, indifferent to dimensions, speak of distances as the crow flies (the Latin reads "per aerem," or cutting through air). Clement IV's bull further specifies that the measuring should be done "by straight line, even where the configuration of the place would not allow it."[25] In other words, the dimension should be calculated abstractly, as if two buildings were united by an imaginary line. And this is precisely what friar Giovanni, the arbiter of the case, did. He summoned a team of experts: the astrologer Benintendo, the professor of grammar Pencio, and the teacher of mathematics (*abbaco*) Paolo. They measured the distance between the two closest points: the southeast corner of the Servite church and the northwest corner of San Fermo.[26] Yet, as numerous buildings stood in between, they followed the course of the road leading from the Servite church to the square of San Fermo, measuring distances in an arrangement that resembled an L shape (figure 18.1). And then, from what were basically two sides of a rectangle triangle, they calculated the hypotenuse (that is, the crow's flight), which resulted in 152 *canne*. The Servites were safe.

But we are not finished yet. There is more to say, as the historical records also mention the measuring tool and how the judge obtained it. As Pilio da Medicina made clear, measuring was tied to the time and place of its deliberation, and friar Giovanni had requested a copy of the *canna* from the Franciscans of Assisi, the original addressees of Clement IV's 1268 bull. The standard arrived. The bishop of Assisi Tebaldo enclosed it as a hemp string in the letter he sent to Verona, which Giovanni opened on a bridge in the presence of both Servites and Franciscans.[27] The string today is gone, but the communal records state that it corresponded to eight palms—that is, Clement IV's definition of the *canna*.[28] This was eventually translated as three *braccia* of Verona, an equivalence that was, however, rejected by the spokesman of the Servites and which reminds us that dimensional translations were never taken for granted and were often antagonized.

18.1 | Map (scale of 1:1000) showing the measurement in 1327 of the distance between the churches of Santa Maria della Scala (top left) and San Fermo (bottom right) in Verona. The dotted line indicates the distance "per aerem" between the two churches; one *canna* (*c* on the map) is 1.9465 meters. Drawing by Emanuele Lugli and Fernando Lugli.

We do not know the dimension of the Veronese *braccio* until later in the century, when it corresponded to about sixty-four centimeters. (I remain deliberately imprecise, as we know that standards oscillated over time.)[29] Yet, even after considering a 5 percent error—and that is a significant margin—the range of 58–70 centimeters is in no way proportional either to the Assisi *passetto* or to eight Roman palms.[30] So the standard seems to have been a papal invention. It would make sense. As in the Middle Ages each trade or material had its specific standards, why should religious distances work otherwise? Space was not the homogeneous, imaginary, and infinite vacuum we have come to regard it as with René Descartes. Rather, space was physical. Measuring made it concrete, differentiating between communal and religious space. Clement IV's *canna* constructed spatial relations of a different order than those produced by communal standards. Urban space was transformed according to the tools and norms of the ruling institution. Yet both church and commune used measurements to link their power to the ground. This is what measuring does: it translates power to the ground, turning it into a physical component of the city.

19

Invisible Boundaries

Let us return for a moment to Ottaviano, the bishop of Bologna. You may have barely noticed him, but he was the one who, in 1268, ignored the three-hundred-*canne* ruling and decided that no new religious house could settle down within Bologna's southeast sector, the Dominicans' quarter. As we saw, his approach was the losing one: popes and cardinals before and after him preferred to define the no-building area in relation to the radius from a church, and not—as he did—in relation to the perimeter around it. And yet Ottaviano's choice was not unpopular. Quite the opposite, it was the way by which many communes protected space.

In his directive, Ottaviano described the urban circuit with topographic precision.[1] The perimeter closely followed the southeast sector of the latest city walls, an enormous palisade that, because of its quasi-circular shape, came to be known as the *circla*.[2] To keep it clear of the private dwellings that kept growing against it like moss, the commune enacted an ingenious system of control. Its officials staked it out and the notary Ranieri of Perugia recorded the distances from one stone peg to another.[3] His lists, known as the *libri terminorum* ("the books of stakes"), still survive today.[4]

Author of the famous law treatise *Ars notariae* (1216), Ranieri was not just any local notary but a celebrated jurist, probably the best available in Bologna at the time.[5] In his measuring task he was assisted by the *scarii*, the public engineers, who discussed with him where to place the stones and carried out the laborious operations to cover the seven-kilometer walls. (The difficulty was not just an effect of size, as the *circla* was a complex of structures that included ramparts, seven-meter-wide ditches, and at least two rings of streets, internal and external to the city.) It may be worth reminding that at the time the *scarii* also supervised the *pietre di paragone* (plate 1) and brought to the project a deep knowledge of measuring.[6]

Staking out the *circla* was onerous, as the pegs were not simply hammered into the ground but were entirely buried to prevent their arbitrary relocation.[7] With their disappearance, the measurements recorded by Ranieri were the only means to retrieve them.[8] Significantly, it was the distances between the stakes that were listed, not their topographical locations: the physical stones were only used to confirm the recorded data. Space thus became immanent to measurements, not to physical objects. If, with the crow's flight norm, measurements reduced the architectural landscape of the city to particular points of departure, the staking of the city lessened the importance of those reference points even more. The city's built environment lost its significance to an independent treatment of space with its own forces and logic.

While increasing popular participation—the Bolognese *pietre di paragone* are first documented in 1250, five years after the staking out of the *circla*—the commune also withdrew from it, leveraging the immateriality of measurements to maintain the effectiveness of the stakes. (This wayward move is analogous to the *pietre*, crafted as if the standards had been removed and substituted by the cast of their dimensions. It is also similar to the 140- or 300-*canne* norm, which could only be imagined abstractly, never experienced.)

The staking of the *circla* had precedents, as it can be considered a form of *terminatio*, the Roman practice of marking boundaries that was still used in medieval times.[9] In 1186, Genoa employed it for its three city markets. The surveyors did not stake the areas out, but measured and recorded the distances between the corners of the buildings overlooking the squares.[10] And after a fire devastated the city of Padua in 1174—a chronicle reports that more than 2,600 buildings went up in flames—Bishop Gherardo also appointed three surveyors (that is, three priests) to retrieve the perimeters of the parishes and mark them with stakes.[11]

Genoa and Padua—I could mention other examples—relied on the architectural fabric of the city to set spatial boundaries. As buildings were identified by their proprietors and ownership changed over time, however, the identification of buildings was not a straightforward process.[12] Moreover, buildings expanded and shrank in ways that medieval officers did not notice. In every passing year the city risked outgrowing the records, if the officials could still remember how to interpret them. And it was to avoid such problems that Bologna constructed an architecture of invisible stakes parallel to those of the city. By corroborating each other, the two intangible systems—the network of underground boundaries and the lists of distances—produced a closed, self-sufficient network that squeezed the real, physical city out of itself.

The *libri terminorum* were an important political document, inspected by the *podestà* himself. The *Sacramentum potestatis*, the list of duties that every incoming *podestà* promised to fulfill, states that he ought to perform the checks twice a year when inspecting the *circla*, in April and August. He also checked the stakes that dotted Bologna's market squares. (This was a more pressing duty, as the inspection took place every other month.) Indeed, the *terminatio* of the *circla* proved so effective that in July 1286 it was extended to the perimeter of Bologna's market areas, today known as the Piazza Maggiore (figure 19.1) and the Porta Ravegnana square. The areas had been surveyed before—the *libri* record that stakes were already in place, thus hinting at some missing documentation—but the 1286 *terminatio* increased the number of pegs, and thus spatial definition. The surveyors also registered the distances between *termini* and street corners, gutters, and waterspouts, thus moving away from a two-dimensional registration of plans for a three-dimensional imaginary network of lines, which enveloped a building in all directions. This practice seems to have closely followed the principles of Robert Kilwardby's *De ortu scientiarum*, written around 1250, which assiduously applied the science of measuring not only to bi-dimensional magnitudes but also to those cutting through air. Kilwardby found that a skilled measurer could

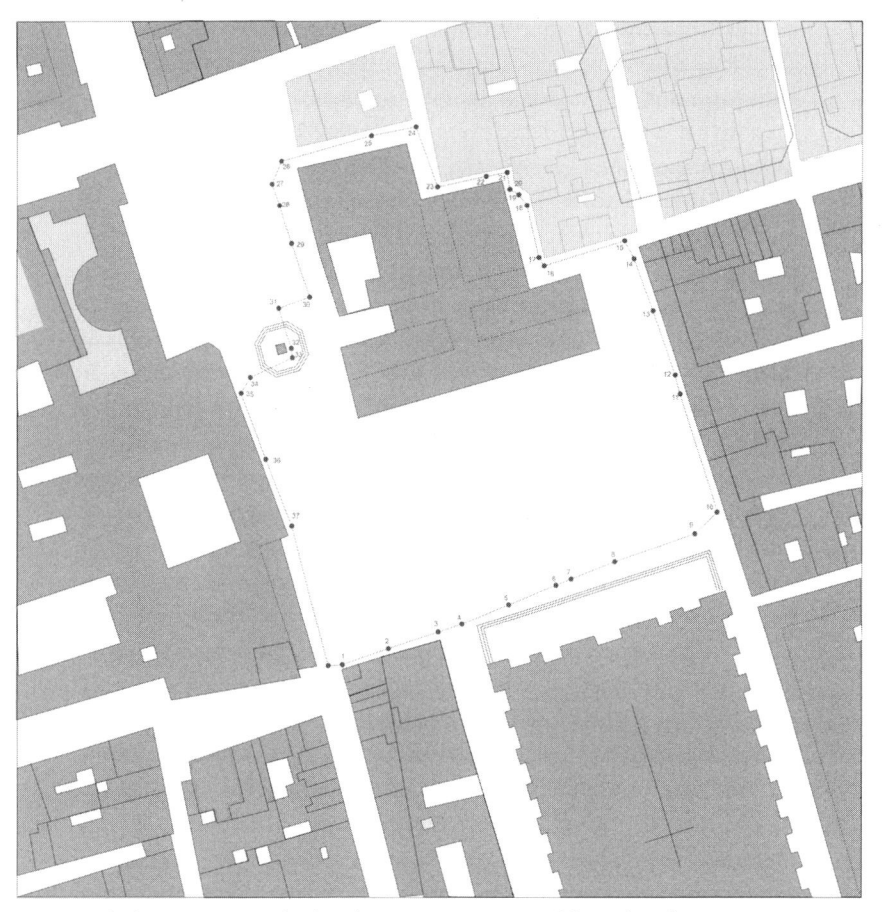

19.1 | The 1286 *terminatio* of Bologna's Piazza Maggiore, one of the city's market squares. Each dot indicates the position of a stake as described in the *Registro grosso* of Bologna's Archivio di Stato. Drawing by Fernando Lugli.

scan space in all directions and thus precisely locate inaccessible points in midair, even intangible ones.[13]

In 1294, following a drastic urban expansion and reconfiguration—the square of the Ravegnana Gate was enlarged and turned into the cloth market, pushing all other trades to Piazza Maggiore—Bologna commissioned a new survey. This included a series of previously uncharted public sites: the *trivium* of the Stiera Gate, the lots near

Santa Maria Maggiore, the surroundings of the Santo Spirito church, and the Carmelite possessions in San Martino. I know that these areas may mean nothing to nonspecialists, but even as a list of toponyms, they give a sense of the commune's increasing control over space. Even small areas, such as a square, church, or crossroad (*trivium*), were then precisely charted. The 1294 version of the *libri terminorum* even included a new section titled "De designatio bancharum mercatorum," which regulated the position of market stalls. As stallholders took their goods out of the square every night, the piazza returned to its original, pristine state as an empty space, thanks to the establishment of intangible boundaries.[14]

Bologna's vast critical survey of 1294 was put in the hands of Giacomo, the son of Benvenuto of Santa Maria in Duno, who took part in the staking out of the *circla*. Giacomo had inherited his father's tools and operative methods, as artisans did in the Middle Ages. Yet the Bolognese surveyors and officials did not keep their practices only within their families, but spread them transregionally. In 1249 Brescia completed a *terminatio* similar to the *circla* after summoning a Bolognese *podestà*. And when disputes flared up in Padua about its parish boundaries, it was a jurist trained in Bologna, Martino Gosia, who drew up the final conciliatory contract.[15]

Brescia's official inventory can be found in the *Liber potheris*, a detailed description of all the city's estates by position, surface, and economic value.[16] The *terminatio* itself appears under rubric 123 (titled *Liber viis*, or the "Book of Streets"), and was undertaken in conjunction with the laying out of a grid of twenty new roads. The streets were staked out at crossroads, and each *terminus* was recorded as to the property in which it was buried and its distance from other pickets. A cross-analysis of the distances between stakes shows that the *Liber viis* is dimensionally consistent. It allows for a re-creation of the street network of the time (figure 19.2).

Differently from previous surveys, which charted only small squares and private properties, Brescia's *Liber viis* and Bologna's *Libri terminorum* regularized enormous urban areas.[17] Other cities, such as Lucca in 1284–87 and Siena in 1315, even charted territories outside their walls.[18] Yet what singles Bologna out from other plans is its precision. For the segments of the *circla*, the notary recorded dif-

19.2 | Reconstruction of the "new" western and southern sectors of thirteenth-century Brescia, according to the *Liber potheris*. The dots indicate the position of the *termini*. Drawing by Fernando Lugli.

ferences to the degree of one *oncia* (around three centimeters). Space was not broadly defined but scrutinized, appropriated, and controlled minutely. The spatial texture of the Bolognese operation was extremely fine, even more meticulous than today's urban planning. And it was so because spatial precision stood for political rigor.

20

Imposing Self-Control

At the Second Council of Lyon, discussions of dimensions revolved not only around space but around clothes. Pope Gregory X attacked "women's immodest appearance" (*immodica feminarum ornamenta*) and appointed Cardinal Latino Malabranca Orsini to put an end to what he saw as conspicuous consumption. Cardinal Orsini did so in 1279, with the *De habitu mulierum* norm, which, besides imposing veils for married women and banning plunging décolletage for single girls, set the dimensions of dress trains (one palm) and plied mantles (two palms).[1]

As usual, there were precedents. Cardinal Orsini co-opted norms that had become rather common in Lombardy, Tuscany, and Romagna, the regions under his direct supervision. Reggio had set lengths for the trains of women's dresses since at least 1242, Siena in 1249, San Gimignano in 1251, and Parma in 1258.[2] In 1250, Bologna forbade women who did not work as prostitutes to wear clothes that touched the ground or use ribbons longer than one *braccio* and a half.[3] In many centers in northern Italy, tailors swore to respect communal protocols and accepted having to pay the same fine as their clients.[4]

Those norms, however, were limited to the city; and by appropriating their logic, Orsini legitimized them over a much wider area. He explicitly asked bishops and vicars to implement his norm speedily and effectively, using the network of religious authorities to spread it as far as he could.[5] He also preached it himself. In Florence, for instance, he caused an outcry among women for reducing the length of the gowns dragging on the ground from one *braccio* and a half to one palm.[6] (Info about the uproar reached Salimbene de Adam in Parma, who, even living in northern Italy, remained informed of Tuscan happenings thanks to the connections he had forged while living there.)[7]

It is difficult to track the effectiveness of the *De habitu mulierum*, given that sumptuary laws, like all laws that directly affected citizens, were renegotiated at the arrival of any new government. In 1288, for instance, the norm was relaxed in Bologna and women's trains grew back to three-quarters of a *braccio*. Shortly afterward, they even stretched to one *braccio*.[8] Yet the importance of those norms lies less in the dimensions they set than in cementing the idea that clothes could be measured. In 1286 a Bolognese woman named Francesca was taken to court because she refused to have her green dress measured. She was celebrating the festivities of Saint Dominic near the homonymic church, and later replied that she did not oppose the control itself, but only its implementation under those circumstances, as the place was too crowded to permit measuring ("*propter tumultum gencium*"). The impracticability (perhaps she meant the indecorousness or even the inconvenience; measuring a dress usually entailed leaving it with the notary) was confirmed by two eyewitnesses. Francesca was released.[9]

Francesca's case is illuminating, as it gives us information about the circumstances of the search. She was checked during a festivity, and in front of the Dominican church. As stated for other cities such as Florence, the captain of the *popolo* or the *podestà* would search for unlawful clothes inside churches and through the squares during sermons, "especially on Sundays and during festivities" (*maxime diebus dominicis et festivis*).[10] We know that people sported their fineries on Sundays. But searching for clothes on holidays was less a safe bet than a way to reinstate the moralistic value of the quest, in line with the denunciations pronounced from the pulpit, around which the quest often took place. It is unsurprising that friars were often appointed as dress checkers. What is remarkable is that sometimes we know their names. In 1343 in Florence, the *religiosi viri* in charge were Brother Marco and Brother Cristoforo da Settimo.[11]

Francesca pointed to the crowded square as a pretext for her release. She may have been trying her luck, but her excuse gestures toward the lack of space, the ultimate justification in the city. The fact that the communes defined women's tunics in relation to the ground—specifying whether they could touch it, or how much of it they could cover—did not so much accommodate women's different heights as it tied sumptuary laws to the preservation of public space. Clothes embezzled space. They impeded traffic and occupied what ought to remain free. Francesca's green train was the Sunday equivalent of a protruding sign shop, or of an unauthorized market stall. The association between sumptuary laws and *terminationes* was not accidental but structural. Both practices depended on measuring as power's organizing principle. Both practices intended to keep the ground clear. They both aimed at restricting possibilities, freezing out time, or at least returning to a point of origin (whatever moment the authorities recognized as origin.) Unlike other provisions which forbade materials and decorations, the measuring of clothing implied a journey to the origin of a garment on the tailor's counter, before it was cut. Measuring's promise of repetition was anchored to a vision of time that could fold upon itself, reaching a still point. And the ideal of stillness came with a heavy political as well as moral load.

It is futile to disentangle practicalities from ethical concerns when dealing with medieval measurements. Medieval sermons forged parallels between the two strands to the point of inextricability. The word "mensura" came to refer to both a physical dimension and a process of self-control; and the term "forma" ended up being used for both the *pietre di paragone* and a predisposition of the soul to good. The preacher Giordano da Pisa declared that the artisan's last was to a shoe what charity was to the soul.[12] (And medieval shoe lasts were not cobblers' inventions, but forms approved by the commune and regulated by the guild.)

The commune co-opted the language of religion and its impulses to discipline and sacrifice. Consider the extreme self-control that Penitents had displayed since the beginning of the thirteenth century, and the ways in which it stimulated sumptuary laws. In 1221 the Penitents of Romagna composed a list of intentions, the so-called *Memoriale proposti*, which would serve as general rule of the order.[13] Penitents would pay tithes, abstain from swearing, and make sure that everyone else did the same. To express their piety, they opted to wear a tunic of common, undyed cloth, whose size had to be such that its value did not surpass six *soldi* (the expression of dimensions through pricing is a constant of sumptuary laws).[14] Such norms had been inspired by previous monastic rules, such as that of the Cistercians, an order consecrated to complete standardization, which they accomplished by prescribing specific measurements for clothes, but also of pillows and bread, as the commune would eventually also do.[15]

Or consider the pursuit of the "common good," an ideal that was thought to be achieved through the abandonment of opportunism. Aristotle explained the concept in the *Nichomachean Ethics*, which first appeared in Latin in the translation of the jurist Burgundio da Pisa, in the second half of the twelfth century.[16] Scholars have emphasized the classical fascination that the concept had for communal authorities. Their meddling into all sorts of private affairs has been explained as a reenactment of *aemulatio*, the Roman principle that defined excess as limiting other people's freedoms.[17] Yet, even from its Aristotelian formulation, the concept had an explicitly religious ring to it. Aristotle wrote that yearning for something better than one's good was "divine."[18] It is, then, little surprise that the concept intrigued men of faith such as Robert Grosseteste and Thomas Aquinas (respectively, translator and commentator of Aristotle's *Ethics*).[19] They, as well as others, reflected on the common good and its ties to religion, charity, and faith, their elaborations suggesting that it was, first of all, a religious pursuit.

Franciscans and Dominicans might have had a modest influence in directly shaping late medieval legislation on measurements. (They took up a more prominent role in the fifteenth century.)[20] Nonetheless, they acted through their powerful role as mediators, constructing an audience out of the divided communities that made up the medieval city. Remigio dei Girolami, a Dominican and a pupil of Thomas Aquinas, dedicated an entire sermon against commercial fraud in measuring and weighing, speaking of it in front of an audience not made up only of merchants.[21] This was possible because the orders themselves were composed of people from all social groups. The very first mention of Bologna's Penitents—a group of only four people—reveals a mixed community: a cobbler, a barber, and two noblemen.[22] And the bonds between religious orders and the rest of the city grew stronger over time. Friars mingled with tailors during their processions, as the statutes of Bologna recall.[23] Before the fishmongers of Reggio Emilia requested the *podestà* to relax a decree, they went to speak to the local friars, asking them to mediate.[24] The friar Salimbene de Adam recorded the fluctuations of the market in Parma with the precision of a broker.[25] In cities so profoundly invested in economic and financial dealings, friars did not condemn trade, as the sacred texts of the church fathers had done before, but tried to pragmatically mediate between civic and religious ideas. In their sermons to a population largely made of traders, and as personal confessors, they dodged questions about merchants' predisposition to cupidity, and instead strengthened their importance as agents of charity.[26] Their revaluation of mercantile competence even led them to denote it as *industria*, a term previously reserved for monastic activism.[27]

So the growing importance of measurements in shaping urban life was not just an effort of the commune or of the guilds. It was not just a matter of training eyes to the new standards, constructing protocols and timing controls, and appointing officers. If measurements had such defining social and cultural roles, as they did in the thirteenth century, it was because religious orders cared about them. For them, measuring became a powerful vehicle of conversion and political participation. And they succeeded in advocating its efficacy because they themselves became models of self-measure.

Most communes appreciated the religious orders' involvement. They put *religiosi viri* in charge of measuring not only because of their numeric competence or honesty, but because their practices triggered certain techniques of the self that appealed to the powerful elites of Italian cities. What such elites wanted to achieve through measurements was probably not just fairness and commercial stability, but also wisdom and piety. If this is true—and I am aware that there is no way to prove it—measuring could have offered not only a way to discipline society but, thanks to its capacity to shift ideas from practice (standards) to morals (self-measure), an operative category to approach life at large.

21

The Ideology of Order

"Why are cities weak?" the friar Giordano da Pisa asked the audience gathered in the cathedral of Florence on the evening of Sunday, October 4, 1304. "Because things go disorderly there. If things went orderly for all citizens, the city would be very strong, whereas cities are ruined and strifes come out because of disorderly things, and the city is made weak."[1]

Giordano's oppositional reasoning of success and failure is worthy of today's populist politics. The city he depicted was on the verge of constant collapse: it was an ever-changing place where things moved (*"le cose vi vanno"*), disagreements were born (*"nascono"*), and citizens were hesitant about their future. To prevent uncertainty, he insisted upon operating on the city materially, by ordering its "things." Order precipitated strength, as it was its material counterpart.[2] Ordering physical things was the first step towards success. Giordano repeated the point often. During a sermon for Lent, he spoke of the "goodness [that] has the power to straighten things." And, after asking his audience, "What are the straightened things?" he tautologically replied, "All those that are in order," concluding his circular thinking with an exhortation to "behave straightly and without deceit: your tunic, your pursue, and everything else: please make them straight."[3]

Giordano took precise, straight borders as an indication of moral conduct, translating a physical quality into an abstract value. He had undertaken similar permutations before, we have seen, as had many other friars and writers. Thomas Aquinas, for instance, spoke of *termini* as a material necessity, part of the ineluctability of living in the secular world.[4] He gave the term a moral tinge, removing it from the technical meaning of the legally binding staked boundaries created by communes. And both Dante and Petrarch employed the word *norma* to refer both to a mason's set square and to a moral ruling.[5] Giordano's words may read today like classic self-righteous preaching, but in the context of the urban life of the time, their meaning reached into the political. Setting an equivalence between the physical world and social order, taking one as a mirror to the other, was in fact a principal move of communal rhetoric. By pairing straightening and acting "without deceit"—a recurring formula in the statutes, in which it often accompanies norms about measuring—Giordano reveals himself to be an accomplice to the commune.

The earliest record of straightening the streets comes from 1208 Vicenza. The commune then specified that all the facades of buildings were to be aligned along a straight line, so as to produce "fairness" (*aequalitas*).[6] Almost a century later, Florence was still rearranging some of its streets so that they could become "beautiful, wide, and straight" (*pulchre, amplae et rectae*). Although no material evidence can be comfortably linked to such undertakings, the statutes confirm that the straightening was obtained by pulling ropes along imaginary lines and rotating them crossways to check the perpendicularity of the sides, a process that we will explore in more detail in part 4.[7] Straightening, however, did not just mean removing the bends in the contours of the roads, but also paving them with polished stones, something that Florence started doing as early as the 1260s. It also meant clearing the roads of dirt and garbage, like in Siena, where the friars made sure that the front of each house was unobstructed for at least one *passetto*.[8]

Friars cleared not just the ground, but also the air. In 1231, Treviso required all streets to be free of signs or other elements protruding up to fifteen *piedi* from the ground. The height of towers was restricted in Bologna, Prato, and Volterra, where they could reach thirty *braccia* at most, and culminate in a canopy of an extra five *braccia*.[9] Dominicans also announced that their churches would not rise over thirty *piedi*, a measure of modesty according to them.[10] But then air and earth were connected, as the measuring of the city's vertical planes was calculated in reference to the ground, as we saw was the case for tunics.

I speak of vertical planes in the plural, because buildings were deconstructed into various elements, each of which seemed to a follow an independent logic. Staircases could go higher in Bologna than in Vicenza.[11] The windows of Siena had to be one *passetto* wide, regardless of the overall size of the building. And it is worth pointing out that records show that since 1309 the task of measuring the windows had been given to Penitents.[12] (Many overlook the impressive architectural knowledge of which religious orders were depositary. And while some scholars have started pointing out the roles of those orders in erecting monumental buildings, there is still much work to do to reconstruct their part in weaving the urban fabric.)[13]

Many historians have detailed that the commune's idea of spatial order was produced through straightening, paving, and measuring.[14] Values such as justice, goodness, and beauty were not just celebrated abstractly but constructed collectively and laboriously. And while order was often presented as a state of being, it should instead be seen as a constant process.[15] Buildings, large and small, expanded in all directions, frequently swelling beyond their consented sizes; and they needed to be brought back into shape. By clearly articulating their possibilities of growth through measurements, communes made sure that cities could always return to order. Measurements were wielded as tools not for making, but for remaking.

The commune, apparently, was not alone in this endeavor. If citizens came to actively seek the commune's urban regulations, it was because preachers had pointed out that God pursued order. And if citizens associated order with quantification, it was because the friars reminded them that God created the world according to measure, weight, and number. The triplet—and here I repeat myself—is deceiving. Counting, weighing and measuring are not similar forms of quantification, and the commune, as well as religion, singled out measuring above the other two because only measurements produce the delusion of participation on which political ideology rests. Measurements are tools of dissimulated activism: they give the credulity of quantifying objectively, while in reality they serve to legitimize the power that, through them, takes hold of space and limits movement. Measurements turn space and bodies into extensions of power. The seizure of measurements is merely descriptive (as proved by the fact that any dimension can be articulated through different standards). And yet, at every act of measuring, measurements contribute to extending the hold of the power that legitimizes them, while producing dimensional relationships that eventually affect ideas and judgments, and trigger actions. Measurements are sly tools of subjugation because, at every use, they slowly turn the world into a place that continues to make sense as long as the power that legitimizes the measurements rests in place.

As gestures both coming from above and supported from below, measurements enact that process of self-corroboration in which lies their ideological efficacy. From such a circular endorsement derives much of their cultural significance: both standards and ratios, both bodily parts and abstract dimensions, both causes and effects of power. As ideological tools, measurements are taken as self-evident. They need no explanation, as the reasoning surrounding their origins is ultimately subordinated to their political legitimization. While counting more easily transcends power and culture, measuring is a form of quantification whose origins and practices are more difficult to remove from the grip of specific, localized authorities. And such a political charge is clear once we realize that the ultimate purpose of measurements is not truth but dissemination, for only if accepted by everyone can measurements effectively operate. Measurements can even be grossly imprecise and uneven, but if endorsed collectively they succeed in their goal. The mathematics of measurement is predicated on the social, and so is their epistemology.[16]

The dependency of measurements on politics comes into a better focus once we realize that measuring norms often stemmed from response to circumstances. In the opening years of the 1090s, the bishop of Pisa Daiberto was asked to settle a dispute over the height of the towers in Pisa, which were growing at a fast pace. He defined the miximum allowed height as thirty-six *braccia*, as this was the size set by Henry IV in a diploma of 1081.[17] The tremendous variety of dimensions and distances set by emperors, communal governments, popes, and their legates is an effect of the lack of awareness about the political expansion of measuring: there was only management. The height of towers, the width of streets, and the length of gowns were produced in response to specific conflicts which, however, could end up taking the role of archetypes.[18] As we saw in chapter 18, the original distance between mendicant churches had been set up at three hundred *canne* because of the looser urban fabric and politics of Ascoli, where the norm seems to have been produced. It was later retouched to 140 as it adapted poorly to denser settlements such as Assisi or Bologna. Starting as impulsive measures for damage containment, dimensional norms did not undergo any systematic review. The retouching of the distance from 300 to 140 *canne* was an exception.

Still, the immense amount of money and energy put into the checks, and the self-consciously monumental character of so much of this control, persuaded citizens of the sincerity and seriousness of the commune's commitment to order.[19] Also, because measurement standards were perceived as neutral tools that would survive any current government, and were endorsed by religious orders who cast their benefits as divine, the dimensional prescriptions of the commune did not produce a sense of domination, but instead emanated the idea of rightness: absolute, enduring rightness.[20]

Such a positivity reflected the public nature of the norms. The commune's quest for legitimacy occured not only by setting boundaries and making people ensure that they were maintained, but by recalling the existence of such boundaries with hammerlike insistence. The regulations of the statutes were proclaimed at the inaugural meetings of artisans, merchants, officials, and Penitents. In Bologna in the 1280s, trumpeters declared every communal act at 204 clearly defined points throughout the city to make sure that everyone would hear.[21] They themselves were embodiments of the measuring craze of the time, as they were obliged to live within fifty *pertiche* of the communal palace so that they could be summoned at short notice. Religious orders moved more subtly but just as effectively in advertising communal goals and policies, mostly because they were victims of the same ideology. They themselves presented measuring as an unbiased search for truth, while using it to implant in subjects a strong sense of identity between them and the government.[22]

For a telling example, reconsider Bonvesin de la Riva's celebration of Milan. To his sleepy and pessimistic citizens, he presented his goal as revealing "the purest truth" (*veritate purissima*), adding that he was set in motion not by interests—he specified that he did receive no compensation or intimidation—but by divine illumination.[23] Like Giordano da Pisa, he saw his city threatened by lack of "unity" (*civilis concordie*), but also thought that the menace was a consequence of ignorance about the city's workings.[24] He thus set on a quantitative exploration of Milan around its "carefully measured" (*diligentissime mensuratus*) city walls. Bonvesin, who defined himself as both friar and Milanese citizen (*frater . . . civis Mediolani*)—a double identity that does not go unnoticed here—mastered measuring. His survey of Milan's was consistent. He clung to dimensions, as well as numbers, as markers of objectivity, because they cast his discourse as accurate. Still, he coupled quality and quantity so frequently in his urban descriptions as to produce a virtual statement about their interdependency. Or, in other words, Bonvesin deployed measurements topologically to such an extent that, instead of constructing a perceptible city, he built an imperceptible order, which may have been invisible but was just as practical as walls and friars checking scales in assuring citizens that a plan was in place.

22

The Height of Christ

Measuring did not find its objects solely in the bricks and the spaces that made up a city. Measuring was also practiced in the solitude of a friar's cell, or revealed to a confessor when admitting a sin. Friars even invited readers to measure religious matter, even if today only a few objects let us glimpse what must have been a not uncommon practice.

One such object is a sumptuous manuscript in Florence's Laurentian Library. It was made in Genoa around 1293 for a wealthy patron of the Friars Minor, as evidenced by the inclusion of numerous Franciscan prayers and saints.[1] One of its first pages represents Christ in a gold-trimmed tunic, clad in an oak-green mantle (plate 8). He triumphantly blesses the viewer while holding a crossed staff, which stands, as he does, on a richly decorated band: a *linea*, according to the accompanying text. "If prolonged twelve times," the text continues, "this segment shows the height [*mensura*] of the body of our Lord."[2]

The illumination tackles the enigma of the intangible, elusive body of Christ. Because Christ's ascension to heaven was both a problem and the ultimate breathtaking miracle, what he left behind was carefully scrutinized and worshipped.[3] Literary references to Christ's stature and appearance received a great deal of attention. How tall was he? How muscular? How long was his beard?[4] And his descriptions in the Gospels and its commentaries were evaluated through the physical evidence of the Holy Land. As a crucial tool for corroboration,

145

measuring played an important role in the process of retrieving evidence about Christ's body.

Yet measuring came in handy for another reason. As a practice that constantly flips the plane on materiality—abstracting the specific and, at the same time, making the metaphysical palpable—measuring served as an ideal interpretative channel to recuperate the body of Christ. Measuring was, after all, substantially equivalent to Christ's ambivalent body. It invoked proximity while maintaining distance.

The Genoese manuscript did not reproduce Christ's actual height, but only a portion of it. It was up to the reader to multiply and imaginatively reconstruct his body at full scale through other media: a string cut to size, two notches on a wall, a series of mental comparisons. In other words, the segment worked only if it was worked through. Measuring was, then, not unlike praying—that is, a formula activated through recitation, whose reiteration produces an alternative to reality from which praying ultimately draws its force.

Likewise, the point of reconstructing Christ's height was not to remake Christ. Rather, it was to activate a process of assembling imaginative elements that enabled believers, for a moment, to occupy a space removed from their own environment and linked to Christ through real conjunctures. In the Genoese illumination, such a space is represented by the thin red square, whose sides match both Christ's height and the *linea* that works as its base. Its simple construction comforted readers about the accuracy of the translation from the vertical, figurative representation of Christ to the horizontal, abstract band. Significantly, the worshippers were not asked to measure the height of Christ directly. It must have been disrespectful to place a ruler on Christ's face. So the square maintained deference while providing an intermediary plane in abstracting Christ's body. Yet the red line was less a middle stage between figurative body and abstract segment than the visualization of the process that the prayer triggered. It was something neither here nor there: not even in between the two, but a whole different thing. The thin red line brought believers to that viscous level of material uncertainty on which the devotion of Christ's height eventually rested. It was a state where it was unclear whether it operated in a world of things or in a world of ideas.

The caption to the illumination continues: "The *linea* has been taken from the golden cross in Constantinople, which was shaped after Christ's body."[5] This cross, which was kept in the round treasury chamber (the *skeuophylakion*) of Constantinople's Hagia Sophia, is documented as early as the tenth century in a chronicle that discloses further facts.[6] The *Diegesis peri tes Hagias Sophias*—this is the title of the chronicle—adds: "The cross had been precisely measured by faithful and praiseworthy people in Jerusalem and for this reason Justinian covered it in silver and precious stones and had it gilded. Still today it heals from diseases and avert demons."[7]

Because of its healing properties, the fame of the cross spread throughout the medieval world. Numerous pilgrims, traveling from as far as Iceland, mentioned it in their accounts of Constantinople.[8] And, like their predecessors in the Holy Land, they, too, reproduced its length on pieces of parchment, some of which were eventually incorporated into prayer books once they returned home.[9] The last record of the cross comes from around the year 1200, when the pilgrim Anthony of Novgorod saw it outside of the Hagia Sophia sacristy.[10] Four years later it was gone, together with the many treasures that disappeared in the sack of the Byzantine capital during the fourth crusade.

Its loss, however, did not damage the cult. In most of Europe, after all, the cross existed only as an idea. If anything, its disappearance urged a continuing process of substitution based on measuring, one that allowed for the production of lengths that could perfectly replace the original. Many of the manuscripts reproducing such a length called it *mensura Christi*, as it is in the case of the Genoese volume. In an oration opening a manuscript in Cambridge's Trinity College Library, however, it is called *longitudo Christi*.[11] The rest of the text—"Hec linea sedecies ducta longitudinem Dominici corporis ostendat"—is the formula that returns most frequently in medieval documents, most of which are lists of prayers.[12] A Florentine manuscript, which represents Christ's height as a blood-red segment (plate 9), copies it at the beginning of a series of invocations for salvation.[13] In a volume in Munich, it figures among charms against worms and diseases that become effective after drawing geometrical figures.[14] And a second Florentine manuscript—which represents the *mensura* as thick as a linear standard—inventories all calamities from which it shields.[15] "Those who wear this measure or keep it in their houses or see it every day cannot die of sudden death on that day," it starts. "And they cannot be harmed by fire or water, nor by the devil, nor by a storm. And the pregnant woman who wears it cannot die of childbirth if she trusts in the name of our God, Jesus Christ."[16]

As in Florence and Cambridge, the devotion of Christ's height often appears at the opening of manuscripts. It is a way to conjure up the physical presence of Christ right at the start, keeping the following pages holy, free from evil. It is particularly fitting as the opening of the Cambridge manuscript, a copy of Saint Augustine's reflection on dogmas, the *83 Quaestiones*. A few pages after the prayer, the text defines evil as something that lacks measure: "Everything which exists is not without some form. But where there is form there necessarily is measure, and measure is something good. Absolute evil, therefore, has no measure, for it lacks all good whatsoever."[17] As an opening, the *longitudo Christi* thus constructs a threshold to goodness, protecting a place of precisely pondered meaning, from which evil and disorder recede.

And it is to keep evil away that a Florentine manuscript suggests keeping the *mensura Christi* in the house, nailed to the wall.[18] The oration stayed in homes as much as it was carried around. Some of the scrolls that reproduce it are so tiny that they could fit in a piece of jewelry.[19] The *mensura Christi* was also worn as a ribbon around the neck or around wrists and ankles. The tradition was more than half a millennium old. Sellers of tapes and ropes cut to the size of Christ's body parts are recorded in Jerusalem since the sixth century.[20] They reproduced the length of his footprints in Pilate's palace and the impressions of his chest and fingers on the column of the Flagellation.[21] Often barely noticeable—like the depressions on the rock of the Ascension (plate 10), in which pilgrims could not tell whether they could discern one footprint or two—measuring served to substantiate something that did not seem to be there.[22]

It is against haziness that measuring plays one of its most important roles. It takes faint traces and turns them into definite dimensions, turning what may just look like a stone into a perfectly outlined footprint. Measuring is not a process that simply reproduces dimensions. Rather, it constructs a precise outline from imprecision, in order to make that precision its own certification. In so doing, it throws the observer into a mode of being in which things lose their unclear, uneven contours and take up a precise mathematical outline. And because precise outlines do not belong to this world, the moment you start measuring, you already are in the realm of ideas.[23]

The *Diegesis peri tes Hagias Sophias* explains that Justinian's cross did not carry traces of Christ's body. Instead, it was a new object constructed from the dimensions that some of Justinian's officials had retrieved in Jerusalem centuries after Christ's ascension.[24] The cross was not a relic—it had never entered into contact with Christ—and yet it was venerated as such.[25] It was seen as a point of origin, as it was the object that certified Christ's real height. Without it, there was no stature: only some indistinct marks on stone.

The process of constructing a relic out of a length is peculiar, but it can be explained on the basis of medieval law. The famous jurist Baldo wrote that "where the form of a thing does not change, the thing itself is said not to change."[26] His maxim was in response to the issue of identity-within-change that pervades the *Corpus iuris civilis*: How could a community maintain its rights if its members kept changing? More than a century before Baldo, Accursio had solved the problem by referring to the image of the continually repaired ship of the Argonauts.[27] "Even if the ship has been partly rebuilt, and even though every single plank may have been replaced, it is nonetheless always the same ship."[28] Accursio affirms that there are objects, like the *Argo*, that have no identity other than their form (and "form" here points to a meaning so large as to account for naming).[29] This group included the *pietre di paragone*, slabs that were regularly substituted and yet regarded as if equivalent to what they replaced. It also included the series of objects of the *mensura Christi*.

Accursio's reasoning supported Thomas Aquinas's interpretation of being as inherently split. Following Aristotle, Aquinas saw two forms of being as coexisting in any object: the *esse naturale*, which was immanent to matter, and the *esse spirituale*, which transcended matter but preserved the form of its being.[30] Spiritual being could be perceived through the senses and translated across space, and also translated the miraculous power it enshrined. Aquinas explained the process through the metaphor of the seal, whose wax "received" (*accepit*) the image of a stamp, but not the metal of which the stamp was made.[31] Carved to appear as negative ruts, the *pietre di paragone* can also be regarded as stone seals (and the checks for which they were used can be seen as a form of stamping, since the standards in need of validation fit their recesses, touching the bottom and extremes of their incisions). Better than Accursio's gloss, which still refers to material traces, Aquinas's wax metaphor explains how the spiritual being of Christ could enter a cross by means of measurement alone.

Let us recapitulate. The impression of Christ's body in Jerusalem was reproduced on a cross in Constantinople, and such a doubling invited pilgrims to measure the cross themselves, sustained by the belief that measuring was an activity so removed from materiality that it came close to a form of spiritual impression. Through measuring, the process of substitution could continue without risking the dilution of its devotional content. Even if Justinian's cross went missing, worshipers could reconstruct Christ's height, as long as they were in possession of the dimension. Some Florentines believed they could do so at the end of the thirteenth century, as they did two hundred years later when, in January 1477, the Dominican convent of San Jacopo di Ripoli printed broadsides with the *mensura Christi*.[32] And even if it went missing, the cross could be reconstructed; it can be seen today in the Church of San Nilo in Grottaferrata, south of Rome, where it is documented from the seventeenth century.[33] It could even originate an altogether different object, such as the fifteenth-century canopy that currently stands in St John Lateran, Rome's cathedral (plate 11).

In the Lateran canopy, the height of Christ is said to be reproduced by the loft of the four columns. They hold a squarish stone slab, which the architect Andrea Palladio described as the table on which the Roman soldiers cast lots for Jesus's robe.[34] The humanist Andra Fulvio, however, took it as the stone on on which Judas's thirty pieces of silver were counted.[35] That is also what a 1518 inventory of the Lateran relics reports, and it differentiates between the slab at the top (*supra*: Judas' table) and the columns supporting it (*sub*: Christ's height).[36]

The canopy is today in the church cloister. Yet, until the end of the sixteenth century, it was in the council hall of the adjoining Lateran Palace. The palace received a great deal of attention from the papacy of Nicholas V (1447–55), who started energetic campaigns to make Rome both the center of the Christian world and the capital of an absolutist state. Rome had never really enjoyed such a dual role

22.1 | Pompeo Ugonio, sketch of the council hall of Rome's Lateran Palace (ca. 1588). Cod. Barb. Lat. 2160, fol. 157v, Biblioteca Apostolica Vaticana, Vatican City. Photograph © Biblioteca Apostolica Vaticana.

before.[37] In particular, Nicholas and his immediate successors redesigned liturgies and refurbished venues to strengthen the pope's role as Christ's vicar on earth. The Lateran council hall was one such space.[38] It was destroyed at the end of the sixteenth century to make room for a new palace.[39] During the demolition, however, the antiquarian Pompeo Ugonio recorded its appearance in a sketch that enables us to reconstruct it with a good degree of precision (figure 22.1).[40] The hall was accessed by three gates. Across the hall, in the main apse, a mosaic represented Christ charging Saint Peter and Saint Paul with the power to carry his word across the world, a theme continued in the ten lateral apses, which depicted the mission to the other apostles.[41] At the center stood the *mensura Christi* canopy, which served as the backdrop for the pope's throne, the *falstidorium*.[42]

The most detailed description of the canopy comes from the organizer of Pope Innocent VIII's investiture on August 26, 1484. Johann Burchard, the master of papal ceremonies, wrote that the pope stood in front of it with a "rod" (*ferulam*) in his hand and "without wearing the mitre" (*sine mitra*), his traditional headgear.[43] All around him was a crowd of cardinals and deacons, who, while staring at the pope, must have seen a sort of three-dimensional rendering of the Genoese *mensura Christi* (plate 8).[44] There, the pope recited *Exaudi Christe* ("Hear, Christ"), the hymn of the most solemn events. As he sung about the need to materialize Christ's words on earth, he was transfigured into his representative on earth. Framed by a structure that reproduced Christ's bodily presence, he became Christ's double. The structure, like the song, substantiated the translation of Christ's mission and miraculous power to the pope.

The connection between the Roman structure and the illumination may be even closer, as the Burgundian pilgrim Georges Lengherent, who was in Rome in 1485, wrote that the Lateran canopy also served as the portal for papal ceremonies. Participants in the rite went through it as a way to join the procession.[45] If this were true—and it would be beneficial to have at least a confirmation of this practice—the Lateran canopy articulated the threshold to moral perfection much as the *mensura Christi* served as the protective frontispiece of manuscripts.

23

The Touch of Measurements

People believed in the healing powers of the *mensura Christi* because measuring was a privileged way to reach the miraculous source of a holy body.[1] Legend has it that a lame Florentine took the measure of his own body, went to Siena's church of San Domenico, and placed the dimension over the tomb of the famous preacher Ambrogio Sansedoni. After the contact, he was able to run and jump again.[2] A second story recalls that a suffocating man was healed by praying on the tomb of Saint Umiliana de' Cerchi in Florence's Santa Croce and, after his prayer was answered, asked a notary to record the size of the bone that the saint had helped him cough up.[3] Such procedures, which mix the legal and the devotional, were common throughout the fourteenth century. The influential theologian Thomas Ebendorfer validated such practices in his work *De decem praeceptis* of around 1438, where he wrote that in order to cure a sufferer, it would be enough to present a priest with "a rod replicating her or his height."[4]

While focusing on measuring, these stories are ambivalent as to what ought to be measured. Taking the dimension of the aching limb, the saint's body, or even the perimeter of the reliquary all seem to have led to the same result. Even within a single episode, the description of the measuring is elusive. Did the Florentine record the size of his height or of his shin bone, which the tale explicitly identifies as the cause of his limping? And how did he reproduce such a dimension? By drawing it, or by cutting a ribbon to size? Such a number of possibilities was less a sign of confusion than an effect of the commutative thinking fostered by measuring. Measuring was not only a contemplative device, a way to mentally linger over the relics, but an indexical process of translation that enabled a living body to enter in contact with a spiritually enhanced one.

The body of Saint Rose of Viterbo—a prodigious child venerated since 1258—had been measured for centuries.[5] Her monastery still owns some of the ribbons that reproduce the height of the saint's diminutive body (plate 12).[6] Every twenty years—the practice is recorded since the early seventeenth century—the nuns in Viterbo took the corpse of Saint Rose out of her coffin and undressed her. They tore her tunic and veil into pieces to be distributed as relics, before garbing the body in new clothes. While undressed, her body was also measured with silk and brocade ribbons, which the nuns then sold to the public.

This devotion is documented in parallel to another more ancient one, according to which people pressed belts against Saint Rose's body and made pregnant women wear them while giving birth.[7] The passage from belts to metric ribbons is indicative of the ambivalence between touching and measuring. Many legends report about it. Another miracle attributed to Ambrogio Sansedoni's sepulcher in Siena, for instance, involves a girl who was healed by simply touching it.[8] The story, which presents the girl as affected by ambulatory problems, is a doubling of the tale of the Florentine paraplegic with which I opened this chapter, and is another indication that measuring and touching were regarded as equivalent gestures.

For Thomas Aquinas, a fellow student of Sansedoni (they went to university together in Paris), touch was the prime sense for healing. Christ could heal a leper through touch, as French kings were also believed to do.[9] If metric ribbons restored health, it was not just because they reproduced the form of miraculous objects. It was also because by entering in contact with a holy body, they became its spiritual extension. An anonymous pilgrim of the early Middle Ages reports that the ribbons that reproduced the size of Christ's feet had been produced by pressing them against the relic of his footprints in Jerusalem.[10] They were thus worn by pilgrims, who felt safe by imagining Christ's body pressed against theirs.

Measuring is, as it was, an effect of touch, since you cannot measure something unless you have placed the measuring tool against the object to measure.[11] Or better, measuring is both a confirmation of touch and a demonstration that contact has been carried out attentively.[12] In the *Golden Legend*, the famous collection of saints' miracles composed by the Dominican Jacopo da Varagine, there is a story in which healing is the consequence of consistent measuring. "In Bologna there was a scholar named Nicholas who suffered such great pain in his kidney and in his knees that he could not leave his bed. He then avowed to God and Saint Dominic, and took a thread to make a candle of his length, and measured his own height, and then his body, his neck, and his chest. When he put the thread around his knees and called, as he did at every measure, the name of Jesus Christ and of Saint Dominic and felt relieved of his pain, he exclaimed: I am freed."[13]

The story presents Nicholas's measuring as the reverse of an act of binding. Nicholas is freed after wrapping the string around many of his limbs. The survey of his body is both faithful and mechanical. He repeats the invocation at every measure, before moving to the next. At the end, it is unclear if he gets healed because he eventually wraps the string around his aching knees or because he has covered and purified his whole body. The story, in other words, dramatizes the indexicality of measuring.[14] The healing potential of measurement is a consequence of the promise of sameness that only mechanical touch and faith guarantee. Sameness is but an act of faith in the carrying out of the reproduction without human intervention. Like faith, sameness collapses spatial, temporal, and material differences, producing a magical chain of bodies and objects that are dimensionally and thus spiritually linked. This process is not only from body to body; as in some of the examples I have mentioned, they also deal with objects. Buildings participate to this process too—which is important, as it shows the scale the measurements traverse: from the intimate to the monumental, and from body parts to urban structures.

Besides housing the tomb of Saint Dominic, Bologna attracted pilgrims because of the Benedictine monastery of Santo Stefano (plate 13). A sprawling cluster of churches, cloisters, and chapels, the complex is famous for replicating many of the holy sights of Jerusalem, such as the Holy Sepulcher and the courtyard where Pilate washed his hands. The Santo Stefano shrines share some architectural elements with their originals, as well as some dimensions.[15] The distance from the Bolognese reproduction of the Holy Sepulcher to the chapel of the Holy Cross, for instance, reproduces the same distance in Jerusalem—a way, scholars say, to participate in the holy perfection of the original (figure 23.1).[16] It has, however, passed unnoticed that the Santo Stefano site was described as a feast of measurements in a twelfth-century manuscript, the *Vita vel acta Sancti Petronii*, which attributes the transformation of the Bolognese site to the activism of the city's patron saint, Petronius.[17]

An aristocrat from fifth-century Constantinople, the man who came to be known as Saint Petronius, was sent as an ambassador to Rome, where he was appointed bishop of Bologna by Pope Celestine I. Once in Bologna, Petronius started a vigorous campaign of church restorations. In particular, he focused his energies on Santo Stefano, which he decided to model after the Jerusalem sites. The *Vita* describes the outcome not only as architecturally dazzling—Petronius brought in a forest of expensive porphyry columns, and had the capitals nested with animals and birds—but also as metrically rigorous. Indeed, Petronius measured the Jerusalem sites himself when he went there as a tax collector. (It was his success at bookkeeping that earned him the promotion to ambassador.)[18]

More specifically, the legend speaks of three moments of measuring. In Jerusalem, Petronius took the distance between the Golgotha and the Mount of Olives, which, his biography explains, he then reproduced between Santo Stefano and a Bolognese hill nearby, on which the church of San Giovanni in Monte was built. He also took the dimensions of the Pool of Siloam, where Jesus healed the blind man. The narration specifies that he took them with a "rod" (*calamo*), a detail that not only conjures up accurate measuring, but also reveals that Petronius entered in physical contact with the site of a miracle. Finally, Petronius commissioned a replica of the Holy Cross "in height and width."[19] Three references to measuring are an exorbitant number, given that the whole narrative of the journey to Jerusalem is less than a page long. Their frequency serves to convey Saint Petronius's level of engagement and, at the same time, to validate the regenerative effects that his dimensional translations soon produced.

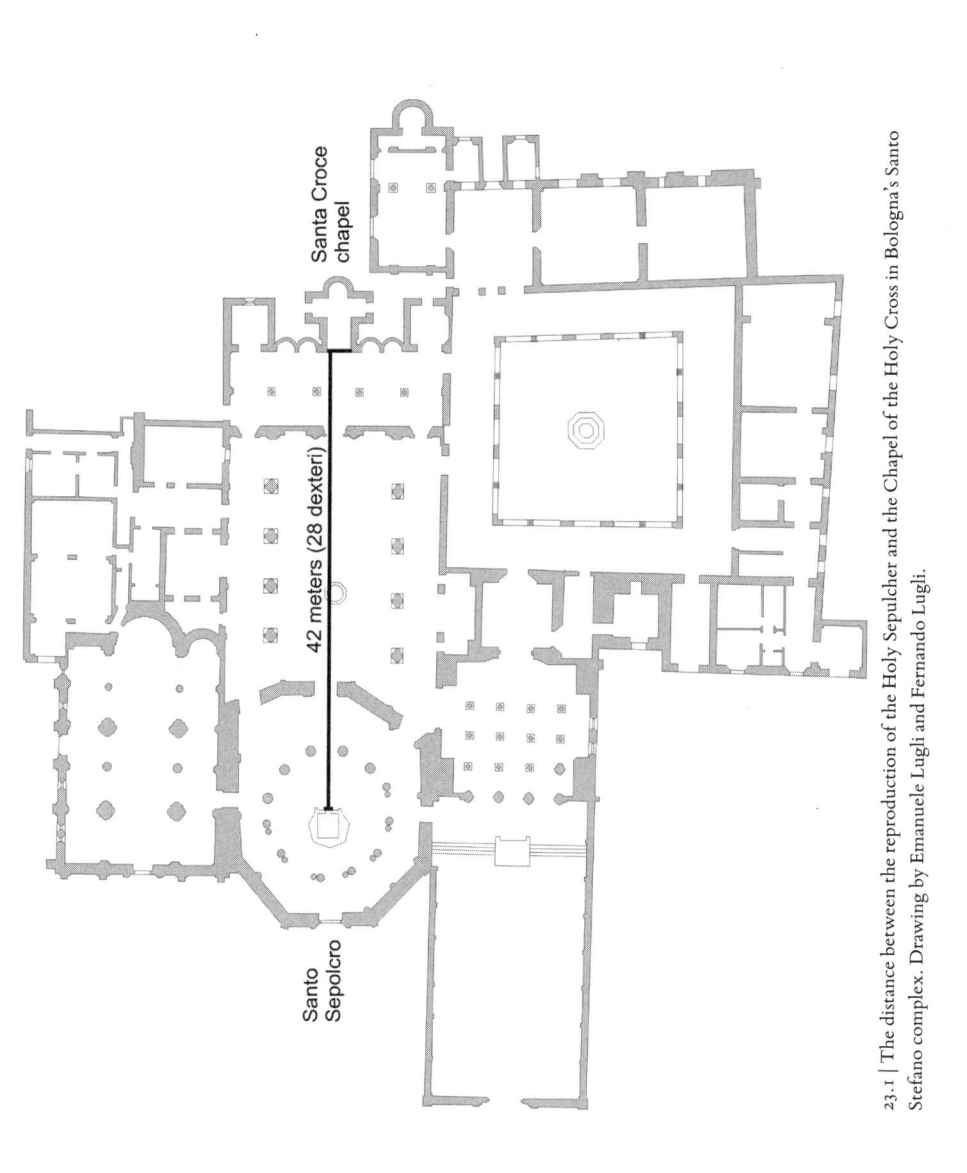

23.1 | The distance between the reproduction of the Holy Sepulcher and the Chapel of the Holy Cross in Bologna's Santo Stefano complex. Drawing by Emanuele Lugli and Fernando Lugli.

The *Vita* plants the idea that thanks to Petronius's measuring, Bologna bloomed. The city became an impressive "garden" (*ortus*) whose abundance fulfilled the premises of its name *Bononia*: "replenished with all sorts of goods" (*omnibusque bonis ad plenum refecta*).[20] The panegyric may be as conventional as the idea that replicating Jerusalem turned Bologna into an evergreen earthly paradise. Yet, if the transformation felt real, that was because measuring was thought to provide access to the miraculous source of eternal life.

The many floral references also turned Saint Petronius into a farmer. The *Vita* explicitly says that he guarded his city "as a farmer protects his gardens."[21] But, given that he is described measuring the distance between Jerusalem and a mountain, it would be better to call him a land surveyor. The image of Saint Petronius climbing a hill with a rod resting on his shoulder, or measuring the borders of a pool located outside the walls of Jerusalem, must have strengthened many listeners' belief in his deeds. Indeed, no other figures in the Middle Ages had a stronger connection to measuring than land surveyors, as part 4 of this book will argue.

PART IV

Fields

24

Dividing Up the Land

Ovid's *Metamorphoses* starts as a poem of natural deterioration. Rather than presenting humanity as laboriously making its way through pain to reach prosperity, Ovid imagined the opposite process. The world was born self-sufficient. Seedless flowers bloomed. Tree branches were constantly loaded with fruit. Streams of milk flowed through glistening crops in an eternal spring. And humans, for Ovid, were completely extraneous to such abundance. They did not even seem to appreciate it. Perhaps they did not understand where they were, or at least this kind of puzzlement is what Ovid hinted at by introducing them through negation: no human ever raked the ground, no human ever felled a tree, no human ever kneeled in front of another.

Such a list makes little narrative sense. Our progenitors cannot possibly have experienced any of those feelings. Whoever lolls in abundance has no sense of its lack. So Ovid was speaking not of the world's first inhabitants, but of his Roman contemporaries, whom he simultaneously cast as readers and as the people about whom they read. He encouraged them to imagine what it must have been to live in paradise and, at the same time, he frustrated this very possibility by recurring to the language of ineffability. By speaking of the archaic through the contemporary, Ovid set the mythical pace for a collection of stories that, while remote, were meant to be relevant. And by presenting perfection as flawless, the *Metamorphoses* played on the fundamental tension in the ways that happiness can only be thought of through the negative.

Ovid's inversions cast a shadow over the glorious opening scenery of his poem. They also prepared his readers for the cracks that were about to come. After Jupiter overthrew Saturn, under whose watch the earth prospered, the unchanging year broke into seasons. No longer protected by the mild stillness of the sky, people fought for abodes. Wars broke out. Many sailed away, enduring lives of peregrination. Those who stayed dug the earth, searching for treasures like mad men. Greed expanded into mining. The age of gold opacified, silver dimmed into bronze and then faded into iron, the matter of swords and drills, which fractured the earth into pieces.

The first book of the *Metamorphoses* is largely a poem about the earth, whose status it constantly checks. And the figure who gives the earth its fatal blow is the surveyor. Ovid describes him splitting the ground into fields and marking the boundaries that divide people and regions. His operations have nothing to do with justice. Rather, Ovid presents him as a violator of freedoms and a constructor of inequality, a creature seized by fear (he is "cautious," not careful).[1]

In the Middle Ages, the *Metamorphoses* was mostly encountered through moralized abridgments, which shrunk the poem to only a selection of its many godly transformations.[2] Still, its hexameters were transcribed from at least the eleventh century, and those copyists learned from Ovid's rich vocabulary of borders and his explorations of limits and their transgressions.[3] To them, the beginning of the *Metamorphoses* must have looked very much like the book of Genesis in the Bible, which also linked the fall of humankind to Cain's plowing of the earth, soaked with the blood of his brother Abel.[4] Both texts presented the furrowing of the ground and the marking of boundaries as acts of violence against the earth, in need of expiation. For Isidore, the bishop of early seventh-century Seville who was much read throughout the Middle Ages, such a sacrilege was punishable: "They say that the first person to yoke oxen to a plow was a private citizen named Homogirus. He was struck by lightning."[5]

Isidore's take sets the theme of the last part of this book, in which I investigate measuring in relation to agriculture, architecture, and land surveying. I started this chapter with a poem, as I will mostly deal with texts—and, in particular, texts that, like Ovid's, metamorphosed over time into texts that had little to do with the originals.

This chapter is largely about the ways in which ever-changing books perpetuated knowledge of measuring, and how those books generated ideas about the earth as the origin of measurements.

There is one main reason for speaking of beginnings at the end of this book. Reversing the traditional chronology is a way to avoid repeating the historicist narrative that is so typical of the study of measurements. The history of measurements usually presents a false sense of progress, and I am keen to break such a pattern, since an origin is often constructed in relation to the needs of whoever sets to search for it. So, inspired by Ovid's folding of his time onto a mythical past, this section rereads measurements of the late Middle Ages through the modality by which a sense of origins attached itself to them.

Such temporal dynamics are of prime importance when discussing measurements. If measurements came to be trusted, it was also out of respect for their old age. Their perdurance over long stretches of time heightened the sense of truth that measurements inspired. But people also believed in measurements because they thought they worked as homogeneous connectors between any two periods. Justice and fairness are predicated, not only in medieval times but today too, on the illusion that measurements stand still through time, cutting through it as an arrow pierces the air. If this concept never came under scrutiny in the age of the communes, but rather was celebrated as a way forward, it was because measurements were said, over and again, to derive from the immobile earth.

Measurements entered history through documents concerning fields. Fields were described by their size and, as tithes and taxes were paid in either money or products without any one system prevailing over the other, measuring also facilitated the calculation of what was due.[6] Even acts of donation or contracts that established feudal bonds articulated the dimensions of fields, as size revealed the value of the commitment.

Measuring was so fundamental to the identity of a field that it yielded the concept of the *modius/moggio*, the field that produces the homonymous amount of grains.[7] It also led to the *mansus*, the field that does not change in size. It has been speculated whether *mansi* came to refer to allotments of identical extensions or, given that the ground could turn rather dramatically from crumbly dry to moist, whether it designated fields of similar productivity.[8] A contract from 1118 specifies that all the *mansi* in the territory of Reggio Emilia corresponded to twelve *iugera*.[9] Yet the evidence stops with documents like that, and after the year 1000 the term is taken to have been more or less synonymous with *pecia* ("piece of land").[10]

The *moggi, mansi,* and *pecie* that made a landowner's estate were scattered across vast regions, often days of travel from the residences of their possessors, who went to visit them only occasionally if at all. But when they did visit, they could identify them through the measurements recorded on their notary acts. Consider a document from 898—a *permutatio* of five pieces of land owned by the monastery of Sant'Alessandro in Albino, in northern Italy. There, each field is identified by three characteristics. First, its location—piece #4, for instance, is to be found in the area called "Pradello." Second, its neighboring properties: piece #4 borders on a field owned by Inselberto Paulone to the east, on those administered by the church of San Giuliano and the king to the south, and on fields belonging to the monasteries of Sant'Andrea and San Giuliano to the west and north. And, finally, a field was described by its dimensions: piece #4 corresponded to 103 *tabulae*.[11]

As medieval land economy made use of documents that often had been drawn up decades if not centuries before, such a tripartite procedure remained useful for a long time.[12] It returned, for instance, when a notary enlisted the land that Countess Rolinda donated to the bishop of Lodi Ambrogio in 1044. This amounted to one-third of fifty-eight fields, which the notary described following this format: "The first piece of land is in the area called Ponticello. It is surrounded by roads east and north, on the other sides it joins the rest of the fields. Its third portion precisely corresponds to nine *perticae* (of a *jugerum*) and sixteen *tavole*."[13] The continuity of the notary practice over centuries witnesses to the time-defying capacity of measuring, which could be carried out in the way it had been carried out for generations.

Both these documents, after all, rely on measurement standards constructed in the eight century by the king of the Lombards, Liutprand. Liutprand devised them probably in preparation for carrying out an extensive survey of the royal land around 733.[14] Yet his chronicler Paul the Deacon gave a different explanation. He wrote that after entering the territory of Milan, Liutprand realized that it did not have a measuring standard. So he resolutely placed his foot on a flat rock outside of the city walls and transferred its length. He then declared that from then onward, everything should be bought and sold according to that dimension.[15]

There is a lot to unpack from this anecdote. It offers the idea that measurements are effortless and straightforward, like scratching a rock with a stone. It is paradigmatic of standards as gestures of power. The new Milanese rule was identified with the king: it was Liutprand's foot, a trace of his body validated by his word. Nothing else mattered: the king did not even spend a word on the way in which it should be preserved. The episode was providential in the way it quickly supplied what the city lacked, and its prodigious speed had something divine about it. Paul the Deacon, who wrote the chronicle at the end of the eighth century, more than a generation after Liutprand's death, seems to have channeled an episode of the Apocalypse.[16] There, the archangel of justice places a foot on the earth and gives St. John the book of wisdom, after which he can measure the Temple of Solomon in Jerusalem. The measuring starts after the ultimate judicial authority stomps on a rock, making his presence reverberate through the earth.

In Liutprand's measurements, twelve of his feet (*pedes*) made one *pertica* and 12 × 24 square *perticae* made an *jugerum*. The square *pertica* could also be divided into twenty-four *tabulae*.[17] Such a system was in vigor throughout the kingdom of the Lombards, which at the height of its expansion occupied almost the whole of the Italian peninsula, minus a stretch of land that linked Rome to Ravenna and other Byzantine outposts in Naples and the south. Because of the system's pervasiveness, it never completely disappeared from Italy. As we saw in chapter 6, it was still employed in Napoleonic times in districts as distant as those governed by Lucca and Turin. Eight hundred years earlier, in the eleventh century, a document about the properties of Countess Beatrice of Canossa—who administered a patchwork of fields stretching from Tuscany to Mantua—specifies that the fields should be measured in "Liutprand's foot and the legitimate *pertica* of twelve feet" (*a pertica legitima de pedibus duodecim a pede Liprandi mensurata*).[18] It is a formula that, besides specifying the system of reference, makes a point about the Canossa's lineage. A family from

Lucca that followed Lombard law, the Canossa increased their power from around 940, when the Marquess of Provence, Hughes of Arles, was elected king of Italy and started giving land to his supporters, including the Canossa family.[19] By the time Beatrice was in power, the Canossa had greatly expanded their properties; but, by officially defining her fields in Liutprand standards, Beatrice emphasized the antiquity of her possessions.

25

The Origins of Medieval Measurements

In the mythologizing of her family's possessions, Beatrice of Canossa's reference to a particular measuring system was a stern necessity. At the time her notary drew up the contract, Italy was going through a phase of political frailty, and measurement standards with them. By the time Beatrice died, in 1076, northern Italy had been governed by a string of weak kings for almost two centuries. After Charles the Fat died in 888 and the Carolingian dynasty came to an end, several military families disputed rule over Italy.[1] Among them was Hughes of Arles, whom the Canossa backed. While the royal chancellery in Pavia—the seat of power since Liutprand's times—continued to function, it did not steer the situation in any clear direction by issuing norms of general validity, but instead granted privileges only to specific addressees.[2] Power then splintered, and fragments of royal rights passed into the hands of important military lords who expanded their estates, as they were the only ones able to protect them.[3] Control also passed into the hands of numerous bishops, supported by military groups. They, too, managed to receive numerous royal privileges for the areas that they administered. A startling example is that of the bishop of Cremona, who in 916 received all duties, from the city and for a territory of five miles around it, in perpetuity. In an age when a king never alienated his privileges but only occasionally granted them (as even their momentary loss could compromise the royal finances), the concession of rights in perpetuity was astonishing.[4] It was even more astonishing in that such an unheard-of concession ought to be considered among a plethora of others, such as that of 998 to the bishop of Pistoia, who obtained the rights over the city's market, which also included the taxes for the grain standards.[5]

Measurements were doubly invested in replenishing a lord's funds. By quantifying productivity, they determined taxation. Yet as unique, privileged objects, they were also subject to taxation. Whoever checked the correctness of standards and pocketed the revenues such control generated paid a fee to the rights holder, who was usually the king or a bishop who could also concede other privileges, such as the management of a road or the digging of a mine. The Canossa family continued to refer to royal standards in deference to the ultimate authority. But with the breaking down of privileges, other landowners constructed their own sets of standards, hoping that the switch would result in a lengthier concession of the right. And as the king and bishops passed the right to them, so they passed it to subordinates, such as ford patrollers, who also developed their own tools over time, forgetful of higher authorities.[6] In the tenth and eleventh centuries we thus witness a process of metric multiplication: the number of measures increased while the territories of validity contracted. From 1066, the cathedral chapter of Verona referred in its contracts to many different units of measurement, one for each party with which it dealt.[7]

In a 996 diploma to grant the use of a few fields in Pisa to the lord of Ripafratta, Manfredo di Roncione, the chancellery of the king of Italy, Otto III, defined their size by referring to a local unit of measurement: "secundum mensuram legitimam et nominationem patrie."[8] By then, *mensura legitima* was a common expression, which appeared in numerous documents.[9] Beatrice of Canossa's notary used it too. But *nominationem patrie* is not a recurring formula. *Patria* unmistakably refers to the territory of Pisa and not to the Kingdom of Italy, for which the royal scribes would have employed the word *regnum*.[10] This linguistic usage is confirmed by several other charters in which the word *regnum* is opposed to *patria*. This diploma, in other words, indicates that by the end of the tenth century Pisa was using local rather than royal measurements.

Pisa was not alone in slipping from the royal grip. Around that time, Bolognese notaries started speaking of *tornature*, a unit that is not part of Liutprand's system.[11] Other cities designated royal standards by local toponyms: the Treviso *jugerum*, the Bergamo *sextarius*.[12] In 1060, Milan constructed its bronze standards for salt and grains.[13] It did so at the end of a series of clashes between the Count of Milan, an appointee of the king of Italy, and the citizens, after which the city took up a new form of organization that, modern historians explain, led to the formation of the commune.[14]

Documents make clear that the multiplicity of measurements continued for decades, producing a jagged metrological landscape, which is all the harder to map out as the terminology produces a sense of homogeneity that did not exist. In 1173 a transfer of properties between the San Cipriano di Murano monastery and the Marquess d'Este specifies that a *mansus* equals a surface of 40 × 100 twelve-foot *pertiche*. But a document from around the same time described a series of *mansi* as forty-eight *biolche*.[15] A slightly earlier document, from 1140, specifies instead that the *iugerum* was made of six *perticae* rather than the traditional twelve.[16]

While clinging to a standardized vocabulary, these documents also reveal the greater care with which notaries specified measurements. And it is also as a way to eliminate ambiguity that the first lengths appeared in stone. The *passus* of Salerno, for instance, is documented as inscribed on a marble column inside the cathedral atrium in the 1070s.[17] But let us be clear about the general validity of such specific cases. The Salerno *passus* did not have a public dimension. It stood inside the bishop's courtyard, protected by tall walls whose gates were locked at night. It did not function amid a network of gazes. In a way, it was not a *pietre di paragone*. More interesting are the mentions of the Florentine standard. Sale documents from 1083 refer to fields around the Tuscan city as "measured with the foot that is designated in the stone placed next to the Gate of San Pancrazio."[18] The gate, named after the eponymous monastery outside of Florence, was the western opening of the city's most recently constructed circle of walls, built in 1078 under Matilde of Canossa, Beatrice's daughter.[19] It stood where today via della Spada meets via della Vigna Nuova, whose name still conjures up a thoroughfare cutting through the monastery's young vineyards, which led to Florence's main market after passing through the San Pancrazio gate. I imagine that the stone incised with the foot

25.1 | The *canna* incised in the San Francesco gate, Volterra. Photographs by the author.

was placed on the inner side of the gate, like the standard still on the gate of Volterra (figure 25.1).[20] While its precise location is unclear, its dimension is less so. It corresponded to the length of a line traced along a 1094 document that traded fields owned by the monks living in Luco di Mugello, some thirty kilometers north of of the San Pancrazio gate (plate 14). The parchment is preserved at Florence's state archives and the the segment is labeled *pes portae* that is, "foot of the gate."[21] The Florentine standard was still incised on the gate in the year 1200, after Florence had constructed a new, wider ring of walls (1173–75) to protect the San Pancrazio monastery and other neighboring possessions from the attacks of the new king of Italy, Frederick Barbarossa.[22]

Rather than zooming out to deal with larger politics, however, I want to stick with the Florentine gate for a bit longer. If we reread the records for Faenza, Bergamo, and the other early adopters of *pietre di paragone*, we can see that the standards were always fixed near doors, mostly cathedral doors (see endnote).[23] Standards were translated from city gates to church doors, shifting their role from marking the threshold of a judicially privileged space to indicating the place of justice. On the San Pancrazio gate, the Florentine *piede* advertised to incomers—market sellers, buyers, visitors, diplomats—the existence of a set of rules. It served as a practical mold, but it also signaled the city's metric sovereignty and economic independence. Gates and doors are regular presences in the history of measurements. *Fratres penitentie* impatiently waited at city gates to weigh incoming products: we saw that in chapter 17. But it is also on a rock at the gate of Milan that Liutprand stamped his foot, thus providing the *pietre di paragone* with an archaic model.[24]

This specific legend was known by Sicardo, the bishop of Cremona who petitioned Frederick Barbarossa to confirm the incredible privileges obtained by his predecessor in 916.[25] Sicardo succeeded, and when compiling his chronicle of the city, he included the episode.[26] So did the jurist Alberto Milioli when composing the history of Reggio Emilia.[27] Before becoming a writer in his retirement years, Alberto was a jurist for the commune. He oversaw cases such as those contained in the Pratissolo file. The copy of the 1265 Statutes of Reggio, from which we have the earliest mention of the city's measurements, is in his hand.[28] It is then not surprising that he recognized the value of Liutprand's story. Yet for us, his inclusion of the anecdote has another validity, as it provides an argument for the ways in which communal directives and legends could blend into the psychological landscape of medieval measurements.

26

Geo-metria

Most of the information about measures comes from the many manuscripts of the libraries of abbeys and cathedrals. It is not surprising to find them there, given that measuring was essential when administering land. Bad land management had led many religious complexes into debt, and sometimes to the brink of poverty.[1] It was not just a matter of collecting tithes and hoping for good weather and industrious workers. It was, first of all, a matter of knowing how much ground they had, and how to move, work, and water it. Since at least Charlemagne, when his cousin Adelard managed the properties of the monastery of Corbie, abbots had searched for information on land management in classical works.[2] At the library of Corbie, monks could learn engineering, agriculture, hydraulics, and surveying from Boethius's translation of Euclid's geometry, Columella's treatise on farming, and Cassiodorus's *Institutiones*, which included a few words on the importance of measurements.[3]

While succinct, these texts were nonetheless critical in establishing geometry as the field for the studying of measurements. *Geo-metria* was, literally, the measuring (μετρέω) of the earth (γῆ)—that is, land surveying.[4] Measuring was conceptualized less as a set of abstract proportions than as operations directly connected to the ground.

The history of such a connection was explained in two ways. Cassiodorus assigned it to the dawn of time, explaining it in terms similar to Ovid's: measures were created to keep wandering humankind still and at peace.[5] The most popular account attributed the invention of measuring and geometry to the Egyptians. Isidore of Seville—just to mention a popularizer of the account—stated that the Egyptians employed it as a measure to restore the size of the fields flanking the Nile banks after the river's regular floods.[6] By returning the land to its original divisions, geometry made the floods inconsequential.[7] It turned inundations from disaster to resource. And it was because of such a transformative capacity that geometers applied the same geometrical principles to surveying the sea and the sky. Geometry, the discipline that erased time by promising a constant return to origins, metamorphosed into astronomy.

The Egyptian story was much repeated. And yet I find that the closeness of the association between geometry and land is difficult to feel today, even when going through its countless occurrences in medieval literature. It can, however, be intuited easily enough when looking at the belfry of the Florentine cathedral (figure 26.1). On its east side, an allegory of Geometry—a woman holding a compass and a quadrant—sits above a scene depicting the plowing of a field (figure 26.2). The pairing is part of a series in which each allegory of the liberal arts is coupled with the inventor of one mechanical art.[8] And here the mechanical art is agriculture, identified as Homogirus, whom Isidore designated—we saw it in chapter 24—as the first man to yoke two oxen together. In the Florentine relief, Homogirus thrusts a triangular blade into the ground so energetically that his oxen's heads are each driven in a different direction, up and down, moving away from their yoke like the two arms of a compass from their hinge. His machine is not just plowing: it is also surveying.

Called *agrimensura* in Latin—a term that scales down the focus of *geometria* from globe (*geo*) to fields (*agri*)—land surveying was made of practices handed down through a heterogeneous body of late Roman works, the *Corpus agrimensorum romanorum*.[9] Modern specialists speak of a body (*corpus*) of works because it is not one text but a series of them. It includes a legal treatise about land divisions, a compilation of geometrical problems, and a list of the names of the Roman regions, plus memos about how to set up a military camp, how to recognize boundary stakes, and how to calculate taxes. And yet, despite its heterogeneity, even the term *corpus* is somewhat misleading. It gives the

26.1 | East side of the bell tower of Florence's cathedral. Photograph by the author.

idea that these texts were assembled following a consolidated rationale and that they worked together, as bodily organs do. So modern scholars set out to reconstruct that reasoning, privileging a few compilations over others, and thus crystallizing the *corpus* into distinct yet interconnected traditions, without realizing that such traditions are only cursorily useful in describing a hodgepodge of texts.[10]

26.2 | Andrea Pisano and workshop, *Invention of Agriculture* (or Homogirus plowing the field) (1334–36). Museo dell'Opera del Duomo, Florence. Photograph by the author.

Consider the manuscript composed in central Italy in the eleventh century and now housed in Brussels's Royal Library.[11] It includes fragments from Euclid as well as another anonymous geometrical treatise, which—it has been recently discovered—is also a reworking of Euclid encrusted with Arabic commentaries. An excerpt on measurements from Saint Augustine's *De quantitate animae* follows, even if it is a treatise on the immortality of the soul that does not belong to the tradition of the *Corpus*. And then the manuscript includes customary texts, such as a section on triangles by Vitruvius Rufus, and a letter on land surveying by Trajan's engineer Balbus.[12]

When dealing with measurements, medievalist copyists joined information as they saw fit, without assessing the interconnection of the sources.[13] They made numerous choices that scholars now regard as mistakes. Medieval scribes worked their texts freely, cutting and joining them in ways that may seem surprising today. Many manuscripts include excerpts not only from St. Augustine but also on astronomy, or from Isidore of Seville's listing of field sizes.[14] It is not infrequent to see Isidore's passage in connection to Varro's *De re rustica*, even if the latter is the very treatise from which Isidore took most of his information about agriculture. Sometimes the two got mixed with other texts that also reworked Varro, giving rise to compilations that repeated the same information again and again.[15] Reading such a book is a bit like being served a meal of gnocchi, french fries, and mashed potatoes.

The *Corpus* is thus less a distinct body of works than a textual avalanche, whose sprawling complexity (despite giving relatively weak information) is the reason why these texts have been considered second-string.[16] This is true across the board. The *Corpus* is devalued by philologists, who smirk at the simplistic, corrupted Latin of its texts. Landscape and architectural historians overlook it, as they do not see direct references to the built environment. Historians of mathematics neglect it since its treatment of geometry is rather childish in comparison to the contemporary achievements of Arabic science.[17] The *Corpus* thus has suffered the same fate as measurements, of which it represents the primary source. I would like to reflect on this state of things for a moment. The failure of measurements to produce sameness is not just an illusion of reproduction. It is also an effect of the disorderly status of textual transmission. The *Corpus* often confuses rather than informs. It scatters the material in numerous directions without offering ways to connect its many parts. So it is hardly surprising if modern historians are frustrated and dismiss both measurements and *Corpus*. Scholars take the *Corpus* as a set of theoretical exercises with little or no practical value.[18] And yet land surveying was of fundamental importance to the Middle Ages. We have evidence for it.

Towards the end of the tenth century, the bishop of Cremona regularly employed three surveyors.[19] And Gerbert of Aurillac, the future Pope Silvester II, read treatises by Roman *agrimensores* at the monasteriy of Ripoll, in Catalonia, and at the monastery of Bobbio, in Italy, whose vast estates he eventually managed as its abbot.[20] In a letter he sent in 983, Gerbert mentioned a manuscript that described methods and tools for measuring. And in a second missive, sent to Adalbold of Utrecht in 998, he referred to an manuscript on land surveying he had previously sent.[21] Monks did not just read books on land management; they sent them to one another. There was a demand for information. People wanted to know.

When Gerbert wrote to him, Adalbold was at Liège, the seat of a famous school for the transmission of surveying knowledge. Some of Gerbert's pupils, like the monk Fulbert, ended up teaching geometry there.[22] Fulbert taught Adelman, who became bishop of Brescia in 1057. We will encounter him again in a moment, but I mention him here to show how information about land surveying moved in and out of Italy through the large networks of cathedral schools and traveling scholars.[23]

The three surveyors working for the diocese of Cremona were trained in law—the document says it explicitly—and indeed many *agrimensores* texts detail rules to recognize boundaries and settle disputes. Boethius even dedicates a section of his book on geometry to such controversies.[24] Land surveyors served to identify divisions that, not unlike Bologna's subterranean *termini*, as we saw in chapter 19, had sunk deep into the ground, swallowed by marshlands and wild forests. By studying the patterns that the Roman and Lombard boundaries were likely to follow, medieval surveyors could restore them or, at least, something that could have been there. Manuals about land surveying thus included descriptions of the most common types of boundaries, most of which were defined geometrically. They often also contained illustrations, which expedited comprehension and increased the surveyors' competence. The so-called *Codex Arcerianus*, a manuscript completed in the sixth century, is filled with illuminations of land scored by lines and geometrical shapes (plate 15). At the time Gerbert became abbot of Bobbio, it was either in Verona or Treviso, from where it circulated through the hands of some of the most venerated scholars of the late Middle Ages. It was commented upon in Petrarch's own hand and then it moved, amid the booty of armies, from Padua to Milan, before landing in the library of Erasmus of Rotterdam.[25]

In another manuscript from the ninth century, stored today in the Vatican Library, an illumination shows a mathematical landscape (figure 26.3).[26] At first the image gives the impression that a chessboard of fields has been constructed just outside the city. But with a closer look, the illustration registers the way land division sinks into the ground and disappears. On the right of Minturnae—the Roman city is identified by a caption—is a grid that has been recently traced to divide the fields. It is new because it says so: it is labeled *assignatio nova*. On the left, the blank area indicates the fields assigned by professionals. The area is not at all blank. The original division may be gone, but its isolated architectural elements—a hexagonal structure, a vaulted mausoleum, a sculpture on a podium—enabled a professional surveyor to recover it since, as the text explains, they served as boundary markers.[27] The miniature image, in other words, visualizes the natural disappearance of physical borders. At the same time it also advertises the surveyor's reconstruction as possible. By placing the freshly traced grid and the invisible boundaries on the same level, the illumination produces a visual argument for their recreation, similar to the ways in which the Egyptians restored their fields after the Nile floods.

The *Corpus* categorizes field shapes according to location and number of sides. It also shows that the Romans preferred charting areas as large as possible through regular chessboards of roads and canals (figure 26.4). Called *centuriatio*, this grid was laid down in imperial times at the end of a military conquest, of which it marked the ultimate act. Through it, surveyors, who formed a contingent of the Roman army, transformed the land subtracted from the enemy, producing something new. Its benefits were hard to underestimate. The regularity of the centuriation facilitated the calculation of distances and the assignment of fields to settlers, as well as the calculation of taxes.[28]

Roman authors such as Varro describe its basic unit as the *centuria*, a square measuring twenty *actus* per side (one *actus* was around 35.5 meters, or 120 Roman feet of roughly 29.7 centimeters).[29] Two *actus* made one *iugerum*, and two *iugera* (or one *heredium*) usually constituted the size of the average plot assigned—by lot—to a family of colonists.[30] Hence comes the name "centuria," which referred to the one hundred plots of two square *iugera*—etymologically. the area that can

quas bellorum ciuilium interuentus exhauserat.
D ito rerum coloniae numero nomine ciuium ampli
aut. quasdam &finibus ideoq. multis regionibus an
tiquae mensurae actus indiuersum nouis limitibus
inciditur. nam t&tantum u&erum lapides adhuc
parent sicut incampania finibus mynturnensium
quorum noua assignatio transfluuium lirem limiti
bus continentur cetra lirem postea assignatum per pro
fessiones u&erum possessionum ubi iam oportunaru
finium commutatione relicus primae assignationis
terminis more arcifinio possid&ur.

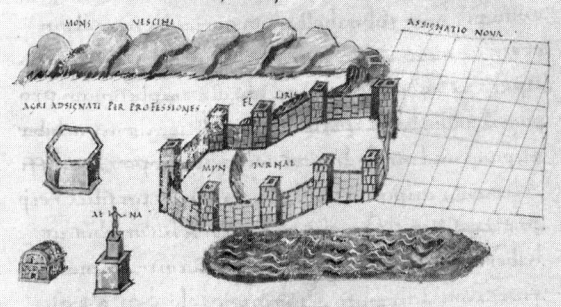

Multis ergo generibus limitum constitutionesin
choatae sunt. quibusdam colonis. k.m. &.o.m. non
longe a ciuitate oriuntur. nam &inproximo esse
deb&. immo sifieri potest exipsa colonia inchoari
sed quotiens u&usta municipia iniuscoloniae trans
feruntur. stantibus iam muris &c&eris moenibus

26.3 | Methods for marking the centuriation around the Roman city of Minturnae, from a manuscript dated 820–40. Pal. Lat 1564, f 88r. Biblioteca Apostolica Vaticana, Vatican City. Photograph © Biblioteca Apostolica Vaticana.

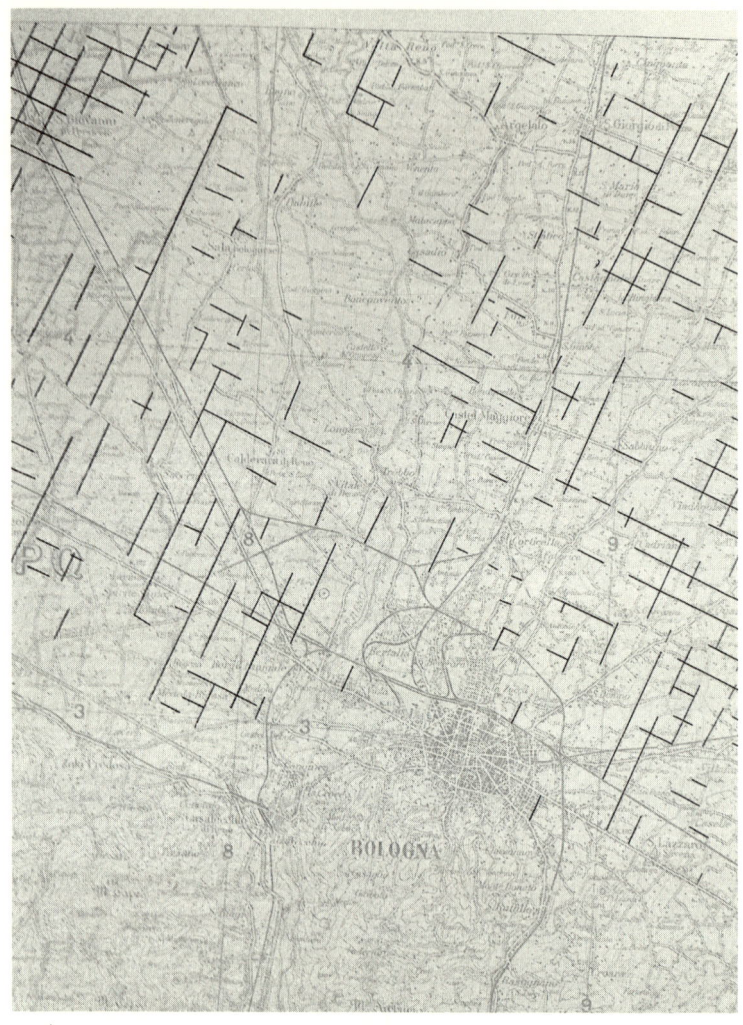

26.4 | Extant centuriation reticulum near Bologna. Drawing on a map in the public domain by Fernando Lugli.

be plowed by a yoke of oxen (*iugum*). Figure 26.5 makes the dimensional relationships easier to grasp.

The centuriation's main axes, either parallel or perpendicular, were traced on the ground using an established procedure. Surveyors relied on the *groma*, a wooden tripod supporting a horizontal cross from

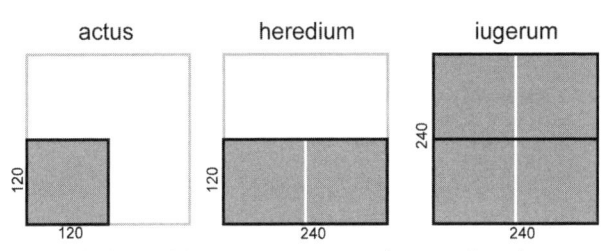

26.5 | Proportional scheme of the centuriation. Diagram by Fernando Lugli.

whose arms plumb lines hung down (figure 26.6).[31] The *agrimensor* established the direction of the new axis by aligning one plummet with its opposite, and shouted to a collaborator where to position the stakes (in Latin, *dictare metas*).[32] To our eyes it may seem a primitive technique, but it was a very reliable instrument, especially for distances encompassing a few *actus*. Only in this way could the Romans manage to circumvent the complex calculations that the curvature of the earth and the slopes of the ground would have required.[33] The correct alignment of the stakes was automatically double-checked while the surveyor enlarged the centuriation grid, but the real authentication came when the sides of each *actus* were measured by rods or ropes.

When designing a square, surveyors could be sure that its angles were at ninety degrees by measuring the diagonal, as it has a fixed relationship with the sides. "Fixed" does not mean "straightforward," though. In his *Metaphysics*, Aristotle defined the ratio between the side and the diagonal of a square as something so complicated that it may break a man's confidence. "All men," he states, "begin by wondering if things are as they are [. . .] as in the case of marionettes, solstices, or the incommensurability of the diagonal of a square with respect to its side."[34] Today we represent their proportion as $\sqrt{2}$. But before the introduction of irrational numbers, the diagonal of the square was expressed using different approximations. A common one was $17/12$.[35] Since the side of the *actus* measured 120 Roman feet, a surveyor could then verify that the quadrangle he had just traced was a perfectly squared *actus* (120 × 120 feet) by checking that its diagonal measured 170 feet (figure 26.7). If the diagonal did not correspond, one of the sides of the *actus* had to be rotated to fulfill that condition.

The surveying continued this way, as a strenuous process of trial

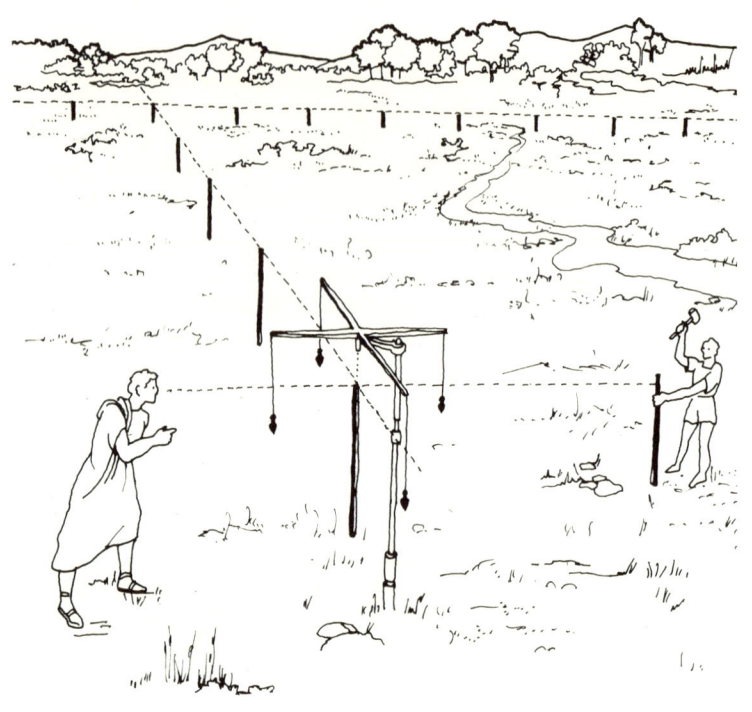

26.6 | A Roman surveyor looks through the plummets of a *groma* and instructs a collaborator in where to place stakes. Drawing by Fernando Lugli.

and error, with stakes repositioned countless times, checked over and over. It was also a delicate matter; its pace was determined by the sudden disappearance of the tiny silhouette of a man behind two out-of-focus plumbs, its operations jeopardized by sudden gusts of wind. Still, the *agrimensores* painstakingly managed to partition Rome's enormous conquests, an extraordinary undertaking which is today admired only by a few specialists.

Although designed to encompass slopes and even mountains, the centuriation found an ideal terrain for application in Italy's flatlands. Many Roman *centuria* are well preserved in the territory surrounding Verona, Padua, Pavia, and Bologna (see figure 26.4). The surviving roads and canals that form their perimeters are still in use.[36]

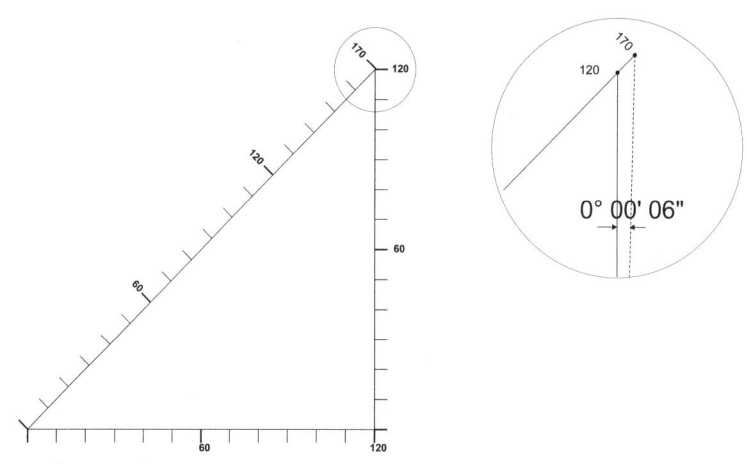

26.7 | Diagram showing the relationship between sides and diagonal in a square whose side is 120 units. If the diagonal is rendered as 170 units, the discrepancy between the angle at the base and the ideal value of 90° is 0°00'06". Diagram by Fernando Lugli.

Near Modena, the sides of the *centuriae*—blocks of twenty *actus*—measure 706.1 meters on average. If we consider the Roman foot as 29.6 centimeters—and I want to add that this is conjecture, as we know nothing about how standardized its dimensions were—the side of the Modenese *centuriae* are four meters short of their ideal dimensions.[37] In other words, the margin of error is negligible.

If the portions of the centuriation have been preserved so well over time, this is mostly an effect of scale. The centuriation stretches for such vast territories, and its meshes are of such a size, that it is impossible for humans to erase it completely. It is even impossible to encompass it by gaze, since its boundaries are too far from one another. In a way, the centuriation exists mostly as an imaginary construction, an idea triggered by the experience of synecdoche: you know it shapes the territory, even if you can only see one of its stones marking a corner.

Yet the centuriation survived, not only because of its inhuman scale. Rather, it was actively maintained. Archaeologists demonstrated that near Faenza, the centuriation was even expanded: in the ninth century, surveyors created new streets and canals in continuity with the old Roman grid.[38] This enlargement may come as hardly surprising, given the hundreds of manuscripts on land surveying written between Charlemagne and Frederick Barbarossa. Still, the archaeological evidence opens up new ways of thinking about the ways in which *agrimensores* manuals spurred a reshaping of the territory. It also reviews data we already have.

When Liutprand took the *jugerum* as the unit for his measurement reformation, did he do so to make use of a popular system? When he reigned, in the first half of the eighth century, land contracts referred to Roman standards.[39] With a decreasing population, though, estates were growing larger. So, he divided the short side of the *iugerum* by 72 rather than 120, as the Romans did, producing a foot of around forty-nine centimeters, which was longer than that of the Romans, and named it after himself. It was employed throughout the Middle Ages in the territory of Pavia, where he set his court and where was eventually buried.

Liutprand's redivision of the Roman system sets up the possibility of a new way of looking at medieval measurements. Italian cities may not have created their standards from scratch. Instead, they could have taken them from their territory. The Bolognese surveyors took the side of three Roman *actus* as 28 Bolognese *pertiche*, thus producing an equivalence between one square Bolognese *pertica* and 60 Roman square feet. Mantua turned 120 Roman feet (the side of one *actus*) into 76 local feet. By interlocking "modern" and classical systems, those cities met contemporary needs without losing their past.

This is, of course, only a hypothesis. It needs to be corroborated by other evidence, as at the moment it stands on dimensional correspondences alone. (And they are insufficient, as dimensions can be easily manipulated.) Such a possibility, however, tends towards probability the moment we consider the plan of cathedrals, as we will see in the next chapter. It also makes sense upon realizing the tight relationship of measurements and land as articulated in the surveying manuals of the time, which point at field division as the source of metrological knowledge.

27

Thinking through Squares

When master builders constructed a church in the eleventh century, there was a moment when they put down their ropes and measuring tools, and stepped aside.[1] Enter the bishop, with his entourage. He walked along the perimeter, sprinkled it with perfumed holy water, and traced letters in the ground near its corners. The ceremony purified the foundations of what was set to become God's house on earth. After the bishop's blessing, a stone incised with a cross was placed somewhere around the perimeter, usually at a corner.[2] It was supposed to never resurface again. But the ways of God are inscrutable, and some of them have reappeared. The diocesan museum of Mantua, for instance, displays the foundation stone of the parish church of Formigosa, a small town north of the Po River. It is a reworked Roman tomb slab, which the Formigosa parishers recarved and inscribed with the date 1064, the foundation year of their new church.[3] Those were the heydays of the consecration ritual I have quickly summarized. The ceremony took a somewhat standardized form only under Pope Gregory VII (r. 1073–1085) and was eventually confirmed in 1095, under Pope Urban II.[4]

The bishop's blessing marked a break in the architectural conceptualization of the church, separating the plan of the building from its elevation. The holy water also made the plan sacred and binding, leaving the latter open to transformations.[5] It was in this tradition that the ground of the cemetery of Pisa Cathedral, the so-called Camposanto, was covered with sacred earth taken from Jerusalem.[6] The separation, however, is better articulated by the many European legends about the foundations of churches. The perimeter of the Fraumünster in Zurich, for instance, was immutable because it was thought to have been demarcated by a rope that angels brought from heaven, which was kept in a chest near the altar until the Reformation.[7] The plan of Santa Maria Maggiore in Rome was thought to have been defined by a miraculous snowfall in the summer of 358. A late thirteenth-century mosaic on the church's façade (figure 27.1) shows Pope Liberius fighting a blizzard to record with a stylus the outline made by the snow, lest it melt away in the heat (as evidenced by the towering stalks in the foreground).[8]

Folklore aside, these legends articulate the quasireligious commitment of medieval architects to planning, which resulted from a rigorous application of geometrical schemes.[9] How wide should a church be? How long? Where to erect a wall? The tracing of squares, the rotation of their diagonals, and other geometrical schemes provided the rationale for the spatial definition of boundaries. It improved communication, too. How could architects convince a group of patrons that a wall needed to stop here and not there, or that the building had to cost such and such and not a different amount? The geometrical schemes proved persuasive: the debates about constructions unfolded along their lines.

Modern architectural historians label these schemes "ad quadratum." Yet they do not really need a name, since they are as pervasive as they are simple. They could just be called "geometrical," as they follow the basic proportions between the elements of the square and its compounds—that is, the most basic figures of Euclidean geometry as well as land surveying. To keep the practices separated does not help in this case. The centuriation was, after all, also a source of basic geometrical proportions. It was Gerbert of Aurillac who said so in his treatise on geometry, the first one not to be a direct translation of Euclid, and one that explains many of the geometrical schemes that can be retrieved in the churches built after the year 1000.[10] Despite Gerbert's words, this convergence has not been stressed enough.[11] But this is what I set out to do in this chapter, where I read the plan of some of the most famous buildings of the time in light of the geometrical structuring of the centuriation.

27.1 | Filippo Rusuti, Pope Liberius recording the boundaries of the miraculous snowfall of 358 in Rome (ca. 1295). Facade of Santa Maria Maggiore, Rome. Photograph: Fondazione Federico Zeri.

To start, many of the churches built between the eleventh and twelfth centuries followed the orientation of the centuriation rather than the solar course.[12] Some adapted to preexistent buildings of early Christian origins, such as the Faenza baptistery, which followed the reticulum by default.[13] And some notable cases—such as the abbey of San Zeno in Verona, or the abbey of Nonantola, simply did not.

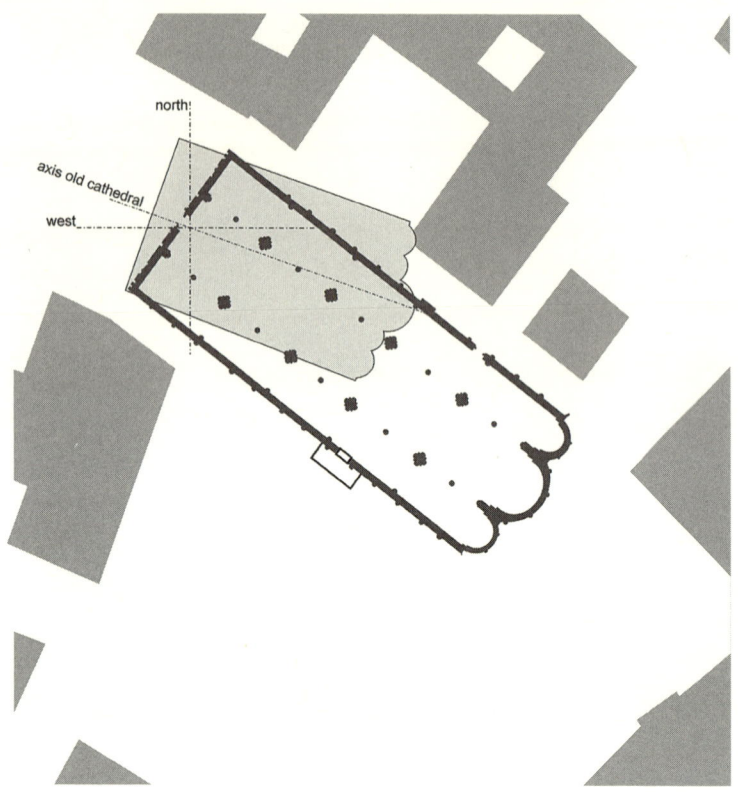

27.2 | Diagram showing the rotation of Lanfranc's cathedral in relation to Modena's preexistent cathedral. Drawing by Emanuele Lugli and Fernando Lugli.

Yet we also have cases for which a stronger association can be argued. The bishop of Brescia Adelman, the one who studied the treatises of the *Corpus* at the cathedral school of Chartres under Gerbert's pupil Fulbert, was probably the patron for the new round cathedral, oriented according to the centuriation.[14] And from 1099, the architect Lanfranc demolished the cathedral of Modena to construct a new building, whose plan was rotated just a few degrees from the previous one, so as to perfectly match the centuriation reticulum (figure 27.2).[15] Far from being a feature only of the rural landscape, the centuriation entered the city, shaping its most important buildings, sometimes in rather subtle ways.

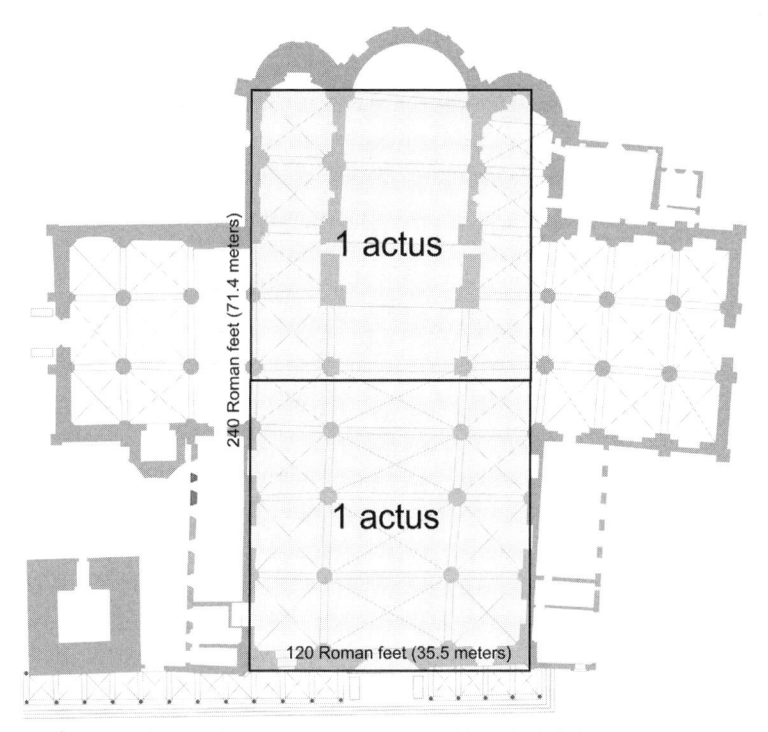

27.3 | Diagram showing the centuriation proportions of the cathedral of Cremona. Drawing by Emanuele Lugli and Fernando Lugli.

While Modena erected Lanfranc's cathedral, Cremona constructed its own.[16] The architects followed the Roman reticulum; yet there, the grid saturated the building at a deeper level. The analysis of the brickwork has in fact revealed that the church was planned without the transept as a two-square rectangle (figure 27.3), a rather common scheme for churches, also recommended by Vitruvius for temples.[17] It may be difficult to see it in modern plans, as the south wall of the nave collapsed near the apse. It is unclear when that happened.[18] Still, the geometrical relationship between the north side and the west facade is perfect—the latter measures 35.5 meters and, with its 71.4 meters, the north wall doubles it—and there is no doubt that the church plan was structured around a two-square rectangle.[19] What is more interesting to note is that each square can also be read as one *actus* (35.3 meters are one *actus*), which means that the Cremona cathedral was designed to fit into a *iugerum*. This seems all the more likely because the so-called first cathedral of Pisa—not the one we see today, but the one before its mid-twelfth-century enlargement—was also built as a two-square *actus* rectangle.[20] Archaeologists found its foundations more than a century ago.[21]

The square proportions of the centuriation could, however, also yield other effects. Lanfranc planned the cathedral of Modena as a square joined to a rectangle whose long side matched the diagonal of the square. The external perimeter of the cathedral (minus the apse, which was added later) measures 24.94 and 24.81 meters on its west and east facades respectively, and 60.65 and 60.38 meters on the north and south sides. In Modena's medieval measurements, for which one *braccio* is around 52.3 centimeters, this makes a rectangle of 48 by 116 *braccia*. Or, to keep distinct the two geometrical shapes that make its plan, I would call it a square of 48 *braccia* plus a rectangle built on its 68-*braccia* diagonal (figure 27.4).[22]

As we saw, the diagonal was of fundamental importance for surveyors, who measured it to check the orthogonality of the square. Medieval commentators understood it. Gerbert of Aurillac discussed its proportion in his treatise on geometry, and his pupil Francon expressed the relationship between side and diagonal of a square as 17/12 in his 1050 treatise, *De quadratura circuli*.[23] This proportion can be retrieved in the Modena cathedral, as its short 48-*braccia* side is four times 12 *braccia*, while the diagonal that adds up to the long side is four times 17 *braccia*.[24] Such a ratio is not purely abstract. The Pisan mathematician Leonardo Fibonacci praised the duodecimal nature of Pisan measurements because they facilitated the computation of the relationship between the side of a square and its diagonal.[25] Indeed, while 12 is 5 + 7, 17 is 5 + 5 + 7.

But there is more, as the dimension of the Modenese *braccio* seems to also have been extrapolated from the Roman centuriation. Indeed, 68 *braccia* correspond to 120 Roman feet. This consonance means that Modenese surveyors saw the side of one *actus* as 68 Modena *braccia*. They could reinterpret the Roman centuriation using the Modenese measurement system without altering any of its boundaries. And because 68 is four times 17, not only was the Modena *braccio* proportional to

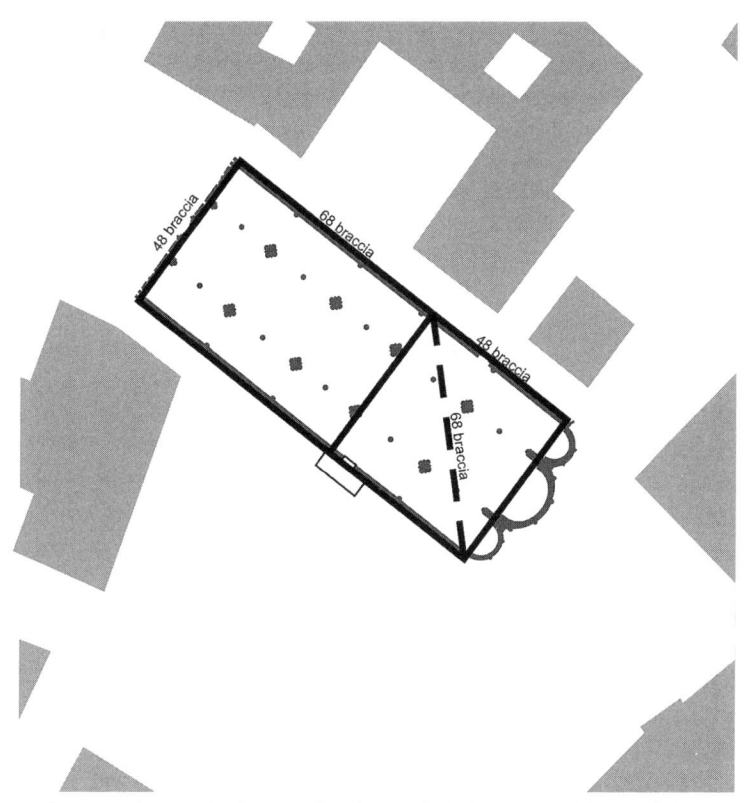

27.4 | Diagram showing the planning of Modena Cathedral as a square of forty-eight *braccia*, plus a rectangle built on its sixty-eight-*braccia* diagonal. Drawing by Emanuele Lugli and Fernando Lugli.

the key unit of Roman centuriation, but it acknowledged in its structure the key proportion to Roman land surveying (that is, the diagonal of a 12-unit square). Lanfranc must have known those proportional correlations, as he purposely designed the cathedral around a diagonal of 120 Roman feet, or 68 Modenese *braccia*. Such a dimension turned the cathedral into both a spatial and a metrological module for the rest of the territory. Even before it received on its facade the *pietre di paragone*, it embodied the city's standards at a structural level. And if either standard or geometrical practices were lost, they could be restored by measuring its sides. The cathedral, in other words, was not only in direct spatial continuity with the territory, it was also its compressed metaphor.

But maybe I am excessively emphasizing intentionality. The dazzling correspondences between the Modena cathedral and the centuriation may point less to Lanfranc's personal learning than to widespread, interlocked practices. The spatial forms that organized architecture and the territory clung to one another so perfectly because they were extensions of the forms employed to interpret the real. Surveyors and architects shared the thought schemes of their patrons—they had to—and all of them relied on geometry to produce a sense of greater perfection.[26] The story of geometrical appreciation, from Boethius in the sixth century to John of Salisbury of six centuries later, is rather repetitive, after all. It says, over and over again, that "mathematics needs nothing else," that "it rests on itself," and "its results are indubitably true and cannot be otherwise."[27]

Medieval texts never questioned the idea that the character of order is geometrical. Even the Bible repeats that point. So it is rather easy for modern scholars of medieval architecture to explain the geometries of buildings by pointing to the apocalyptic visions of Saint John or the prophecies of Ezekiel, who saw God measuring the heavenly Jerusalem with a linen string and a brass rod.[28] Those passages glorify building with dimensional care, and present the sacred in the function of measuring.[29]

Yet, as those biblical images make clear, measuring did not communicate order itself. Measuring was not a means to an end, but the process of making order visible as such. Balbus's letter to Celsus, one of the fixtures of the manuscripts belonging to the *Corpus agrimensorum romanorum* tradition, states that every "observation of measuring starts and ends in a sign, which is that part made of nothing."[30] Geometrical forms have no substance, but the surveyor can make them manifest through measuring. Measuring is the gesture that identifies the point in space at which a dimension starts and ends. It is the way by which abstract geometry becomes, if not tangible, at least intelligible. And it can deconstruct an imposing cathedral into a set of rational procedures that reveal the building's adherence to God's order. In so doing, measuring materializes the conditions for architectural appreciation, which is predicated on the general pre-appreciation of geometrical forms. Measuring does not enact order, but it is the process that interlocks various practices, enabling forms to exist across them and through society. Measuring links ideas and institutions in a seamless way. And it does so because it constantly joins the source of legitimization to its destination.

28

Frustrating Bodies

When, in the eighteenth century, savants debated whether or not to derive the metric system from the meridian, their discourses contained an echo of medieval geometry, which hinges on the earth from its first syllable (γῆ). Yet it is more difficult to speak of a continuity when they argued that measurements came from the body, as medieval scholars did not agree on the role it played in relation to measurements. Not all of them considered, as Isidore of Seville did, the earth and the body as part of the same system. When listing standards, Isidore linked the former to the latter in a progression that blurred the end of one and the beginning of the other. He wrote: "Our ancestors with great ingenuity divided the world into parts, and the parts into provinces, provinces into regions, regions into districts, districts into territories, territories into fields, fields into centuries, centuries into acres, acres into zones, zones into *actus*, then rods, paces, steps cubits, feet, palms, *unciae*, and digits." He concluded: "A digit is a finger."[1]

The friar Bartholomeus Anglicus copied this passage verbatim in his encyclopedia, a fixture of almost every Franciscan library.[2] So did Uguccione da Pisa, in the dictionary we encountered in chapter 14. Uguccione, however, also showed himself to be receptive to a second way of relating bodies to measurements. He linked the verb *metior* ("to measure") not only to *membrum* (a "bodily limb") but also to *methodus* (a "remedy," even in the medical sense of "cure"), thus explaining that measuring could heal because it revealed the natural point of health.[3]

"Everything in nature rejects either lack or excess and is preserved through the property of measure," wrote Tertullian in a sentence that sums up the understanding of nature as incessantly tending towards the average.[4] Medicine, the science of the means (*modus*), operated on external and internal factors (the dryness of the air, the amount of food) so as to preserve and restore health, also a mean.[5] Medieval doctors chased, as much as they propagated, an idea that the natural body ought to inhabit the middle ground. It is not by chance that "normal" in medieval Latin is often referred to as *aequalitas*, or *mediocritas*, a term that denoted a natural state of the body rather than what was unremarkable.

And such a body was thought to be as internally balanced as it was dimensionally proportionate. Boethius opened his book on geometry by speaking of how geometry is useful to physicians.[6] And Galen, the greatest classical physician, was thought to have believed that measurement standards were in the body, which could thus serve as a measurement system.[7] "It is necessary that the proportion of the foot corresponds to those of the hand," Galen also wrote, constructing a notion of body limbs as mutually commensurate.[8] The thought cemented ideas of the existence of an average, normative body of clearly defined measures.[9] And even if the treatise in which he mentioned those points, *De usu partium*, only arrived in Italy in Nicholas of Reggio's fourteenth-century abridged translation, *De juvamentis membrorum*, there were many other ways by which to make the connection.

The sizes of limbs were heavily scrutinized in the Middle Ages, as they opened or closed possibilities. To enter the military or the ecclesiastical hierarchy, men had to be of average stature, which the Roman writer Vegetius identified as around six feet.[10] Disproportionate hands and feet were tolerated, but only to an extent. Pope Gregory IX's Decretals (1230), which entered Canon Law, prescribed that a priest without two fingers or one who had lost a part of his hand could not celebrate the mass. Many asked for exemptions—the mutilated clerics do not seem to have been few—but the norm remained valid, as public exposure of the priest's deformity would have invalidated the sacrality of the rite.[11]

So profound was the belief that a healthy body produced correct proportions that it is not uncommon to encounter episodes such as that of Giliolo Leonardi, who, in the absence of standards, surveyed a street with a foot he constructed from his own palm ("ad pedem manus dicti Gilioli").[12] Medical knowledge supported the idea that Leonardi's palm could replace the local standard. The mathematician Leonardo Pisano, better known as Fibonacci, observed that the ways in which his contemporaries measured could be divided into two groups. There were "those who measure with their arms, forearms, or steps, while others use *pertiche* or another measurement standard."[13] This remark reveals a split between those who thought of measurements as equivalent to limbs and those who realized that standards were artificial tools. Still, Leonardi's operation—which took place in Assisi in 1304—was conducted in the presence of two eyewitnesses. Idealized notions about human body did not remove the political need for certification and control.

But the interdependence of measuring and bodily health was not expressed only in terms of size, since measuring extorted considerable degrees of physical labor. The *Corpus agrimensorum romanorum* abounds with verbs of movement. Surveyors walked and climbed. They shouted to one another. Above all, they spent a lot of time seeing. The centuriation is largely a product of vision: every stake was placed by someone who was instructed, remotely, by a colleague operating the *groma* (see figure 26.6). There was nothing curious or imprecise about this practice since, as ancient writers wrote, seeing followed straight lines, like rays of light or the threads of a spider's web.[14] Lines entered the pupil from the outline of an object as the catheti of a triangle reached their vertex.[15] Avicenna's *De anima* explained how the rays of vision that meet in the pupil are similar to the tip of a pyramid, a correlation that must be taken seriously, like all those made by the detailed, authoritative Arabic scientific treatises that from the mid-twelfth century were translated into Latin.[16] Some thinkers reversed the relationship of eyes and objects, insisting that the visual rays left, rather than entered, the pupil.[17] Still, they too agreed that optics followed the rules of geometry, and considered seeing and surveying as related triangulatory activities.[18] Optics relied on intangible rays to construct a geometry that surveyors eventually mapped onto the territory.

The unyielding connection of optics and geometry was at the core of what modern scholars have named the increasingly optimistic belief in vision as capable of producing certitude.[19] Its apex came with Roger Bacon's treatise on optics *Opus maius* (completed in 1267), in which he provided a comprehensive study of the geometricization of vision in a section—the fifth, called "Perspectiva"—which started circulating independently.[20] Bacon also provided a list of nine circumstances that could impede vision and compromise its results, such as the distance of the observer or the quality of the air. "It must be understood that when these nine conditions do not depart from moderation—that is, when they are met, neither in excess nor in deficiency—then vision achieves certainty."[21] It is interesting to see how this sentence predicates visual truth on a sense of *mediocritas*. While recognizing that certitude was an effect of the body (its position and visual capacity), it points again to balance as a precondition of correctness.

Bacon's list was important given that, in surveying, vision overcame the limitations of the body. In *De ars mensoria*, one of the treatises often encountered in the *Corpus agrimensorum romanorum*, Frontinus asked what to do when a surveyor found himself in need to measure things that were physically inaccessible. He defined the problem as one of *altitudo*: How to calculate the height of towers? How to survey a mountain that soars on the other bank of a river? Frontinus provided a solution that surveyors and architects adopted for many centuries. The key, he explained, was compression.[22] The unfathomable dimension could be calculated if its characteristics were reproduced at the smaller scale of the surveyor's body, so that he could measure them directly.

Frontinus's text stops here, without explaining how to proceed. It is so fragmentary that critics wonder if what survives is an abridgment of the original treatise. Such doubts intensify once we consider an anonymous eleventh-century text that specialists call the *Geometria incertis auctoris*, which supplies the missing information. To measure a faraway mountain, he writes, take a wooden stick and place it in front of you. Then lie on the ground and look at the top of the mountain so that the pole crosses in front of it. Then, without moving, ask someone to mark on the pole the point where you see the mountaintop. Stand up and measure two things: the distance between your eye and the pole, and the height of the point of intersection from the ground. The problem of *altitudo*, in other words, is solved by recurring to similar triangles. The height of the mountain is to the height you marked on the pole as its distance is to the span between your pupils and the spot where you stuck the pole.[23] The body has dissolved. *Geometria* reduces it to a series of specific points on the ground and in the air. Such a solution remained valid for many centuries to come (figure 28.1).

But let's stick to the making of this medieval treatise for a moment, so to clarify its enmeshment with the historical threads that I have spun. As I have said, the *Geometria* is anonymous. It was recently suggested as the work not of a sole author, but of a group, perhaps researching at one of the large cathedral schools in northern France. The authors could have been pupils of Gerbert of Aurillac, the well-traveled mathematician whom we saw managing the estate of Bobbio monastery in chapter 26, and to whom the text was attributed at the end of the nineteenth century. More likely, they were students of his pupil Fulbert of Chartres.[24] Reflections on the identity of the authors are welcome news for what is a relatively unknown text. Scholars of medieval science prefer studying treatises of abstract mathematics by renowned authors rather than, as is the case with the *Geometria*, anonymous works of applied geometry.[25] But, despite its enigmatic origins, the *Geometria* is important. It stands at the very roots of the so-called *geometria practica* tradition, a series of texts whose origins are usually found in the *Didascalion* of Hugh of Saint Victor, a monk from northern Europe who, however, drew much of his information from the *Geometria*. And the treatises of *geometria practica* led to early modern books on surveying and measuring, up to Cristiani's treatise of chapter 6. The *Geometria*, in other words, made engineering textbooks possible.

After all, the *Geometria* is a manual for the manipulation of the world through geometrical schemes. Their range of application is impressive. The *Geometria* takes its readers from rivers to mountains. It uses mirrors, quadrants, and astrolabes.[26] In one case it recommends tying strings to two arrows and shooting them at the top and bottom of a tower.[27] But the variety of its examples eventually does not matter, as they are all solved by resorting to similar triangles and some proportional principles. The *Geometria*, in other words, compresses the variety of the world to a few simple rules, and in that maneuver it

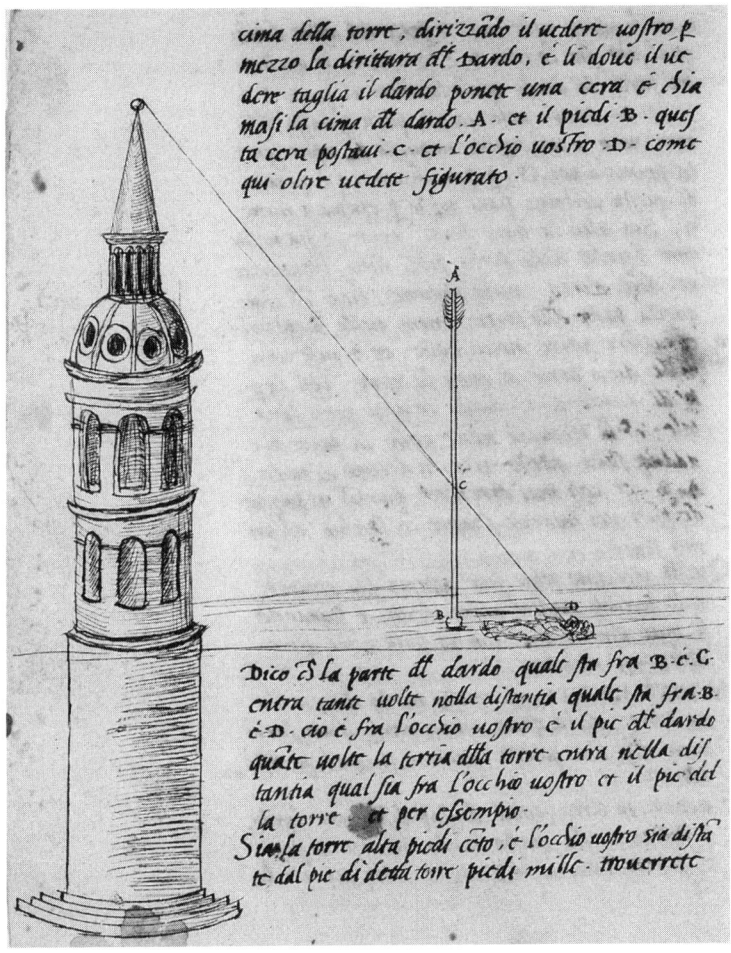

28.1 | The measuring of a tower by means of a rod stuck in the ground. Exercise from an early-sixteenth-century copy of Leon Battista Alberti's *Ex ludis rerum mathematicarum*. Ms. 2942.1, f. 2v, Biblioteca Riccardiana, Florence. Photograph courtesy of Ministero dei beni e delle attività culturali e del turismo.

synthesizes the body to geometrical elements. At one point it recommends placing one's eyes at the level of the ground, which means putting a side of one's face in the dirt and pushing aside concerns about the thickness of one's cheekbones.[28] The body is reduced to its eyes, and the eyes to pupils—or, better, to only one pupil. The body is compressed to a point in space and is considered a holder of straight lines. It can even be substituted with flying arrows, whose trajectories so closely resemble human vision. Measuring, then, may not need the body after all. The vocabulary of feet, palms, and fingers invokes a physical engagement that may once have existed; but in the eleventh century, measuring was already detaching itself from it.

29

Fibonacci's Standardizations

The Pisan mathematician Fibonacci reworked much of the material contained in manuals of practical geometry in his 1220 treatise, the *Practica geometriae*. The continuity is not just an effect of its title. In the treatise, he describes surveyors as geometers of vision and repeats the example of the double arrow, even if Fibonacci's darts target a tree rather than a tower.[1] Because his material was not original, Fibonacci is today heralded more as a popularizer than as an innovator.[2] Scholars have identified many of his sources, pulling them apart and tracing their paths from Asia to Tuscany.[3] But then, Fibonacci never claimed to be original. Quite the opposite. In his first treatise, the *Liber abaci*, he describes his work as a compilation of Indian mathematics, mixed with some Euclid.[4] In the prologue he recalls how he learned to solve mathematical problems while staying in Béjaïa, Algeria, at whose customs house his father Guglielmo worked as a public scribe for the city of Pisa.[5] Fibonacci claims that as a kid he noticed the difficulties of Pisan merchants: how clumsy they were in front of their confident Arab counterparts, and how many mistakes they made. So he set out to study the computational and geometrical practices of the Indians, and composed his book as an aid for inexperienced Latin speakers ("et gens Latina de cetero sicut hactenus absque illa minime inveniatur").[6] If we are to believe what he wrote, he traveled through the Mediterranean to familiarize himself with mathematical problems and refine his solutions. His journeys present his mathematical knowledge as comparative. But they also serve to appeal to his readership of merchants, presenting Fibonacci as one of them.[7]

While Fibonacci was more a translator than an original writer, a compiler rather than an innovator, this is not how he was perceived at the time. His work inspired a plethora of translations, abridgments, and reworkings which all refer to him as their primary source and from whose esteem they hoped they would profit.[8] Vivified, updated, and locally adapted by countless other texts, Fibonacci's treatises lived up to the end of the fifteenth century, and their proliferation put a stop to other traditions.[9] The waning of *agrimensores* texts, whose content Fibonacci absorbed and rewrote in an uncomplicated prose, can be explained by his success.[10]

Fibonacci's success was, however, an effect less of clarity than of timing. His first book, the *Liber abaci*, composed in 1202, hit the market at the beginning of Italy's currency and measurement revolution. It directly addressed such a transformation rather than being, as it is often said, another element of the general commercial growth of Italy. The real innovation of the book, after all, comes in chapter 8, in which Fibonacci lists the currencies and measurement standards in use in numerous Mediterranean harbors, providing conversion rules from one to another.[11] It is thanks to this compilation that we know many of the measurements in use in the early thirteenth century. That is, that Pisa used a ten-palm *canna*, while the *canna* in Sicily, northern Africa, and Provence measured eight (Pisan) palms. (The three regions, for Fibonacci, were metrologically one: a result of the intensity of Arabic trading.) Fibonacci also discusses standards of capacity, specifying that at home the *cantaro* was made of 100 *rotuli*, and that the weight for cheese was 220 *cantari*. The latter, Fibonacci explains, could also be considered as 2,200 *libre*, which simplified its relation to the corresponding measures in Messina, where one *rotolo* was 300 *libre*.[12]

Such clarifications must have been greatly praised, because in his following book, the *Pratica geometriae*, Fibonacci included an even longer section on measurements, dedicating most of it to surface units, which he had left out of the *Liber abaci*. And his treatment was more detailed. Among its revelations, Fibonacci showed what a sophisticated system of measurements Pisa had developed. We can speak of a "system" because Pisa did not just cobble together the standards employed locally; it constructed them as a sequence. It can in fact be simplified as a duodecimal progression of standards: one square *denaro* (1 × 1 foot), one square *pertiche* (6 × 6 feet), one square *scala* (12 × 12 feet), one *stioro* (sixty-six square *pertiche*), and one *moggio* (twenty-four *stiori*)—with the insertion of one extraneous element, the *pano* of five and one-half *pertiche*. Its presence hints at a complicated history, whose shaping forces were not yet finished in Fibonacci's times. Rather than erasing the *pano*, the Pisans kept it, retouching the value of the *stioro* so that it could simultaneously accommodate twelve *pani* and sixty-six *pertiche*. Its inclusion was an opportunity to shift between a system in base twelve and one in base eleven, enabling a significant degree of refinement in the computation of land.

But Fibonacci's sense of detail came as part of an across-the-board shift, a new cultural awareness about measurements that is reflected in his linguistic competence. Despite admitting the limits of his knowledge, Fibonacci displays a rich vocabulary of measurements. He explains that each city employed a different standard: "the *iugerum*, or the *aripennio* or the *carruca*, the *tornatura*, or the *cultura*, or other standards they are accustomed to."[13] I am aware of no other medieval writer who mapped the measurements of the medieval world as extensively as Fibonacci.

Fibonacci may profess ignorance, but this is only circumstantial, confident as he is that mathematics provides the means to translate any standard into another. Fibonacci explains that the first step of any equalization is to separate between the "quality" (*qualitate*) and the "quantity" (*quantitate*) of standards, as it is only through the latter that the fundamental qualitative differences can be overcome. Consider, for instance, the following problem of the *Liber abaci*: "If 20 *braccia* of wool cost 3 Pisan *lire* and 42 cotton *rotuli* cost 5, how many *rotuli* correspond to 50 *braccia*?"[14] The qualitative difference between *rotuli* and *braccia* is erased by reducing them to the same currency. That is where the solution lies. To solve the problem, Fibonacci divides 42 *rotuli* by 5 (*lire*) to find that one *lira* would buy him 8-2/5 *rotuli*. He repeats it for the *braccia*, which gives 6-2/3 *braccia* for each *lira*. And, once the two standards are homogeneous, it takes him little to realize that 50 *braccia* (prince: 7-1/2 *lire*) correspond to 63 *rotuli* (that is, 7-1/2 × 8-2/5 *rotuli*).

The graphic system Fibonacci introduces considerably facilitates the computation. In the two boxes that he pens next to his description (right margin of figure 29.1), he separates currency and measurement standards in different columns and proceeds to multiply diagonally, starting from the dimension whose price we do not yet know (50 *braccia*). It then divides the result by the combination of the remaining two. So 50 × 3 × 42 divided by 20 × 5 = 63.

Even if the operation takes quite a bit of energy to process, the merchant is spurred to undertake it by the confidence that a common denominator does exist. Fibonacci often makes money the medium of the conversion. Throughout his text, however, he also tends to reduce every linear standard to the Pisan palm.[15] So, while presenting the tools for standardizing the world, Fibonacci already makes it appear more connected than it is. Fibonacci's *Liber abaci* is thus not just a manual to explain the mechanics of trading to Pisans, but a picture of how a connected world may be. Fibonacci's book facilitates the optimism on which the promise of sameness eventually rests.

In Fibonacci's map, the Mediterranean Sea seems unified by a sort of universal palm. The choice is innocent when we remember that the hand is also the tool by which merchants counted. By standardizing every linear standard in relation to the palm, Fibonacci finds in hands a common denominator for both quantities and qualities. He urges merchants to learn how to count by hand. "Those who wish to know the art of calculating," he writes, "must know computing with hand gestures"—a passage so important that later versions of his book even devoted a full-page illustration to it (plate 16).[16] And there he goes: explaining that curving the ring finger over the middle of the palm signifies six, and that touching the forefinger with the thumb so as to make a circle means ten. But if you stretch the forefinger and thumb, the hand says twenty; and if you keep them in position while lowering the ring finger, you have twenty-six.

Scholars have emphasized how Fibonacci popularized the use of Arabic numerals, which until him had only been sparsely used.[17] But Fibonacci did not only standardize mathematical practices: he also reconfigured bodily movements around those of the educated Muslims. Even if separated by linguistic barriers, Pisan merchants could nevertheless trade with anyone by making use of their bodies. Fibonacci's *Liber abaci* was thus a *chironomia*, a treatise for speaking

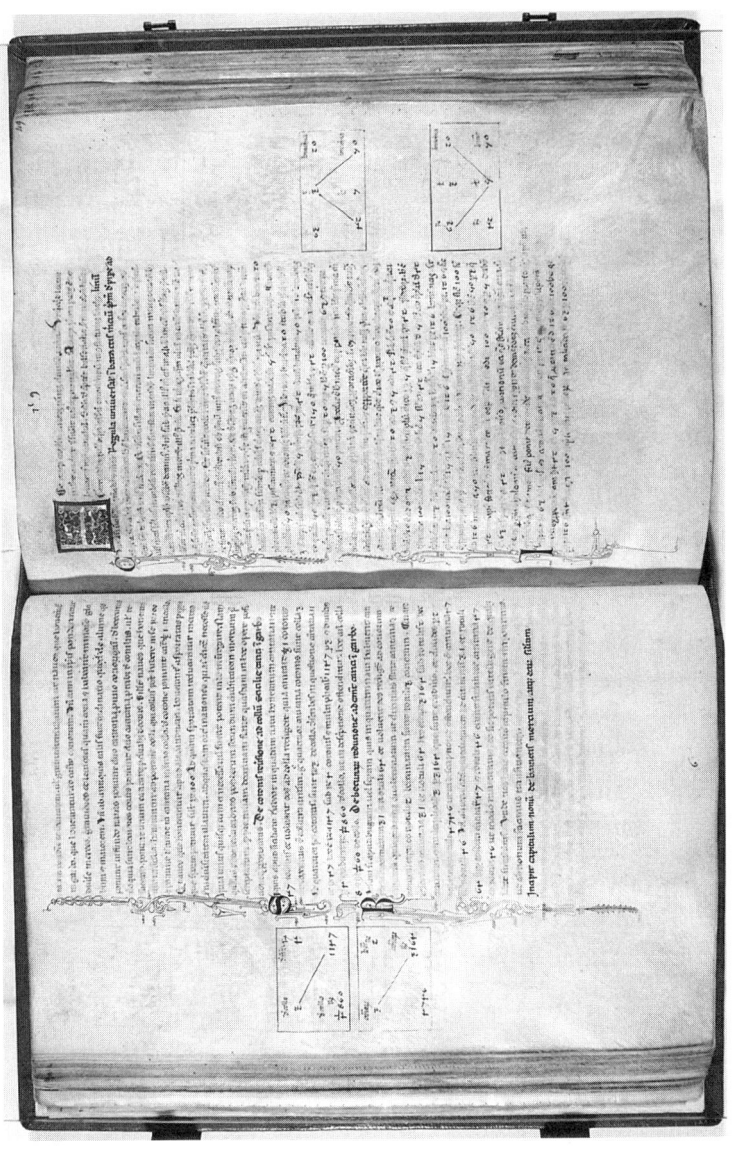

29.1 | A page of Fibonacci's *Liber abaci* (early fourteenth century). In the right margin, two diagrams for computation provide an answer to the conversion problem described in the adjacent text. The manuscript is considered a faithful copy of the second edition (1228) of the *Liber abaci*. Codice magliabechiano Conv. Soppr. C 1, 2616, f. 49r, Biblioteca Nazionale Centrale, Florence. Photograph courtesy of Ministero dei beni e delle attività culturali e del turismo.

through hands that brought merchants close to actors, like the mimes whom the friar Salimbene de Adam saw in the streets of Pisa around 1243, "performing a play with appropriate gestures."[18]

Fibonacci's *Liber abaci* transformed Pisan merchants' way of thinking by operating on their bodies. Through his instructions, Pisa saw the surge of a new, compact class of merchants skilled in Mediterranean counting, who could take advantage of the impressive network of other traders, as they spoke the same bodily language.

Such a shift, however, was not possible without the consent of politics. Fibonacci's success was not an effect of his pacing himself to the rhythm of mercantile needs. Rather, it was enabled by the commune of Pisa, which paid an annual salary of twenty *soldi* to Leonardo for teaching its merchants.[19] The commune even made Fibonacci's *Liber abaci* a requirement for public surveyors. In its statutes of 1286, around forty years after Fibonacci's death, it insisted that surveyors had to be able to read, write, measure, assess, and count through an *abbaco* (*faciendi rationem ad ambacum*).[20]

The school for wealthy merchants Fibonacci set up in Pisa represents another of the centers that diffused metrological culture in the Middle Ages.[21] After the cathedral schools of northern France, the law schools of Bologna, the libraries where Dominicans and Franciscans studied economics treatises, and the communal palaces, where measurement standards were compared on a regular basis, there were mercantile institutions. Scholars often speak of *scuole d'abaco*, in the plural, forgetful of the fact that their number was far from high. The Pisan commune guarded Fibonacci's teaching as a significant advantage in maintaining the city's trading supremacy. But then, secrets were hard to keep in a world as intertwined as that of merchants and scholars. Especially so, as Fibonacci advertised his skills and started a correspondence with the mathematicians of Frederick II's court as well as Cardinal Ranieri Capocci, deacon of the church of Santa Maria in Cosmedin in Rome.[22]

The first recorded *magister abaci* after Fibonacci was Pietro da Bologna in 1265, but he, like his son who followed in his wake, may have taught only privately.[23] For public lectures outside of Pisa, we need to wait until 1279 and go to San Gimignano. The commune of Verona requested a school in 1277, but it is unclear whether it managed to set it up before 1285, the date of a record of a payment to *magis-*

ter Lotto from Florence.[24] But even towards the end of the century, Fibonacci's teaching was ignored, even resisted. In 1299, Florence's guild of wool-cloth traders, the Arte della Lana, forbade its members to make use of Arabic numerals. Instead, it insisted that numbers be written either in Roman numerals or as words. Similarly, in Padua the accountants had to write numbers "in clear Roman letters, not in ciphers" (*non per cifras sed per litteras claras*).[25] If we have a powerful sense that Fibonacci's teaching was successful, however, it does not come from the mushrooming of *abaco* schools, but from the fact that merchants believed in, and themselves continued, his standardization project.

"Continuity," however, is a strong word, as Fibonacci's followers, while mapping out trading links, did not produce the same results. In a mercantile *Rechordanza* composed in Venice before 1311, there are the metrological conversions of Fibonacci's *Liber abaci*. "In Venice, the standard by which they sell grain is called *staio*, and you should know that 3 Venetian *staia* make 1 *qafiz* of Tunis and that for 300 *staia* there are 6 *qafiz* left."[26] This equivalence means that the conversion of three Venetian *staia* to one Tunisian *qafiz* was approximate: the error is of around 2 percent (two *qafiz* for every one hundred *staia*). Like Fibonacci, the anonymous compiler of the *Rechordanza* accepts some levels of imprecision as a condition of measuring. He ignores them until such errors break the system. In Armenia, he recommends that his readers avoid dealing in grains. Soap is fine, but "wheat and barley are sold by a measure that is called *marzapan*, which the Armenians deliberately keep enlarging and shrinking so that no one can tell its real value two months in a row."[27]

At around the time that the *Rechordanza* was composed, another merchant, Francesco Balducci Pegolotti, started keeping a diary of the measurements he encountered on his journeys, dividing them by product. He did so for around thirty years, during which time he traveled on behalf of the Bardi, one of Florence's most prominent mercantile families.[28] His *Pratica della mercatura* covers a much larger area, one that exposes the enormous network of the Bardi business interests.[29] Yet it is precisely the scale and the detail of his descriptions that reveal the fragility of his project. Tracking measurements from various cultural regions was a failed operation from the start, for two main reasons.

First, Pegolotti observes that measuring was hardly up to the merchants; it was instead an activity undertaken by officials who said nothing about their ways of proceeding. In Constantinople, superintendents measured fabric for a fee of half a karat per roll, and charged the same amount to inspect it for manufacturing defects (but traders paid a full karat to get both services, and their fabrics folded).[30] Second, Pegolotti's conversions are questionable. For instance, he makes an enormous error (over 10 percent) when converting the Venetian *libra grossa* to the *qantar* of Pera, the Genoese settlement across from Constantinople.[31] Modern scholars speak of lack of consistency, and it is possible that Pegolotti gathered data from different sources without having a direct experience of all standards. Such a discrepancy, however, shows that his promise of homogeneity holds only at a reduced scale. It works within the rather limited Tuscan network of Provence and the northern coasts of Tunis, which Fibonacci had already presented at an already mature level of standardization. But the moment it encounters profound cultural differences (such as in Armenia) or spreads from London to Pera, his information becomes less reliable.

Increased detail and scale reveal the mirage of standardization, but so does the structure of Pegolotti's book. After all, he does not provide consistent information, either. For example, in Tunis he gives conversions between local standards and those of seventeen other cities. In Tabriz the cities have dwindled to six, and in Feodosia (medieval Kaffa) they are only two.[32] Scholars assume that his information reflects the trading of each harbor, and that his text is a mirror of the unevenness of the commercial world of the time. If so, the Mediterranean of the first half of the fourteenth century was patchy: trading deserts spread between busy ports. The original title of Pegolotti's book, after all, does not speak of mercantile practices (*Pratica* was the name its eighteenth-century editor gave it). He instead called it *Libro di divisamenti di paesi (e di misure di mercatantie)*, which, much like Marco Polo's *Le divisament dou monde*, reveals the act of dividing the world into parts.[33] Pegolotti does not construct sameness. Rather, he maps unevenness, thus helping merchants to identify the busiest axes in a world that, rather than marching towards standardization, was structurally impotent to achieve it.

Conclusion

The Metamorphoses of Measurements

One of the most scorching passages in *Against Verres*—the oration by which Cicero lambasted the Sicilian governor for exaction—deals, obviously, with measuring. Cicero throws us into Rome's enormous temple of Castor and Pollux during one of Verres's inspections.[1] The guardian of the temple, Publius Junius, has died, leaving behind an underage son. Verres sees an opportunity to make a profit out of the boy's inheritance. He asks how much it would cost to restore the temple, without realizing that it is in perfect condition. The ceiling looks as if it has been freshly painted. And even the most scrupulous checkup concludes that there is nothing to do. So one of Verres's henchmen suggests something absurd: to check all the columns with a plumb line, as "there is hardly any pillar that is exactly perpendicular" (*fere nullam esse columnam quae ad perpendiculum esse possit*). Bingo. There is some resistance from the boy's guardian, but he is a soft-spoken man, easily hushed. Some relatives also try to stop the work from going ahead, but Verres is elusive, difficult to approach. He has decided that he will have all the columns of the temple checked, thus making a profit out of a pointless service.[2]

The episode is exemplary of Verres's limitless wickedness. He appropriates something public for private gain. He extorts money for nothing. He abuses a minor. Cicero could not possibly have found another story that would hit so many nerves. The episode is also telling of the reputation that measuring had already created for itself:

that of an activity that can pass for useful even when useless. What is remarkable in Cicero's narration is the quickness of Verres's reaction to the proposal to measure. His impulse is an effect of the facility by which dimensions can be manipulated: he knows he will win in all events. Yet, on the whole, measuring immediately connotes an important adjudicative operation, something technical and urgent.

In this book, measuring has often appeared as a solution: an aid to resolving controversies, a way to investigate the real, a means to get to the bottom of things or to get the hell out of them. And throughout these pages, I have tried to show how problematic the solutions provided by measuring can be. As the columns of the temple of Castor are not perfectly vertical—and not just because of optical correction, as is often said, but because buildings are never the flawless geometries architects create in their minds—so measuring provides results that, more often than not, are simply good enough. When describing the effort of laying down the centuriation, Hyginus Gromaticus, the author of one of the treatises included in the *Corpus agrimensorum romanorum*, claimed that the goal of his profession was to strive for a system that would be sound and consistent.[3] Perfection was not the point. And some of the examples I have examined—the Neapolitan palm in chapter 3, Sacchetti's novella in chapter 14, the *mensurae Christi* in chapter 22, and Fibonacci's conversions in chapter 29—reveal the range in the degree by which measuring operations fell short of the results they were claiming to yield.

In a way, this book is a four-act drama built on broken promises. Like some of the plays that defined that genre, it starts in the eighteenth century. In part 1, "Safes," I explored the construction of modern measurement standards to reflect on how they pretend to be both neutral ratios and objects that dislodged human industry. In part 2, "Squares," I tracked the appearance of the medieval precursors, the *pietre di paragone*, to speak about measurements' pretensions to diffuse power by shifting from royal appanage to public pursuits. In part 3, "Cities," I documented the involvement of people of religion to discuss the ways in which measurements encouraged a belief in justice as well as in an order that linked the self to the universe. And, finally, in part 4, "Fields," I attempted an archaeology of medieval measurements to show how they both defied temporality and emerged, in drawing from time and working against it, as conduits of sameness across the ages.

When the idea of writing a book about measurements started coalescing in my brain, I did not plan its chronological structure so as to proceed backwards: it simply happened. As the research expanded, I found myself retrieving the roots of many of the practices and ideas about measuring between the eleventh and the early fourteenth centuries. Over time I came to see that era as more important than the later metric revolutions. Now that I am writing these last sentences, I still believe this is the case. The enlightening age of measuring reforms is an opacified version of the energies that measurements absorbed in the "dark" Middle Ages.

There is no single explanation for the emergence of such an intense, sustained engagement. In this book I have shifted the viewpoint many times to offer a glimpse of the multiplicity of historical agents. While exploring whether the breakdown of royal privilege, the display of the *pietre di paragone*, and the involvement of religious orders in policing measurements were linked, it became clear to me that none was responsible for the cultural, social, and moral transformation I saw in medieval measurements. No single operation did, let alone those that for centuries have been firmly disciplined. But the circulation of surveying texts, the warnings that preachers yelled from the pulpit, and the staking of public streets all made some difference nonetheless. By way of concluding, I cannot but offer a survey of all the activities that made measurements take the directions they did. The agent of change was not a revolution or an event of a magnitude for which historians coin names. The displays of measurements have no label. Even the expression *pietre di paragone* is one of which I often disapprove, given that in modern Italian it ordinarily refers to the basanite slabs against whose surfaces early modern goldsmiths dragged minerals to check their purity. (While hesitant, I still convince myself it is useful, given that both practices have to do with corroboration and control.) But the historical shifts with which this book is concerned are not associated with a dramatic image. The temporal bracket is loose, as sometimes is the evidence. Medieval chroniclers did not record much about the making of measurements. The textual information is scant—a passage here, a mention there—and the material evidence is hardly more abundant. Still, by putting all those pieces together, I think I start to see some of the directions that things followed.

This book has treated measurements, the quintessential tools for stability, as agents for movement. Even more, it has presented measures as channels to metamorphosis. They worked, as they still do, across scales and forms. They translated dimensions from one place to another, across materials and time. Sometimes the illusion of continuity they forged was taken seriously, as when nineteenth-century politicians from Turin believed they were employing the standards that King Liutprand had devised 1,100 years earlier. That they convinced themselves of the narrative is understandable, given that measurements are often constructed to deceive. In the seventeenth century the Papal States produced slabs with the measurement standards for every principal city. The stones imitated the *pietre di paragone* that had been created for over three centuries, even if today we can hardly tell the early modern displays from their medieval ancestors. But such a distinction makes little sense, since measurements trigger a form of thinking in which time loses importance and becomes secondary. The dating of the *pietre* is quickly forgotten, their genesis cast in a mythical limbo. The chronological contours of measurement need to vaporize, as much of their validity depends on suspending the clock. This willful forgetting is already happening for the meter, despite the fact that it was created only two centuries ago and forced upon people with such tirelessness that detailed records of its enforcement are to be found even in the most remote libraries. When eighteenth-century savants debated about whether to create it out of the earth, some historians, and Voltaire with them, supported the idea by pointing out that in doing so they were reenacting operations of centuries earlier. The metric system, like other forms of measurement, yearns for hooking onto the past even when it claims to produce something utterly new. Little did the French scientists remember how much of the past their royal measurements carried within them. As time-defying, hard-to-see, quickly forgotten conduits, measurements do not just translate past dimensions; they also bring sedimented anxieties and cultural malaises with them.

So when we speak of measurements today, we must also consider those aspects of the past that they carry with them. Much of this past is streaked with political conflict. Measurements, I write one last time, were fields of contention and tools through which power extended its grip. We saw how the Pisan commune antagonized the guild of bricklayers, and how Pope Innocent IV barely defended the Humiliates, tired of weighing goods at the gates of Milan. We also commented on the failures of Barbarossa and Napoleon. Power dictated the ways to manufacture standards and the amount of attention they deserved, as it is through that attention that control takes shape. Such considerations have not gone away. Think of the opposition of Americans to the metric system, or the centrality the London Stock Exchange derives from relying on the meridian zero.

Authorities attend to measurements because they prove useful in maintaining the order that keeps them in power. Size standards are subtly coercive tools that pass for neutral instruments of trade. But measurements do far more than that. They dissolve power into space, shaping ideas about the latter and chaining it irreversibly to the former. Measurements give space to power, and make use of people to tighten such connections. With every measuring act—no matter if approximate or precise—measurers and makers abide by an authority that demands respect because of the measurements it maintains.

Yet success is not just the result of wielding power, giving citizens no choice in the matter, but a question of both sides conspiring to keep the narrative aloft. People refused to reject measurements, because they were gaining something in the process too. And this is why this book has been a collection of stories about not only the winners' successes but also the losers' willingness to accept the state in which they were thrown. This book has often tracked the inertia that carries lives and the stories communities have crafted to console themselves and pull things back, insisting that standards stay as they are. After all, measurements are characteristically reactionary. They hamper possibilities while setting the bar for the amount of critical thinking they can bear. When I measure, my observations depend on the precision of the standards I employ. (A centimeter may matter for shoe sizing, but not as much for land surveying.) I accept the implicit social conventions. (Is it relevant if a butcher gives me a few more grams of ham?) I abide by the relevant political institutions. (In the United States I speak of feet, never of meters.) I may question an aspect of the process, but not the process itself. Don't move past the border, or the whole system may collapse.

I have been moved to write this book by my frustration with a passive acceptance of this border. I have explored the history of medieval measurements to argue that it is important to see what happens above and below it. In a way, this book embraces the well established deconstructionist approach typical of science studies. It is preoccupied with complicating our assumptions, showing how even measurements are things—and, as Heidegger says, "things" are always necessarily gatherings of many different items.[4] So size standards are not mere objects, but objects that come with assumptions, desires, and projections. The history of measurements-as-things is particularly complicated because these things tend to turn into abstract ratios and thus dissolve into a different order. When I think of them in concise form, as I am doing now that I search for precise, conclusive definitions, they often seem to drag the train of the whole world, and it is then that I realize that their cultural validity cannot be explored in full. This book has gathered many strands, showing, for instance, how a change in legal formulations affected both the locations of monasteries and the length of women's dresses. Yet it can only remain fragmentary when compared to the ubiquitous pressure of measurements upon the real.

So measurements are organized into endless and endlessly different sets—they can only exist as such—even if much of the history I have reconstructed depends on their occultism. Certitude is predicated on somewhat simple narratives. The *tavole di ragguaglio*, Paucton's seemingly endless history of the world, Ovid's shattering of peace—they all explain some aspects of the history of measurements in memorable terms. To smooth their construction, measurements must have that natural, innocent look of those familiar objects that do not interfere with what you see or what you may consider right and true. Assurance benefits from unpretentiousness, and measurement standards have taken up the unassuming form of geometrical diagrams, stone incisions, and metal bars. I hope to have shown how much is hidden by such purported simplicity.

In a time where the construction of facts is being disputed and scientific evidence is under constant pressure from political forces that have only to gain from a state of disarray, to express doubts about the validity of measuring is to arm such forces with a fatal bullet. This book, however, is not about the dismantling of scientific facts. On the contrary, it aims at emancipating people from quick assumptions, from the way power manipulates events to naturalize arbitrary choices, and, more importantly, from the way identity is prematurely accepted without scrutiny.[5]

I understand that things may appear strangely disjointed now. Are we supposed to trust measuring or not? Well, this question is not the one to ask, as a yes-or-no answer is what measuring is often required to supply. The only way not to repeat such a simplistic dualism is to explore measuring so as to enforce other critical questions about effects and intentions, modalities and circumstances. In other words, we need to bend measurements to produce alternative results to the black-and-white dilemma in which they cast us.

Measurements can provide more nuanced solutions because they contain multiple elements. They can be turned inside out to reveal the authority that presides over them and the ways they move from the particular to the general, from matter to abstract. Power has largely ignored such complexity, and has cherry picked what was convenient. But once we embrace the complex nature of measurements—the way they hide multiplicities within them and their results—we also realize their inherent fragility. If anything, I would like this book to express the need for caring about them. Their history is as rich as it is fascinating, and if I have tried to uncover some of its aspects, it is not to invalidate them but to understand what conditions make their grip stronger or weaker. So do not ask whether measurements should be trusted. Rather, employ them to reveal the circumstances through which their results produce sense. Use them against the traditional ways in which they have been used, to discover a new range of results.

I have often stated that measuring promises sameness, but sameness is never a product of measuring; you need to believe in it before any measuring is even undertaken. Sameness is constructed through discourses, forms, and practices that lie outside of measuring. Sameness is largely a product of abstractions. It emerges from the intersections between mental operations that are at the core of the construction of forms. And it is only through the link between measuring and forms that measurements come in contact with sameness.

In the late nineteenth century, when measurement standards—even those platinum bars that reflected scientists' complacent smiles—were shown to be inaccurate under the magnifying lenses of microscopes, some people searched for other ideas. In September 1870, one month after the heads of the world's metrological cabinets met in Paris to discuss the metric system one more time, James Clerk Maxwell gave a lecture at the British Association for the Advancement of Science in Liverpool in which he spoke of new ways to define standards.[6] "The discoveries of physics," he declared, "have revealed to the mathematician new forms of quantities which he could never have imaged for himself."[7] Atomic studies showed that objects were composed of bodies that no one could see nor handle, but which were always in motion, impossible to stop or break. "If we wish to obtain standards of length, time, and mass which shall be absolutely permanent, we must seek them not in the dimensions, or the motion, or the mass of our planet, but in the wave-length, the period of vibration, and the absolute mass of these imperishable and unalterable and perfectly similar molecules."

The tone was enthusiastic, the sense of the new palpable. Maxwell pointed to what is today regarded as a breaking point in the history of measurements. It matters little that his wavelength definition of standards continued centuries-old ties between measuring, light, and seeing. What matters is that measurements would no longer be seen like light; they would become light.

In 1960, at the international conference of weights and measures, the meter was redefined in relation to the radiation of krypton 86. More precisely, it was acknowledged as the 605-nanometer wavelength of the orange spectral line. Despite the new declarations of precision, such a definition was abrogated six meetings later, in 1983, when scientists realized that they had made such progress in measuring the frequency of radiations that they had to redefine the speed of light, and the meter with it. The measuring tool and the object to be measured continued to support each other. As the philosopher Boethius wrote in the sixth century, mathematics stands on itself.[8] So it is likely that scientists will need to retouch their current definitions of measurements again with the advancement in their understanding of light behavior. They will need to do so to keep some level of connectivity between abstract and physical forms, as measurements have done for centuries. Yet, in the meantime, while measurements enjoy such a close relationship to light, let us consider them neither as the multiplier of an orange wavelength nor as traces of the passage of light. Let us instead see them for a moment for what they are: bleary, incandescent sunsets that, for as long as they last, engulf everything that surrounds them, as they are both a luminous source and spectacles in themselves.

Acknowledgments

In the endnotes I have thanked all the people who gave me advice for this book, next to the readings they suggested. I am very grateful to all of them, as well as to those who invited me to reflect on measurements at workshops and conferences. Special mentions go to Marina Cotugno, who helped me obtain images for the illustrations at light speed; Renaldo Migaldi, who helped me to rethink what constitutes exactness in writing; and Randy Petilos, my editor at the University of Chicago Press, for being so patient and caring.

This book, however, would not be what it is were I not sustained by the constant encouragement of three people. Marvin Trachtenberg at New York University's Institute of Fine Arts, has shaped my ideas about measurements, space, and medieval architecture to such an extent that there is almost no page in this book where I cannot detect his magical thinking. He was the first person to believe in this project, dispensing advice at every turn and turning my lumpy thoughts about the poorly photocopied images I brought to him into intellectual firebrands. John Gagné has scoured the outcomes of those experiments. He read every sentence, polishing it and sharpening it as the dexterous writer he is. Academic books are, naturally, collective endeavors. As the interminable projects they often feel like, they grow with the writer's surroundings, as hair swells in humidity. Still, the last drafts of this book were very much fueled by an ongoing dialogue between John and me. Finally, Katharine Park is the one who propelled this book into the world. She has been a driving force since I met her, reading drafts, suggesting publishing houses, and promoting this manuscript in New York even before I had completed it. She did all this with the grace and brilliance for which she is rightly famous, never asking for anything in return—not even the bottle of Tuscan red I keep promising her, which still sits on the dining table around which my parents, my brother, and his wife have heard me speaking incessantly about measurements. The members of my family have always been my champions and allies, never flinching from asking questions and helping in ways I could not have foreseen—as has Michal, who has said yes to every journey, physical or metaphorical, I have ever proposed.

Notes

PREFACE

1. The fight of Bologna's administration against graffiti has been recorded by the local press. See, for instance, Gian Luca Terio, "Avanti contro i graffi," *La Repubblica*, Cronaca di Bologna, September 12, 2007; Elisa Anzolin, "Parte il piano contro i graffiti: Via le scritte dal ghetto e Zamboni," *La Repubblica*, Cronaca di Bologna, September 19, 2007; Silvia Bignami, "Lotta ai graffiti, la Cancellieri accelera," *La Repubblica*, Cronaca di Bologna, May 6, 2010; Eleonora Capelli, "La fama internazionale dei writer è iniziata sui muri di questa città," *La Repubblica*, Cronaca di Bologna, May 6, 2010.
2. Paolo Segneri, "Il Confessore istruito," in *Opere*, 4 vols. (Venice: Baglioni, 1712), 4:641.
3. For instance, an instance of this use is in Franco Sacchetti, *Il trecentonovelle*, ed. Emilio Faccioli (Turin: Einaudi, 1970), CCVII, p. 628: "per ovviare alla infamia dell'ordine."
4. Edmund Husserl defines philosophy as the science of the most obvious truth (*selbstverstaendlich*). Edmund Husserl, *Logische Untersuchungen* (Tübingen: Max Niemeyer, 1993), 22. See also Francesca Rivetti Barbò, *Essere nel tempo: Introduzione alla filosofia dell'essere, fondamento di libertà* (Milan: Jaca Book, 1990), 15–16.

CHAPTER 1

1. Wenzel Gustav Kopetz, *Allgemeine österreichische Gewerbs-Gesetzkunde*, 2 vols. (Vienna: Friedrich Folke, 1830), 2:332.

2   Alfred Francis Pribram, *Materialien zur Geschichte der Preise und Löhne in Österreich* (Vienna: Carl Ueberreuter, 1938), 121. Leopold's edicts date to March 13, 1781, and July 11, 1782.

3   Emanuele Lugli, "Cesare Beccaria e la riduzione delle misure lineari a Milano (1771–1789)," *Nuova informazione bibliografica* 12, no. 3 (2015), 579–601.

4   For instance, the king of Sicily, Ferdinand III, introduced new standards with the law of December 31, 1809. See *Codice metrico siculo* (Catania: Stamperia dell'università, 1812), 57.

5   The quest of Florence's standards of measurements is described in Leonardo Ximenes, *Del vecchio e nuovo gnomone fiorentino e delle osservazioni astronomiche, fisiche ed architettoniche fatte nel verificarne la costruzione* (Florence: Stamperia imperiale, 1757), 1–10. For an equivalent quest in Milan, see Cesare Beccaria, "Della riduzione delle misure di lunghezza all'uniformità per lo stato di Milano," in *Opere*, 2 vols. (Milan: Società tipografica dei classici italiani, 1821–22), 2:453.

6   The renowned historian Gaetano de Sanctis is often credited with the aphorism "Metrology is not a science; it is a nightmare" (La metrologia più che una scienza è un incubo; unless otherwise noted, all translations are my own). For an example, see Vincenzo Banzola, "Le antiche misure parmigiane e l'introduzione del sistema metrico decimale negli Stati Parmensi," *Archivio storico per le province parmensi* 18 (1966), 139. I do not know whether de Sanctis ever said that, but in his writings his skepticism is somewhat softer. See Gateano de Santis, *Scritti minori*, 6 vols. (Rome: Edizioni di storia e letteratura, 1966–72), 2:191. Even when de Sanctis is not cited, such a view is pervasive. See Juergen Schulz, "Le antiche misure lineari secondo Sebastiano Serlio e il problema dei loro valori," in *Lezioni di metodo: Studi in onore di Lionello Puppi*, ed. Loredana Olivato and Giuseppe Barbieri (Vicenza: Terraferma, 2002), 363–71; Bruno Andreolli, "Misurare la terra: Metrologie altomedievali," in *Uomo e spazio nell'alto medioevo*, 2 vols. (Spoleto: Centro italiano di studi sull'alto medioevo, 2003), 1:172–73. The earliest definition of measures as an "intolerable nightmare" (unleidlichen Alptraum) that I could find is in Karl Anton Henschel, *Das bequemste Maas- und Gewichts-system gegründet auf den natürlichen Schritt des Menschen* (Kassel: Bertram, 1855), 9. Thanks to a translation published in the same year, the book became very popular also in France.

7   Ugo Tucci, "Pesi e misure nella storia della società," in *I documenti*, vol. 5.1 of *Storia d'Italia*, ed. Ruggiero Romano and Corrado Vivanti (Turin: Einaudi, 1973), 581–612. An exception is Witold Kula, *Les mesures et les hommes* (Paris: Editions de la maison des sciences de l'homme, 1984).

8   The best book on the history of the metric system is Ken Alder, *The Measure of All Things: The Seven-Year Odyssey and Hidden Error That Transformed the World* (New York and London: Free Press, 2002). Also useful are the slim exhibition catalogue *L'aventure du mètre* (Paris: Musée national des techniques, 1989) and Louis Marquet, *L'Epopée du mètre: Histoire du système métrique décimal* (Paris: Délégation à l'information et à la communication, 1989).

9 Maurice Crosland, "The Congress on Definitive Metric Standards, 1789–9: The First International Scientific Conference?" *Isis* 60, no. 2 (1969), 226–31.
10 Alder, *The Measure of All Things*, 301–8.
11 Emanuele Lugli, *Unità di misura: Breve storia del metro in Italia* (Bologna: Il Mulino, 2014), 37–52.
12 See Kathryn M. Olesko, "The Meaning of Precision: The Exact Sensibility in Early Nineteenth-Century Germany," in *The Values of Precision*, ed. Matthew Norton Wise (Princeton, NJ: Princeton University Press, 1995), 103–34.
13 [Arthur-Jules] Morin, "Notice historique sur le système métrique, sur ses développements et sur sa propagation," *Annales du Conservatoire Impérial des Arts et Métiers* 9 (1870), 611–13.
14 Héctor Vera, *A peso el kilo: Historia del sistema métrico decimal en México* (Santa Úrsula Xitla: Libros del escarabajo, 2007), 15–39; Tamano Mitsuo, "Japan's Transition to the Metric System," *Commercial Weights and Measures*, no. 3 of *US Metric Study Interim Report* (July 1971), 97–102.
15 Maurice Crosland, "Nature and Measurement in Eighteenth-Century France," *Studies on Voltaire and the Eighteenth Century* 87 (1972), 277–309.
16 Helen E. Longino, *The Fate of Knowledge* (Princeton, NJ: Princeton University Press, 2002), 97–123.

CHAPTER 2

1 "Nôtre corps n'est pas une régle absoluê, sur laquelle nous devions mesurer les autres." Nicolas Malebranche, *De la recherche de la verité*, ed. Jules Simon (Paris: Charpentier, 1842), 36.
2 "Pour comparer les choses entre elles, ou plutôt pour mesurer exactement les rapports d'inégalité, il faut une mesure exacte, il faut une idée simple et parfaitement intelligible, une mesure universelle et qui puisse s'accommoder à toute sorte de sujets." Malebranche, 487–88.
3 "Non v'ha matematico, che di questo non faccia il maggior uso." Girolamo Francesco Cristiani, *Delle misure d'ogni genere antiche, e moderne* (Brescia: Bossini, 1760), 10. The Académie des sciences had sent French measurements to numerous European institutions, as recalled in *Mathématiques*, ed. Jean-Baptiste le Rond D'Alembert, vol. 13 of *Encyclopédie méthodique* (Paris: Panckoucke, 1789), 130–31.
4 The Royal Society of London asked Christiaan Huygens for a sample of the Rhenish standard in 1664. See Joella G. Yoder, *Unrolling Time: Christiaan Huygens and the Mathematization of Nature* (Cambridge, UK: Cambridge University Press, 1988), 154. Twenty years later, the Florentine scientist Vincenzo Viviani sent the Florentine standards to Pietro Paolo Caravaggio, a math professor in Milan. See Vincenzo Viviani to Pietro Paolo Caravaggio, 31 October 1684, in Florence: Biblioteca Nazionale, Fondo Galileano 256.

5 | Antonio Favaro, *Il metro proposto come unità di misura nel 1675* (Mâcon: Protat, 1901), 17–19 and 108–10; Ken Alder, *The Measure of All Things: The Seven-Year Odyssey and Hidden Error That Transformed the World* (New York and London: Free Press, 2002), 97.
6 | Klaus A. Vogel, "Cosmography" in *Early Modern Science*, vol. 3 of *The Cambridge History of Science*, ed. Katharine Park and Lorraine Daston (New York: Cambridge University Press, 2006), 493; Paolo Casini, *Newton e la coscienza europea* (Bologna: Il Mulino, 1983), 60.
7 | Gabriel Mouton, "Nova mensurarum geometricarum idea," in *Observationes diametrorum solis et lunae apparentium* (Lyon: Mathieu Liberal, 1670), 427–48.
8 | Bruce T. Moran, "Courts and Academies," in *Early Modern Science*, vol. 3 of *The Cambridge History of Science*, ed. Katharine Park and Lorraine Daston (New York: Cambridge University Press, 2006), 251–71; John L. Heilbron, "The Measure of Enlightenment," in *The Quantifying Spirit in the Eighteenth Century*, ed. Tore Frängsmyr, John L. Heilbron, and Robin E. Rider (Berkeley: University of California Press, 1990), 207–42.
9 | Ken Alder, *Engineering the Revolution: Arms and Enlightenment in France, 1763–1815* (Chicago: University of Chicago Press, 1997), 292–318.
10 | Among the first Italian journals to announce the creation of the metric system are: *Giornale de' letterati* 91 (1793), 3–14; *Memorie per servire alla storia letteraria e civile*, 26 July 1795, 22–24; Luigi Brugnatelli, "Sulla metrologia: Avviso istruttivo sulla fabbrica delle misure di lunghezza," *Avanzamenti della medicina e fisica: Opera periodica che serva di seguito al «Giornale fisico-medico»* 9, no. 4 (1796), 58–74.
11 | Pietro V. Baraldi, *La fisica e la meccanica applicate all'industria* (Milan: Gnocchi, 1850), 8. The manual was reprinted in 1864.
12 | Luigi Canina, *Ricerche sul preciso valore delle antiche misure romane di estensione lineare* (Rome: Bertinelli, 1853), 20.

CHAPTER 3

1 | Pierluigi Panza, *Antichità e restauro nell'Italia del Settecento* (Milan: Franco Angeli, 1990), 223–25.
2 | Domenico Comparetti, *Virgilio nel medio evo*, 2 vols. (Livorno: Francesco Vigo, 1872), 1:126–29 and 237–38.
3 | On the popularity of Virgil's funerary site, see Melissa Calaresu, "Looking for Virgil's Tomb: The End of the Grand Tour and the Cosmopolitan Ideal in Europe," in *Voyages and Visions: Towards a Cultural History of Travel*, ed. Jás Elsner and Joan-Pau Rubiés (London: Reaktion Books, 1999), 138–61.
4 | Paolo Antonio Paoli, *Avanzi delle antichità esistenti a Pozzuoli, Cuma e Baia* (Naples: 1768), 13.
5 | Paoli himself hinted at this convention in the commentary to his plan of the Pozzuoli amphitheatre, which, he stated, "revealed all the measurements of

the architectural fabric and its parts" (colla quale si fanno palesi le misure di tutta la fabbrica, e delle sue parti). See Paoli, *Avanzi*, 18. See also Panza, *Antichità*, 227.

6 Sabina Lang, "The Early Publications of the Temples at Paestum," *Journal of the Warburg and Courtauld Institutes* 13 (1950): 57; Monique Mosser and Marie-Raphaël Paupe, "Il dorico senza base in Francia," in *La fortuna di Paestum e la memoria moderna del Dorico: 1750–1830*, ed. Joselita Raspi Serra and Giorgio Simoncini, 2 vols. (Florence: Centro Di, 1986), 2:295–99.

7 Iain Gordon Brown, "The Picturesque Vision: Fact and Fancy in the Capriccio Plates of Robert Adam's 'Spalatro,'" *Apollo* 136 (1992), 76–82. Generally, Arnaldo Momigliano, "Ancient History and the Antiquarian," *Journal of the Warburg and Courtauld Institutes* 13 (1950), 285–315; John A. Pinto, *Speaking Ruins: Piranesi, Architects, and Antiquity in Eighteenth-Century Rome* (Ann Arbor: University of Michigan Press, 2012), 11–50 and 275–77.

8 Johann Wolfgang von Goethe, *Italian Journey, 1786–1788*, tr. and ed. Wystan H. Auden and Elizabeth Mayer (London: Folio Society 1962), 210. See also John A. Pinto, *City of the Soul: Rome and the Romantics* (Hannover and London: University Press of New England, 2016), 16.

9 Renzo Sabbatini, *Le mura e l'Europa: Aspetti della politica estera della Repubblica di Lucca* (Milan: Angeli, 2012), 117–18.

10 Francesco Muttoni, *Architettura di Andrea Palladio: Nella quale sono ridotte in compendio le misure, e le proporzioni delli cinque ordini di architettura* (Venice: Pasinelli, 1741), table 2. The same illustration returns in Francesco Muttoni, *Architettura di Andrea Palladio*, 8 vols. (Venice: Pasinelli, 1740–48), 3:9.

11 Maximilien Misson, *A New Voyage to Italy*, 5th ed., 2 vols. (London: J. & J. Bonwick, 1739), 1: 620–21. Misson took his trip in 1688 and published his account, in French, in 1691. His list of measurements does not appear in the four earlier editions. On Misson's success, see John Towner, "Literature, Tourism and the Grand Tour," in *Literature and Tourism*, ed. Mike Robinson and Hans C. Andersen (London: Continuum, 2002), 228–29.

12 This reflection is inspired by Jonathan Rutherford, "The Third Space: Interview with Homi Bhabha," in *Identity: Community, Culture, Difference* (London: Lawrence & Wishart, 1990), 210.

13 Misson, *A New Voyage* (1739), 1: 620, segment 11. It represents "half of the palm of Naples," and I measured it as 115 mm. The double—that is, the palm—would then be 230 mm.

14 Such custom was common. See Girolamo Francesco Cristiani, *Delle misure d'ogni genere antiche, e moderne* (Brescia: Bossini, 1760), 21.

15 "Un altro punto e ancora più importante dovrebbe esser quello di sapere con precisione l'estensione delle terre del Regno. Non è vergogna che in Europa, dove la geometria ha la sua reggia, vi sian de' paesi ignoti, non essendovene nella China? Non posso adunque dar qui che de' calcoli vaghi, finché il braccio sovrano non ci dia più certe misure." Antonio Genovesi, *Lezioni di commercio* (Milan: Società tipografica dei classici italiani, 1824), 275.

16   Ferdinando Visconti, *Del sistema metrico della città di Napoli e della uniformità dei pesi e delle misure che meglio si conviene a' reali dominj di qua dal faro* (Naples: Stamperia reale, 1838), 38–41.

17   Visconti, *Del sistema metrico*, 38–41. See also Emanuele Lugli, *Unità di misura: Breve storia del metro in Italia* (Bologna: Il Mulino, 2014), 121–37.

18   Saverio Scrofani, *Memoria su le misure e pesi d'Italia, in confronto col sistema metrico francese* (Naples: Monitore delle due Sicilie, 1812), 61–64.

19   Melchiorre Delfico, *Memoria sulla necessità di rendere uniformi i pesi, e le misure del Regno* (Naples: Porcelli, 1787).

20   Joseph-Marie de Gérando, *Histoire comparée des systèmes de philosophie*, 3 vols. (Paris: Henrichs, 1804), 1:599.

21   Cristiani, *Delle misure*, 23. Cristiani gives the palm in Parisian *pieds du roi* as 0 feet, 9 inches, 6 lines, and 4 decimals. At page 21, notice the asterisk following the palm to indicate that Cristiani received it secondhand.

22   Alexis-Jean-Pierre Paucton, *Métrologie ou Traité des mesures, poids et monnoies des anciens peuples et des modernes* (Paris: Desaint, 1780), 777. There, Paucton gives the palm as 0.809 *pieds du roi*. As the *pied du roi* corresponds to 324.839 mm, the *palmo* is then 262.79 mm. Visconti, however, took Paucton's definition of the Neapoltian palm as imprecise. See Visconti, *Del sistema metrico*, 43–44. On Paucton's fame, see Jean-Baptiste Louis de Romé de l'Isle, *Métrologie* (Paris: Printed by the author, 1789), ix.

23   The enormous, thirty-five-sheet *Mappa topografica della città di Napoli e de' suoi contorni* was commissioned in 1750 by Giovanni Carafa III, Duke of Noja, and completed after his death (1768) by the architect Gaetano Bronzuoli in 1775. It was finalized by the professor of architecture Nicola Carletti. To celebrate the completion of the map, Carletti also published some explanatory notes. Nicola Carletti, *Topografia universale della città di Napoli in Campagna felice e note enciclopediche storiografiche* (Naples: Stamperia raimondiana, 1776). On the map, see Giordana Bruno, *Atlas of Emotion: Journeys in Art, Architecture, and Film* (New York: Verso, 2002), 362–63.

24   This observation is not mine. It comes from Visconti, *Del sistema metrico*, 43–44.

25   Feliciano Scarpellini, *Prospetto delle operazioni fatte in Roma per lo stabilimento del nuovo sistema metrico negli Stati Romani* (Rome: Mariano de Romanis e figli, 1811), 41.

26   Alain Badiou, *Being and Event* (London: Continuum, 2006), 38–48.

27   Leonardo Ximenes, *Del vecchio e nuovo gnomone fiorentino e delle osservazioni astronomiche, fisiche ed architettoniche fatte nel verificarne la costruzione* (Florence: Stamperia imperiale, 1757), 321.

28   Cesare Beccaria, "Della riduzione delle misure di lunghezza all'uniformità per lo stato di Milano," in *Opere*, 2 vols. (Milan: Società tipografica dei classici italiani, 1821–22), 2:453.

29   A fascinating recent article on the institutional efforts at standardization in the late nineteenth century is Jonathan H. Grossman, "Standardization (Standardisation)," *Critical Inquiry* 44, no. 3 (2018), 447–78.

CHAPTER 4

1. Ottorino Piangiani, *Vocabolario etimologico della lingua italiana*, 2 vols. (Milan: Dante Alighieri, 1907), 2:1104.
2. [Barnaba Oriani], *Istruzione su le misure e su i pesi che si usano nella Repubblica Cisalpina* (Milan: Comitato governativo, 1801). Oriani is credited as the author in Gaetano Melzi, *Dizionario di opere anonime e pseudonime*, 3 vols. (Milan: Giacomo Pirola, 1848–59), 2:55. The law dating 15 *piovoso* IX (February 4, 1801) is in *Raccolta delle leggi, proclami, ordini ed avvisi pubblicati in Milano dal giorno 13 pratile anno VIII* (Milan: Veladini, 1800), 2:61–62. The expanded version is *Tavole di ragguaglio fra le nuove e le antiche misure e fra i nuovi e gli antichi pesi della Repubblica Italiana* (Milan: Veladini, 1803). A second edition was published in 1806. For all those editions, see Emanuele Lugli, *Unità di misura: Breve storia del metro in Italia* (Bologna: Il Mulino, 2014), 51.
3. Stuart J. Woolf, "La storia politica e sociale," in *Dal primo Settecento all'unità*, ed. Ruggiero Romano and Corrado Vivanti, vol. 3 of *Storia d'Italia* (Turin: Einaudi, 1973), 161–66 and 192–93; Fulvio Conti and Angelo Varni, "L'Emilia Romagna nell'età napoleonica (1796–1815)," in *Legazioni e ducati tra riforme e rivoluzione*, ed. Sante Medri, 2 vols. (Bologna: Analisi, 1989), 2:9–39.
4. Feliciano Scarpellini, *Prospetto delle operazioni fatte in Roma per lo stabilimento del nuovo sistema metrico negli Stati Romani* (Rome: Mariano de Romanis e figli, 1811), 30–32.
5. "Con una paziente operazione, che inutile sarebbe di descrivere, si determinarono l'estremità del Campione della Canna." Scarpellini, *Prospetto*, 33.
6. Paolo Casini, "Ottica, astronomia, relatività: Boscovich a Roma, 1738–1748," in *Newton e la coscienza europea* (Bologna: Il Mulino, 1983), 143–71.
7. Charles Hutton, *A Philosophical and Mathematical Dictionary*, 2 vols., 2nd. revised ed. (1795–96, London: For the author, 1815), 1:400.
8. Ken Alder, *The Measure of All Things: The Seven-Year Odyssey and Hidden Error That Transformed the World* (New York and London: Free Press, 2002), 81–86.
9. Scarpellini, *Prospetto*, 37–38.
10. [Oriani], *Istruzione su le misure*, 43–44.
11. What we know of the local standards from Lucca comes from the *tavole* published in 1809, and the information about those in use in Turin is drawn from the *tavole* compiled in 1818. For both, see *Tavole di ragguaglio dei pesi e delle misure già in uso nelle varie provincie del Regno col peso metrico decimale approvate con decreto reale 20 maggio 1877, n. 3836* (Rome: Stamperia reale, 1877), 379 and 701.
12. Leonardo Ximenes, *Del vecchio e nuovo gnomone fiorentino e delle osservazioni astronomiche, fisiche ed architettoniche fatte nel verificarne la costruzione* (Florence: Stamperia imperiale, 1757), 7–8.
13. The precision of the *tavole*, of the order of a thousandth of a millimeter, is the effect of arithmetic calculations. After all, the *tavole* constructed dimensions by drawing from other textual sources. To calculate the dimension of the Milan *braccio*, scientists read the reports of Cesare Beccaria and Paolo Frisi,

who considered it as 0.5949 of the French *toise*. As the quarter of the Parisian meridian, or 10,000 meters, had been calculated as 30,784,440 Parisian feet, one meter was known to be 3.078444 feet. So as long as any local standard had been measured either in relation to the old French system or in relation to any geodesic dimension (such as the nautical mile), it was possible to retrieve its metric dimension without recurring to any physical object.

14 Vincenzo Banzola, *Le antiche misure parmigiane* (Parma: La Nazionale, 1968), 160–61. On the critique of local presses, see Gioachino Simondi, *Tavole di riduzione* (Turin: Fratelli Pic, 1844), v.

15 Napoleone Reggiani, *Il perfezionamento del sistema metrico decimale in Italia ed i risultati della Convenzione del metro* (Rome: Eredi Botta, 1890), 6.

16 Michael Kershaw, "The Reality of Precision in the 19th Century: Reevaluating the Role of Geodesy," in *Integrating Generations*, Proceedings of the Conference of the International Federation of Surveyors (Stockholm: International Federation of Surveyors, 2008), 2–3; Michael Kershaw, "The 'Nec Plus Ultra' of Precision Measurement: Geodesy and the Forgotten Purpose of the Meter Convention," *Studies in History and Philosophy of Science* 43, no. 3 (December 2012), 563–64.

17 Vladimiro Valerio, *Società, uomini e istituzioni cartografiche nel Mezzogiorno d'Italia* (Florence: Istituto geografico militare, 1993), 243.

18 Witold Kula, *Measures and Men* (Princeton, NJ: Princeton University Press, 1986), 98.

19 Stewart Schapiro, "Vagueness and Abstraction," in *Vague Objects and Vague Identity: New Essays on Ontic Vagueness*, ed. Ken Akiba and Ali Abasnezhad (Dordrecht: Springer, 2014), 220.

20 Giuseppe Parenti, *Prime ricerche sulla rivoluzione dei prezzi in Firenze* (Florence: Cya, 1939), 73; Michael F. Hendy, *Studies in the Byzantine Monetary Economy c. 300–1450* (Cambridge, UK, and New York: Cambridge University Press, 1985), 337; Kula, *Measures and Men*, 111.

21 Michele Luzzati, "Note di metrologia pisana," *Bollettino storico pisano* 31–32 (1962–63), 211–13 and especially 216. The conclusion is worth repeating in full: "In sostanza le modeste discordanze che si hanno fra i vari ragguagli . . . testimoniano tutte di una approssimazione all'origine della quale non doveva esserci tanto una mancanza od una inesattezza di informazione dei mercanti, quanto più una reale differenza fra i singoli strumenti di misurazione in uso. È sforzo vano il cercare di risalire alla capacità di una misura medievale, e dobbiamo accontentarci di indicare una media delle capacità di una determinata misura." See also Enrico Fiumi, "Economia e vita privata dei fiorentini nelle rilevazioni statistiche di Giovanni Villani," in *Storia dell'economia italiana: Saggi di storia economica*, ed. Carlo. M. Cipolla (Turin: Einaudi, 1959), 327–28.

22 François Menant, *Campagnes Lombardes au Moyen Age: L'économie et la société rurales dans la région de Bergame, de Crémone et de Brescia du X au XIII siècle* (Rome: Écoles françaises d'Athènes et de Rome, 1993), 798.

23   "As the study of metrology, standards, empires, has shown so well, smooth continuity is the hardest thing to get." Bruno Latour, "Network, Societies, Spheres: Reflections of an Actor-Network Theorist," *International Journal of Communication* 5 (2011), 802.

CHAPTER 5

1   Ken Alder, *The Measure of All Things: The Seven-Year Odyssey and Hidden Error That Transformed the World* (New York and London: Free Press, 2002), 93.
2   "L'unité de mesure . . . a été prise dans la nature, qu'elle est déduite des dimensions mêmes de la terre, en sorte qu'il sera toujours possible de la retrouver et de la rétablir, si elle était détruite ou altérée." Comité d'instruction publique de la Convention Nationale, *Procès-Verbaux*, ed. James Guillaume (Paris: Imprimerie nationale, 1894), 2:638.
3   On mathematized nature, see Quentin Meillassoux, *After Finitude: An Essay on the Necessity of Contingency* (London: Bloomsbury, 2008), 115. On secretive nature, see Pierre Hadot, *The Veil of Isis: An Essay on the History of the Idea of Nature* (Cambridge, MA, and London: Belknap Press, 2006), 237–46.
4   See, for instance, J. C. Philibert [pseud. of Jean-Baptiste Charles Legendre de Luçay], *Introduction à l'étude de la botanique*, 2 vols. (Paris: Delalain, 1802), 2:501.
5   "Toutes nos connaissances viennent des sens, et particulièrement du toucher: parce que c'est lui qui instruit les autres." Étienne Bonnot de Condillac, *Traité des sensations*, 2 vols. (London and Paris: De Bure, 1754), 256. See also Michael J. Morgan, *Molyneaux's Question: Vision, Touch and the Philosophy of Perception* (Cambridge: Cambridge University Press, 1977), 72–79.
6   "Les nouvelles mesures ordonnées par Décret de la Convention Nationale du . . . 7 Avril 1795 . . . ne sont point des mesures arbitraires, comme le sont toutes celles qu'on a imaginées jusqu'ici: elles sont fondées sur des bases fixes et invariables, et prises dans la nature, et même de manière que l'élément de ces mesures peut être avoué par tous les habitants de la terre; que chacun peut dire, cette mesure m'appartient." Mathurin-Jacques Brisson, *Réduction des mesures et poids anciens, en mesures et poids nouveaux* (Paris: Didot, 1798), 7. Brisson was a professor of physics and one of the member of the Académie. The passage was much copied, even outside scientific circles. For instance, it returns in Claude Mathieu Delagardette, *Les ruines de Paestum ou Posidonia* (Paris: Barbou, 1798), 75.
7   Dan Edelstein, *The Terror of the Natural Right: Republicanism, the Cult of Nature, and the French Revolution* (Chicago: University of Chicago Press, 2010), 8. Gianna Pomata and Lorraine Daston have pointed out that the eighteenth century saw a rise in the number of people who both looked up to nature and felt that they could speak for it. Gianna Pomata and Lorraine Daston, "The Faces of Nature: Visibility and Authority," in *The Faces of Nature in Enlightenment Europe* (Berlin: Berliner Wissenschafts-Verlag, 2003), 1–16.

8 | The expression "crown jewel of metrology" is by Charles Coulston Gillispie, *Science and Polity in France: The Revolutionary and Napoleonic Years* (Princeton, NJ: Princeton University Press, 2004), 469.

9 | The metal bar is one meter long, twenty-five millimeters broad, and four millimeters thick. See Albert Pérard, "The Metric System," *Aeronautics* 15, no. 1 (August, 1946), 27. Pérard observes that today the two surfaces are convex and much distorted. The platinum bar was deposited in the Archives on June 22, 1799. Lenoir was present. See William A. Smeaton, "The Foundation of the Metric System in France in the 1790s," *Platinum Metals Review* 44, no. 3 (2000), 125–34.

10 | Jeremy Jennings, *Revolution and the Republic: A History of Political Thought in France since the Eighteenth Century* (Oxford: Oxford University Press, 2011), 6–9; Morag Martin, *Selling Beauty: Cosmetics, Commerce, and French Society, 1750–1830* (Baltimore: Johns Hopkins University Press, 2009), 90–96.

11 | Harvard University claims that the brass meter in its Collection of Historical Scientific Instruments (figure 5.1) is one of the original models made at the request of the Assemblée Nationale with the law of December 9, 1794. It was bought in 1961.

12 | Karl Ulbrich, "Das Klafter- und Ellenmaß in Österreich," *Blätter für Technikgeschichte* 32/33 (1970–71), 1–34.

13 | Pérard, "The Metric System," 27.

14 | Donald McDonald and Leslie B. Hunt, *A History of Platinum and Its Allied Metals* (London: Johnson Matthey, 1982), 56–57 and 77–80.

15 | Jacques-Cristophe Valmont de Bomare, *Minéralogie, ou Nouvelle exposition du règne minéral*, 2 vols. (Paris: Vincent, 1762), 2:168.

16 | [Arthur-Jules] Morin, "Notice historique sur le système métrique, sur ses développements et sur sa propagation," *Annales du Conservatoire Impérial des Arts et Métiers* 9 (1870), 179–94.

17 | Gillispie, *Science and Polity*, 252–53.

18 | "Le moins susceptible de s'altérer, le moins dilatable par la chaleur, le moins condensable par le froid." Comité d'instruction publique de la Convention Nationale, *Procès-Verbaux*, 2:640 and 643.

19 | Alder, *Measure of All Things*, 265; McDonald and Hunt, *A History of Platinum*, 180; Iris Moon, "Immutable Décor: Post-Revolutionary Luxury in the Platinum Cabinet at Aranjuez," in *Les progrès de l'industrie perfectionnée: Luxe, arts décoratifs et innovation de la Révolution française au Premier Empire*, ed. Natacha Coquery, Jörg Ebeling, Anne Perrin Khelissa, and Philippe Sénéchal (Toulouse: Presses universitaires du Mirail, 2017), 129–35. I am thankful to Iris Moon for sending me her essay.

20 | "Ce singulier métal n'a pas encore dans le commerce de prix déterminé." Alexis-Marie de Rochon, *Essai sur les monnaies anciennes et modernes* (Paris: Prault, 1792), 146. See also Joseph Dombey's letter to André Thouin, March 15, 1785, in Ernest Théodore Hamy, *Joseph Dombey: Médecin, naturaliste, archéologue, explorateur du Pérou, du Chili et du Brésil (1778–1785)* (Paris: Guilmoto, 1905), 138.

21 Charles W. J. Withers, *Placing the Enlightenment: Thinking Geographically about the Age of Reason* (Chicago: University of Chicago Press, 2007), 114–21.
22 Raul Hernandez Asensio, *El Matemático Impaciente: La Condamine, las pirámides de Quito y la ciencia ilustrada (1740–1751)* (Lima: IFEA, 2008), 63–103.
23 Such an hypothesis was first articulated in Pierre Louis Moreau de Maupertuis, *La figure de la terre* (Paris: Impremerie royale, 1738), 179–81.
24 "Une mesure fixe et invariable, à laquelle la suite des siècles ni la distance des lieux n'apporteraient aucune altération." Charles-Marie de La Condamine, *Mesure des trois premiers degrés du méridien dans l'hémisphère austral* (Paris: Impremerie royale, 1751), 1.
25 La Condamine's proposal dates to July 29, 1758. See Charles-Marie de la Condamine, "Remarques sur la toise-étalon du châtelet," *Mémoire de l'Académie Royale des Sciences 1772*, no. 2 (1776), 482–501.
26 The expedition to Lapland took place in 1736 and aimed at measuring the arc meridian between Tornio and Mount Kittis. Pierre-Louis Moreau de Maupertius, *Relation du voyage fait par ordre du Roi au cercle polaire* (Paris: Imprimerie royale, 1738). See also Marthe Emmanuel, *La France et l'exploration polaire* (Paris: Nouvelles éditions latines, 1959), 165–78. On its inferior quality to that of the mission in Peru, see Pierre Bouguer, *La figure de la terre* (Paris: Jombert, 1749), iii–iv.
27 La Condamine, "Remarques," 501.
28 Alexis-Jean-Pierre Paucton, *Métrologie ou Traité des mesures, poids et monnoies des anciens peuples et des modernes* (Paris: Desaint, 1780), 19.
29 Feliciano Scarpellini, *Prospetto delle operazioni fatte in Roma per lo stabilimento del nuovo sistema metrico negli Stati Romani* (Rome: Mariano de Romanis e figli, 1811), 36; [Barnaba Oriani], *Istruzione su le misure e su i pesi che si usano nella Repubblica Cisalpina* (Milan: Comitato governativo, 1801), 42.
30 Together with Christiaan Huygens's experiments on the pendulum, La Condamine's report is considered foundational for the construction of the meter. See *Rapport fait à l'Institut National des Sciences et Arts le 29 prairial an 7 (17 June 1798) sur la mesure de la méridienne de France* (Paris, 1799), 2.
31 "Une pierre si dur qu'on ne peut la briser sue une enclume d'acier, ni la réduire par la calcination, ni par conséquent en extraire le minerais qu'elle enserre, qu'avec un travail infini et beaucoup de fraix." Jorge Juan and Antonio de Ulloa, *Voyage historique de l'Amérique méridionale*, tr. Eléazar de Mauvillon, 2 vols. (Paris: Jombert, 1752), 1, 375. See also McDonald and Hunt, *A History of Platinum*, 15–17; Neil Safier, *Measuring the New World: Enlightenment Science and South America* (Chicago: University of Chicago Press, 2008), 132.
32 "Le peu de connaissances que les chymistes out eues jusqu'ici de l'histoire naturelle de la platine, et la petite quantité qu'ils en ont eue en leur possession, ne leur a pas permis d'y appliquer encore en grand les travaux de la métallurgie." Guillaume Thomas Raynal, *Histoire philosophique et politique des établissements et du commerce des Européens dans les deux Indes*, 3rd ed., 5 vols.

(Geneva: Pellet, 1781), 4:193. On platinum as enigmatic, see also Paucton, *Métrologie*, 633.

33   William Lewis, "Experimental Examination of a White Metallic Substance Said to Be Found in the Gold Mines of the Spanish West-Indies," *Philosophical Transactions* 48 (1753–54), 638–89. See also Claude Morin, *La Platine, l'or blanc ou le huitieme métal* (Paris, Le Breton and others, 1758), 6.

34   Raynal, *Histoire philosophique et politique*, 4:157–58.

35   Jérôme Lalande, "Lettre sur un nouveau métal simple, appelée platine," *Journal des sçavans*, 1758, 46–59. On Lalande, see Alder, *The Measure of All Things*, 81–89; McDonald and Hunt, *A History of Platinum*, 44.

36   "Cette somme assez considérable mettra nos sçavants en état de faire des expériences en grand." Joseph Dombey to André Thouin, April 20, 1780, in Hamy, *Joseph Dombey*, 51 and 55. See also Alder, *The Measure of All Things*, 250. On Dombey's tragic mission to the United States, see Andro Linklater, *Measuring America: How the United States Was Shaped by the Greatest Land Sale in History* (London: HarperCollins, 2002), 147–53.

37   Claude Louis Berthollet and Bertrand Pelletier, "Rapport Fait au Bureau de Consultation, sur les moyens proposés par M. Jeanety [sic] pour travailler le Platine," *Annales de Chimie* 23 (1792), 20–33.

38   Alder, *The Measure of All Things*, 228.

39   Seymour L. Chapin, "'In a Mirror Brightly:' French Attempts to Build Reflecting Telescopes Using Platinum," *Journal of the History of Astronomy* 3 (1972), 87–104.

40   Isaac Newton, *Opticks; or, A Treatise of the Reflections, Refractions, Inflections and Colours of Light* (London: William Innys, 1730), 249 and 258–60. See also Michael White, *Isaac Newton: The Last Sorcerer* (London: Fourth Estate, 1997), 165–71. On Mégnié, see McDonald and Hunt, *A History of Platinum*, 64.

41   Jean-Pierre Claris de Florian, *Oeuvres*, vol. 7 (Paris: Dufart, 1803), 62–63. In the end, the mirror is described as "l'emblême de la société." See also the fable "Le miroir de la vérité" at 126–27.

42   Sabine Melchior-Bonnet, *The Mirror: A History* (London: Routledge, 2002), 153–54. On the craze for mirrors in France, see also Ewa Lajer-Burcharth, *Necklines: The Art of Jacques Louis David after the Terror* (New Haven: Yale University Press, 1999), 136–38. I am grateful to Richard Taws for suggesting this book.

43   "Des glaces représentant des vuides, ne pouvoient raisonnablement être admises qu'au bout d'une enfilade d'appartements pour prolonger son étendue, ou sur des murs opposés à des croisées." Jacques-François Blondel, *Cours d'architecture*, 6 vols. (Paris: Desaint, 1771–77), 5:66. See also Melchior-Bonnet, *The Mirror*, 81–82.

CHAPTER 6

1. "Le pro[to]type ou étalon naturel auquel les Anciens avoient rapporté leurs mesures, est la mesure de la terre." Alexis-Jean-Pierre Paucton, *Métrologie ou Traité des mesures, poids et monnoies des anciens peuples et des modernes* (Paris: Desaint, 1780), 102.

2. "Les Anciens avoient un étalon naturel de mesure, pris dans la grandeur d'un degré du méridien . . . cet étalon immatriculé dans la nature et de la valeur de la quatrecent-millieme partie d'un degré du méridien, étois universel et commun à l'Asie, à l'Afrique ey à l'Europe." Paucton, *Métrologie*, viii. For the definition of *immatriculer* as "inscribed," see *Dictionnaire de Trévoux*, 8 vols. (Paris: Compagnie des libraires associés, 1771), 5, 83.

3. Nicolas Fréret, "Essai sur les mesures longues des anciens," *Memoires de littérature tirés des registres de l'academie royale des inscriptions et belles-lettres* 24 (1756), 433. On Fréret, see Paucton, *Métrologie*, 4.

4. Voltaire, *Essai sur le moeurs et l'esprit des nations*, vol. 16 of *Oeuvres* (Paris: Stoupe, 1792), 353. The story of Al-Ma'mun's expedition was well known. It was described in Al-Farghani's summary of Ptolemy's *Almagest* and repeated in Roger Bacon's thirteenth-century *Opus Maius*. See also Richard Cumberland, *An Essay towards the Recovery of the Jewish Measures and Weights* (London: Chiswell, 1686), 46.

5. Paucton, *Métrologie*, 101–14; Girolamo Francesco Cristiani, *Delle misure d'ogni genere antiche, e moderne* (Brescia: Bossini, 1760), 20–21 and 36.

6. Paucton, *Métrologie*, 110–16; Charles Arbuthnot, *Tables of Ancient Coins, Weights and Measures* (London: Tonson, 1727), 69.

7. Manfredo Tafuri, *La sfera e il labirinto* (Turin: Einaudi, 1980), 53.

8. "Une mesure fixe et invariable, à laquelle la suite des siècles ni la distance des lieux n'apporteroient aucune altération, a été desirée dans tous les temps." Charles-Marie de La Condamine, *Mesure des trois premiers degrés du méridien dans l'hémisphère austral* (Paris: Imprimerie royale, 1751), i–iii.

9. Cristiani, *Delle misure*, 13.

10. John S. Wilkins, *Species: A History of the Idea* (Berkeley: University of California Press, 2009), 47–96.

11. Paucton, *Métrologie*, 138.

12. Cristiani, *Delle misure*, 82; Paucton, *Métrologie*, 136.

13. Cristiani, *Delle misure*, 5.

14. Cumberland, *Recovery of the Jewish Measures*, 46–48.

15. Prospero Balbo, "Del metro sessagesimale, antica misura egizia, rinnovata in Piemonte," *Memorie della Reale accademia delle scienze di Torino* 29, no. 2 (1825), 21. See also Emanuele Lugli, *Unità di misura: Breve storia del metro in Italia* (Bologna: Il Mulino, 2014), 55–56.

16. Two examples should suffice. Giulio Cordero di San Quintino, "Delle misure lucchesi e del miglior modo di ordinarle," *Atti della Reale accademia lucchese di*

scienze, lettere ed arti, 1821, 3–28; Gustavo Uzielli, *Le misure lineari medioevali e l'effige di Cristo* (Florence: Bernardo Seeber, 1899), 13–14.

17 "Furono le primordiali misure de' più antichi disciplinati governi l'unico fonte, da cui ci derivarono tutte quelle che sono in pratica oggidì. Egli è impossibile d'indicarne nominatamente il loro antidiluviano autore, attesa la foltissima oscurità di que' primitivi tempi, benchè appaia dalle sacre carte, che elle ignote non fossero a Noè, mentre eseguì la costruzione dell'arca secondo le misure che gli vennero da Iddio assegnate." Cristiani, *Delle misure*, 3 and xxii.

18 Martin J. S. Rudwick, *Earth's Deep History: How It Was Discovered and Why It Matters* (Chicago: University of Chicago Press, 2014), 55–77; Paolo Rossi, "La storia della terra da Hooke a Buffon" in *Le scienze fisiche e astronomiche*, ed. William R. Shea, vol. 2 of *Storia delle scienze* (Turin: Einaudi, 1992), 290–317.

19 Pliny, *Natural History*, 7.56.

20 Euripides, *Palamedes*, frag. 578 Nauck.

21 "Il est certain que cette invention est beaucoup plus ancienne." Paucton, *Métrologie*, 325.

22 "L'invention des mesures & des poids doit être aussi ancienne que le monde." Paucton, *Métrologie*, 320; Giuseppe Flavius Josephus, *Antiquitates Judaica* 1.2.

23 Paucton, *Métrologie*, 292.

24 Paucton, *Métrologie*, 125–26 (on Hero) and 145 (Vitruvius); Cristiani, *Delle misure*, 5 (on Pythagoras).

25 "Mais le pied géométrique avoit une propriété qui lui étoit plus particulière, les mesures anciennes avoient été reglées sur les proportions naturelles d'un homme de moyenne taille." Paucton, *Métrologie*, 106. See also 126.

26 "La taille des hommes de moyenne stature étoit donc, il y a près de quatre mille ans, 61.632 pouces, comme aujourd'hui." Paucton, *Métrologie*, 145.

27 Cristiani, *Delle misure*, 4.

28 "Dans tous les états du monde, soit que l'on parcoure les annales de l'antiquité, soit que l'on jette un coup-d'oeil sur les gouvernements actuels, on appercevra partout & en tout temps la même vigilance sur la conservation du prototype des mesures." Paucton, *Métrologie*, 10.

29 Such an idea was articulated by Wilhelm von Humboldt in his *Ideen zu einem Versuch, die Grenzen der Wirksamkeit des Staats zu bestimmen*, of 1792. See Wilhelm von Humboldt, *The Sphere and Duties of Government*, tr. Joseph Coulthard (London: John Chapman, 1854), 1–10.

CHAPTER 7

1 Ken Alder, *The Measure of All Things: The Seven-Year Odyssey and Hidden Error That Transformed the World* (New York and London: Free Press, 2002), 391n.

2 Law no. 50, June 10, 1814, in *Raccolta generale delle leggi per gli Stati di Parma, Piacenza e Guastalla* (Parma: Tipografia ducale, 1824), 2:23–25.

3 "Si vous rencontrez un homme qui au lieu d'arpents, de toises et de pieds,

NOTES TO PAGES 50–53

vous parle d'hectares, de mètres et de décimètres, vous avez mis la main sur un préfet." Rome, October 11, 1828. François-René Chateaubriand, *Mémoires d'outre-tombe*, ed. Edmond Biré (Paris: Garnier Frères, 1975), 5:12.

4 | Charles Dickens, *Hard Times* (Boston: Ticknor and Fields, 1867), 375–76.

5 | Cited in Witold Kula, *Measures and Men* (Princeton, NJ: Princeton University Press, 1986), 12. See also Robert P. Crease, *World in the Balance: The Historic Quest for an Absolute System of Measurements* (New York: Norton, 2011), 245.

6 | "Hättest du [Italian antiquarian] mehr gefühlt als gemessen, wäre der Geist der Massen über dich gekommen, dir du anstauntest, du hättest nicht so nur nachgeahmt, weil sie's taten und es schön ist." Johann Wolfgang Goethe, "Von Deutscher Baukunst," in *Schriften zur Kunst, Schriften zur Literatur, Maximen und Reflexionen*, vol. 12 of *Werke* (Hamburg: Wegner, 1953), 8. See also Enrico Castelnuovo, "Il fantasma della cattedrale," in *Il Medioevo al passato e al presente*, ed. Enrico Castelnuovo and Giuseppe Sergi, vol. 4 of *Arti e storia nel Medioevo* (Turin: Einaudi, 2004), 3. Over time, Goethe changed his mind, and in a letter of 1791 he showed some interest in proportional canons. See Erwin Panofsky, *Meaning in the Visual Arts* (Chicago: University of Chicago Press, 1955), 107.

7 | Maximilien Misson, *A New Voyage to Italy*, 5th ed. (London: J. & J. Bonwick, 1739), 1:357. This information is repeated in previous editions. See, for instance, the London edition of 1695, 2:321.

8 | Misson, *New Voyage to Italy* (1739), 1:357. Misson thought that those tools would provide a traveler "with as much exactness as you could desire." He himself measured the columns of the Pantheon portico. See 2:10.

9 | Thomas Nugent, *The Grand Tour*, 3rd ed. (London: T. Evans, 1778), 42.

10 | *The Gentleman's Magazine*, January 1753, 162.

11 | "Du hast dir das herrlichste errichtet; und kümmert die Ameisen, die drum krabbeln, dein Name nichts, hast du gleiches Schicksal mit dem Baumeister, der Berge auftürmte in die Wolken." Goethe, "Von Deutscher Baukunst," 7.

12 | See, for instance, Ugo Foscolo, *Ultime lettere di Jacopo Ortis* (Turin: Einaudi, 2005), 138.

13 | Pietro Selvatico, "Prelezione al corso di storia architettonica (1856)," in *Scritti d'arte* (Florence: Barbera, 1859), 291–307.

14 | Leopoldo Cicognara, "Elogio di Andrea Palladio," in *Discorsi letti nella Real veneta accademia di belle arti di Venezia*, 8 vols. (Venice: Picotti, 1810).

15 | Amico Ricci, *Storia dell'architettura in Italia* (Modena: Regio-Ducal Camera, 1857–59), 1:65, 228, 394, and 443, 2:5.

16 | "Il giudicio di un occhio dotto, ed esercitato sia più superbo quanto più delicato di qualunque compasso e misura." Gian Francesco Galeani Napione, *Monumenti dell'architettura antica* (Pisa: Capurro, 1820), 3:151.

17 | "E misura l'occhio queste quantità con i razzi visivi quasi come con un paio di seste." Leon Battista Alberti, *De pictura* (New Haven: Yale University Press, 1956), 46.

18 Arthur Schopenhauer, *Die Welt als Wille und Vorstellung* (Leipzig: Brockhaus, 1819), 3:382–83; Friedrich Schlegel, *Storia della letterature antica e moderna* (Naples: Marotta, 1838), 197–99.

19 Camillo Boito, "I restauri in architettura," *Il nuovo e l'antico in architettura*, ed. Maria Antonella Crippa (Milan: Jaca Book, 1988), 115.

20 Jean Baptiste Louis Georges Seroux d'Agincourt, *Histoire de l'art par les monuments depuis sa décadence au IVe siècle jusqu'à renouvellement au XVIe*, 4 vols. (Paris: Treuttel et Würtz, 1823), plates 37, 39, and 47. The Italian edition was published in Prato in 1826. See Ingrid Renée Vermeulen, *Picturing Art History: The Rise of the Illustrated History of Art in the Eighteenth Century* (Amsterdam: Amsterdam University Press, 2010), 214–62. Leopoldo Cicognara, *Storia della scultura*, 2nd ed. (Venice: Picotti, 1823–24). Cicognara's second edition differs considerably from the first edition of 1813–18. See Barbara Steindl, "Per un inquadramento della 'Storia della scultura': Impostazione storiografica e rapporto con Giordani," in *Leopoldo Cicognara: Storia della scultura dal suo risorgimento in Italia fino al secolo di Canova*, ed. Francesco Leone, Barbara Steindl, and Gianni Venturi (Bassano del Grappa: Il poligrafo, 2007), 15–62; Elisabetta Rizzioli, *L'officina di Leopoldo Cicognara: La creazione delle immagini per la Storia della Scultura* (Rovereto: Osiride, 2016), 23–41.

CHAPTER 8

1 In the documents, Rolandino is called *"sindaco"* of Reggio. As the duties of this office are unclear, I prefer to translate the term generically, as "representative." In 1260, Brunetto Latini was *sindaco* of Montevarchi and, in that capacity, was sent as ambassador to Alfonso X of Castile. He is called "sindico" in the Florentine register for the 1260 battle of Montaperti, fought against the Sienese. See *Il Libro di Montaperti (an. MCCLX)*, ed. Cesare Paoli (Florence: Vieusseux, 1889), 34 (February 26). For Latini's journey to Castile, see H. Salvador Martínez, *Alfonso X, the Learned: A Biography* (Leiden: Brill, 2010), 156–58.

2 "Gerardus Spadinasacus, mensurator terre perticavit et raxonavit omnes supradictas pecias terre et eas adinvenit et quamlibet earum tot et tantum esse ut dictum est supra et scriptum per me notarium." *Liber grossus antiquus Comunis Regii*, ed. Francesco Saverio Gatta, 6 vols. (Reggio Emilia: Costi/Goretti, 1944–62), 1:144, document 55.

3 "Iuraverunt precepta potestatis et domini Rolandi de Mutis sindici comunis Regii et de veritate dicenda, et extimare bona fide sine fraude omnes supradictas pecias terre consignatas et demonstratas eidem sindico comunis nomine dicti comunis et quamlibet earum per se per dictum dominum Ubertinum tam pro comune quam pro ipso domino Ubertino, remoto hodio, amore timore, prece et precibus. Actum in palatio comunis Regii, feliciter." *Liber grossus*, 1:144–45, document 56.

4 | *Liber grossus*, 1:146–47, document 58; 1:147, document 59. Both acts were drawn in "dicta ecclesia Sancti Iohannis Baptiste."
5 | *Liber grossus*, 1:147–53, documents 50–51.
6 | For another land transaction recording all the measurements of the fields, see *Liber grossus*, 1:187, document 84.
7 | Massimo Mussini, *Una città e il suo battistero: La chiesa di San Giovanni Battista a Reggio Emilia* (Milan: Amilcare Pizzi, 1991); Augustine Thompson, *Cities of God: The Religion of the Italian Communes, 1125–1325* (University Park: Pennsylvania State University Press, 2005), 27–28.
8 | The document is in Andrea Balletti, *Le mura di Reggio dell'Emilia* (Reggio Emilia: Forni, 1984), 110. Generally, on the square, see Gino Badini, "Piazza Prampolini," in *Lo specchio della città: Le piazze nella storia dell'Emilia Romagna*, ed. Francesca Bocchi (Casalecchio di Reno: L'inchiostroblu, 1997), 244–46; Massimo Mussini, "La piazza," in *San Prospero: La basilica del patrono di Reggio Emilia* (Reggio Emilia and Milan: Fondazione Manodori, 2005), 17–20.
9 | Scholars point out that in the Middle Ages it was natural to do business in churches after the morning services, when such buildings reverted to the laity. See Thompson, *Cities of God*, 24, and Michael Camille, *Images on the Edge: The Margins of Medieval Art*, (London: Reaktion, 1992), 91. The practice was so widespread that in 1242 Reggio forbade any business in the cathedral on Saturdays, the market days. "Quod nulla persona in die sabbati emat vel vendat in ecclesia maiori." *Consuetudini e statuti reggiani*, ed. Aldo Cerlini, Corpus statutorum italicorum 16 (Milan: Hoepli, 1933), 13.
10 | In the late eighteenth century, Reggio's *pertica* was recorded as 3.18 meters in length, and the *braccio da legno* (six of which made one *pertica*) as 0.53 meter. See [Barnaba Oriani], *Istruzione su le misure e su i pesi che si usano nella Repubblica Cisalpina* (Milan: Comitato governativo, 1801), 110.
11 | [Oriani], *Istruzione su le misure e su i pesi*, 116. See also *Tavole di ragguaglio dei pesi e delle misure già in uso nelle varie provincie del regno col sistema metrico decimale* (Rome: Stamperia reale, 1877), 621.
12 | Law book 4, rubr. 26–27. *Consuetudini e statuti reggiani*, 253–54. Thompson (*Cities of God*, 24) dates the measurements to 1259, but this is unsupported. For *reza* as door, see Wilhelm Meyer-Lübke, *Romanisches etymologisches Wörterbuch* (Heidelberg: Carl Winter, 1911), 7169a.
13 | Mussini, *Una città e il suo battistero*, 116.
14 | Reggio Emilia, Museo civico, cat. no. 206.
15 | The local saying "Saint John the Baptist reveals deceptions" (*San Svan al far veder gl'ingan*) echoes this popular belief. Cited in Bocchi, *Lo specchio della città*, 57.
16 | The plan of the bishop's palace is still a matter of debate. Miller thinks it extended up to the northern apse of the cathedral. See Maureen C. Miller, *The Bishop's Palace: Architecture and Authority in Medieval Italy* (Ithaca, NY: Cornell University Press, 2000), 64.

CHAPTER 9

1. In Tuscany and Lazio, the substitution took place in the early 1860s, after the unification of Italy, as testified by the dates stamped at the corner of the tablets (see figure 9.2). Their history has not yet been written.
2. Francesca Bocchi, *Per antiche strade: Caratteri e aspetti delle città medievali* (Rome: Viella, 2013), 169.
3. On the history of the statutes, see Andrea Zorzi, "Scrivere le regole: L'Italia degli statuti," in *Atlante della letteratura italiana*, ed. Sergio Luzzatto and Gabriele Pedullà, 3 vols. (Turin: Einaudi, 2010), 1:48–54.
4. Jean-Claude Maire Vigueur, *Cavalieri e cittadini: Guerra, conflitti e società nell'Italia comunale* (Bologna: Il Mulino, 2004).
5. A few examples: Michele Luzzati, "Note di metrologia pisana," *Bollettino storico pisano* 31–32 (1962–63), 191–218; Luciana Frangioni, *Milano e le sue misure: Appunti di metrologia lombarda fra Tre e Quattrocento* (Naples: Edizioni scientifiche italiane, 1992); Maria Giagnacovo, *Appunti di metrologia mercantile genovese: Un contributo della documentazione aziendale Datini* (Florence: Firenze University Press, 2014).
6. A recent, up-to-date history of the Italian communes is Giuliano Milani, *I comuni italiani* (Rome and Bari: Laterza, 2009).

CHAPTER 10

1. "Ante ianuas maioris ecclesiae posuit mensuram pedis ad terram mensurandam." Magister Tolosanus, *Chronicon Faventinum*, Rerum Italicarum Scriptores 121, ed. Giuseppe Rossini (Bologna: Zanichelli, 1936–39), 114.
2. The church is first mentioned in 1143. See Paolo Porta, "La cattedrale paleocristiana nelle fonti storico-monumentali," in *Faenza: La basilica cattedrale*, ed. Antonio Savioli (Florence: Nardini, 1988), 13–25: 15.
3. "Omnes parietes debeant amuelari ad parietem antiquam quae est super rezzios selciati eccl[esiae] S. Mariae Maioris. . . . in muro eccl[lesie] S. Marie Maioris prope parietem textorum signatam in ipso muro." In Luigi Angelini, "Le antiche misure segnate sulla fronte di S. Maria Maggiore in Bergamo," *Atti dell'Ateneo di scienze lettere e arti di Bergamo* 28 (1953–54), 103–8: 104. Bergamo, like the rest of Lombardy, fell to Napoleon in 1796 and adopted the metric system by 1801. See [Barnaba Oriani], *Istruzione su le misure e su i pesi che si usano nella Repubblica Cisalpina* (Milan: Comitato governativo, 1801), 109.
4. *Statuto del secolo XIII del Comune di Ravenna*, ed. Andrea Zoli and Silvio Bernicoli (Ravenna: Ravegnana, 1904), 121–22 and 138; Franco Spinelli, *Gli statuti del comune e delle corporazioni della Brescia medievale: Alle radici dell'umanesimo civile e del razionalismo economico* (Brescia: Delfo, 1997), 236 and 315; Carlo d'Arco, *Della economia politica del Municipio di Mantova e tempi in cui si reggeva a repubblica* (Mantua: Negretti, 1842), 341–43.

5 | *Gli statuti del Comune di Padova*, ed. Andrea Gloria (Padova: Sacchetto, 1873), 269 ("mensura"). In Reggio Emilia the standards are called "tabelle" or "moduli." See Enrico Guidoni, *Il Duecento* (Rome and Bari: Laterza, 1989), 373. In Todi "forma," see Samuel David Gruber, "Medieval Todi: Studies in Architecture and Urbanism" (PhD dissertation, Columbia University, 1998), 402.

6 | *Il Constituto del Comune di Siena dell'anno 1262*, ed. Lodovico Zdekauer (Milan: Hoepli, 1897), 178–79; Giuseppe Abate, *La medievale "Piazza grande" di Assisi* (Assisi: Porziuncola, 1986), 136. In Assisi, standards are also called "forme."

7 | "Statuimus quod omnes et singuli tegularii . . . teneantur et debeant facere tegulas et canales bene coctos, longitudinis et amplitudinis mensure sculte in pariete palatii populi et in porta Sancte Marie." *Amelia e i suoi statuti medievali*, Enrico Menestò, ed. (Spoleto: Fondazione centro italiano di studi sull'alto medioevo, 2004), 479. Today Amelia's *pietre* can be found in the so-called Loggia dei Banditori.

8 | *Statuti di Perugia dell'anno MCCCXLII*, ed. Giustiniano degli Azzi, 2 vols. (Rome: Loescher, 1913–16), 2:437.

9 | "Mensurare ad perticam pedis Ariprandi qui pes et eius forma fiat in porta Sancti Petri et Sancti Andree expensis Communis Mantue in lapide vivo, si factum non est." D'Arco, *Economia politica*, 341–43.

10 | "E ancora si può mettere qui su tutte le misure di pietra . . . e staranno così lungo el muro della chiesa scolpite nelle pietre e cavate." Filarete, *Trattato di architettura*, ed. Anna Maria Finoli and Liliana Grassi, 2 vols. (Milan: Il Polifilo, 1972), 1:289.

11 | Francesco Berlan, *Le due edizioni milanese e torinese delle consuetudini di Milano dell'anno 1216* (Venice: Grimaldo 1872), 265 ("segnati et mensurati . . . ad lapidem"); Cesare Campori, *Statuta civitatis Mutine anno 1327 reformata* (Parma: Fiaccadori, 1864), 251–53 ("passus incisus in petra Bonissime firmus permaneat ita quod non minuatur") and 249–50 ("ad modum antiquum designatum in petra Bonissime").

12 | On *firmitas* as a legal concept, see Ennio Cortese, *Il diritto nella storia medievale*, 2 vols. (Rome: Il cigno Galileo Galilei, 1994–95), 1:321–25; Mario Ascheri, *I diritti del medioevo italiano: Secoli XI–XV* (Rome: Carocci, 2000), 86; Federico Roggero, *"Per guadium et fideiussorem": La wadia germanica nelle glosse alla lombarda* (Rome: Viella, 2003), 53–56.

13 | Glauco Maria Cantarella, *Pasquale II e il suo tempo* (Naples: Liguori, 1997), 170–71.

14 | Adriano Franceschini, *I frammenti epigrafici degli statuti di Ferrara del 1173 venuti in luce nella cattedrale* (Ferrara: Ferrariae decus, 1969).

15 | The inscription reads: "HOC CIRCA TEMPLVM SIT IVS MERCANTIB[VS] AEQVVM PONDERA NEC VERGANT NEC SIT CONVENTIO P[RAVA]." The expression is erudite. It echoes Lucan, *Pharsalia* 8.280 ("mentisque meae quo pondera vergant"). It was translated as "Around this temple, let the merchant's law be just, his weights true, and his covenants faithful" by John Ruskin, who prided himself on having discovered it. See *The Works of John Ruskin*, ed.

Edward Tyas Cook and Alexander Wedderburn, 39 vols. (London: George Allen 1903–12), 33:442. See also David-Everett Blythe, "A Stone of Ruskin's Venice" in *New Approaches to Ruskin: Thirteen Essays*, ed. Robert Hewison (London: Routledge, 1981), 157–73:157.

16 "Quod . . . nec furtum facient. nec treccamentum. nec falsitatem infra curte sancti martini." From Dennis Romano, *Markets and Marketplaces in Medieval Italy, c. 1100 to c. 1400* (New Haven and London: Yale University Press 2015), 217–19.

17 Sarah Rubin Blanshei, "Perugia 1260–1340: Conflict and Change in a Medieval Italian Urban Society," *Transactions of the American Philosophical Society* 66, no. 2 (1976), 1–128:59–60. Also Paolo Serafini, "Antiche misure umbre," *Nuova economia* 85, no. 1 (1973), 76–80.

18 Attilio Bartoli Langeli, *Codice diplomatico del Comune di Perugia*, 3 vols. (Perugia: Deputazione di storia patria per l'Umbria, 1983–91), 1:313–14.

19 "Ad mensuram brachii in campanili Sancti Laurentii designati." *Statuto del Comune di Perugia del 1279*, ed. Severino Caprioli, 2 vols. (Perugia: Deputazione di storia patria per l'Umbria, 1996), 1:417.

20 *Gli statuti del Comune di Bassano dell'anno 1259 e dell'anno 1295*, ed. Gina Fasoli (Venice: Real deputazione di storia patria per le Venezie, 1940), 75.

21 Giorgio Giulini, *Memorie spettanti alla storia, al governo, ed alla descrizione della città, e della campagna di Milano ne' secoli bassi*, 12 vols. (Milan: Bianchi, 1760–61) 6:81–83. On the *pescheria*, see Ada Grossi, *Santa Tecla nel tardo medioevo* (Milan: ET, 1997), 72–90.

22 "Falsa mensura non est mensura. item falsum pondus non est pondus." Alberto Gandino, *Tractatus de maleficiis* (Venice: Francesco di Francesco, 1578), 388. See also Diego Quaglioni, "Alberto Gandino e le origini della trattatistica penale," *Materiali per una storia della cultura giuridica* 29 (1999), 49–63.

23 Giovan Battista Basile, *Il Pentamerone ossia le fiaba delle fiabe*, trans. Benedetto Croce, 2 vols. (Naples: 1891), 1:77–86. Basile's collection of fables was first published in 1634–36. The episode of the foot check is not in Strabo's story of the slave Rhodopis, usually considered the earliest known variant of the tale. There, the Pharaoh identifies Rhodopis because of the identity of the two slippers, not because of the slave's foot size. Strabo, *Geographia* 17.33. See Graham Anderson, *Fairytale in the Ancient World* (London and New York: Routledge, 2000), 26–33.

24 "Venute le femmene tutte . . . lo re . . . provaje lo chianiello ad una ped una a tutte le commitate, pe vedere a chi jesse a capillo ed assestato, tanto che potesse conoscere da la forma de lo chianiello chello, che jeva cecanno." Basile, *Il Pentamerone*, 1:85–86.

25 Aristotle, *Physics* 213b32–213b34, trans. R. P. Hardie and R. K. Gaye, in *Complete Works of Aristotle*, ed. Jonathan Barnes, 2 vols. (Princeton, NJ: Princeton University Press, 1984), 1:363. For a critique, Alain Badiou, *Being and Event* (London and New York: Continuum, 2005), 56.

26 "Linearis pes est, per quem lineas vel longitudinem aliquam metimur nihil interim de altitudine vel latitudine curantes." Gerbert d'Aurillac, *Opera math-*

NOTES TO PAGES 70-73   241

  ematica, ed. Nikolaj Michajlovič Bubnov (Berlin: R. Friedländer and Sohn, 1899), 56.
27 "Quae ita intellectu capienda est." Gerbert d'Aurillac, *Opera mathematica*, 53.
28 "Et nota quod cubitus . . . duorum pedum in longum et totidem digitorum in latum . . . longitudinem contingit." Bonvesin de la Riva, *De magnalibus Mediolani*, ed. Paolo Chiesa (Milan: Scheiwiller, 1997), 2:5.
29 Guidoni, *Il Duecento*, 368.
30 *Statuta Communis Parmae*, ed. Amadio Ronchini, 4 vols. (Parma: Fiaccadori 1855–60), 2:75. See also Giuseppe Micheli, *Le corporazioni parmensi d'arti e mestieri* (Parma: Battei, 1899), 54–55.
31 *Statuti pistoiesi del secolo XII*, ed. Natale Rauty (Pistoia: Società pistoiese di storia patria, 1996), 311. See also Natale Rauty, "Appunti di metrologia pistoiese," *Bullettino storico pistoiese* 10 (1975), 3–48:13. In 1265 the measurement checks passed to the Opera di Sant'Iacopo. See Natale Rauty, "Intervento del Comune nel controllo delle misure a Pistoia (secoli XII–XV)," in *Pistoia: Città e territorio nel medioevo* (Pistoia: Società pistoiese di storia patria, 2003), 227–46: 239. See also Lucia Gai, *L'Opera di S. Jacopo in Pistoia e il suo primo statuto in volgare (1313)* (Ospedaletto: Pacini, 1994), 199–200.
32 Guidoni, *Il Duecento*, 376. The norm comes from the statutes of 1265.
33 "Ad modum designatum sub voltis comunis." *Statuti del Comune di Bologna dall'anno 1245 all'anno 1267*, ed. Luigi Frati, 3 vols. (Bologna, Tipografia regia, 1869–84), 2:153. The information is confirmed at 1:189. Frati specifies that the standards were carved on the *podestà's* palace and not on a generic governmental building.
34 Frati, *Statuti del Comune di Bologna*, 2:149–54. On the numerous Bolognese statutes, see Anna Laura Trombetti Budriesi, "Gli statuti di Bologna e la normativa statutaria dell'Emilia Romagna tra XII e XVI secolo," in *Mélanges de l'École française de Rome: Moyen Âge* 126, no. 2 (2014), https://doi.org/10.4000/mefrm.2396. One *oncia* is around 3.1 centimeters, or one inch and a quarter.
35 Modena, Archivio storico. Camera segreta, Cartulari 1, *Statuta Comunis Mutine spectantia ad officium judicis super victualibus et extraordinarii deputati*. See Achille Lodovisi, "In Domo Bone Opinionis," in *La bona opinione: Cultura, scienze e misure negli Stati Estensi, 1598–1860*, ed. Debora Dameri, Achille Lodovisi, and Giulia Luppi (Campogalliano: Museo della bilancia, 1997), 21–62: 21. The statutes of Modena were approved in 1327 by Marquess of Ferrara Obizzo III d'Este, but the office of the *iudices super victualibus*, who were in charge of measurement standards, dates to 1278. See Marco Folin, "Il governo degli spazi urbani negli statuti cittadini di area estense," in *Signori, regimi signorili e statuti nel tardo medioevo*, ed. Rolando Dondardini, Gian Maria Venticelli (Bologna, Pàtron, 2003), 337–366, note 5.
36 I reconstruct the various stages in Emanuele Lugli, "Hidden in Plain Sight: The *Pietre di Paragone* and the Preeminence of Medieval Measurements in Communal Italy," *Gesta* 49, no. 2 (2010), 77–95: 86–90.
37 See Lorella Grossi, "Pesi e misure del Ducato Estense nel Museo civico d'arte medioevale e moderna di Modena," in *La bona opinione*, 193–202: 197.

38 | Tommasino de' Bianchi, *Cronaca modenese*, 12 vols. (Parma: Fiaccadori, 1862–84), 4:170.
39 | Lodovisi, "In domo bone opinionis," 23.
40 | "Un solo de' fornaciai, de' tavernieri, de' beccai, de' mugnai, de' pescatori, de' rivenditori nostri, che tenesse misura e bilance giuste, né che il giusto vendesse a' compratori." Angelo Pezzana, *Storia della città di Parma*, 5 vols. (Parma: Forni, 1837–59), 5:210.
41 | Roberto Navarrini, "I pesi e le misure a Mantova in Antico Regime," in *I Gonzaga: Moneta, arte, storia*, ed. Silvana Balbi de Caro (Milan: Electa, 1995), 112–23: 118.
42 | Navarrini, "I pesi e le misure," 119–21. Measuring aroused suspicion also in Venice. See James Shaw, *The Justice of Venice: Authorities and Liberties in the Urban Economy, 1550–1700* (London: British Academy, 2006), 78–79.

CHAPTER II

1 | *Statuti del Comune di Bologna dall'anno 1245 all'anno 1267*, ed. Luigi Frati, 3 vols. (Bologna, 1869–84), 2:149–54. On Reggio, see Enrico Guidoni, *Il Duecento* (Rome and Bari: Laterza, 1989), 376.
2 | Manuela Bernardi, "Misure lineari e per laterizi," in *Raccolte comunali di Assisi: Monete, gettoni, medaglie, sigilli, misure e armi*, ed. Maurizio Matteini Chiari (Perugia and Milan: Electa, 2000), 183–86.
3 | "Haec itaque regalia esse dicuntur: moneta, vie publice, aquatica, flumina publica, molandina, furni, forestica, mensure, bancatica, ripatica, portus, argentarie, pedagia, piscationis redditus, sestaria vini et frumenti et eorum, que venduntur ad mensuram, placita, batalia, rubi, restituciones in integrum minorum et alia omnia, que ad regalia iura pertinent." *Die Urkunden Friedrichs I. 1158–1167*, ed. Heinrich Appelt, in *Monumenta Germaniae Historica: Diplomata regum et imperatorum Germaniae*, vol. 10, no. 2 (Hannover: 1979), 65. Standards are not mentioned in what is taken today as the most eminent of Barbarossa's *diplomata*, promulgated at the Diet at Roncaglia in 1158. Such absence may help to explain why scholars of imperial edicts have paid little attention to measurements.
4 | Renato Bordone, "L'amministrazione del regno d'Italia," in *Federico I Barbarossa e l'Italia*, ed. Isa Lori Sanfilippo (Rome: Istituto storico italiano per il medio evo, 1990), 133–156: 135–39.
5 | "Et damus eis [to the consuls and the commune of Genoa] quod in terris, quibus negotiatum iverint, homines eorum habeant unum vel duos vel plures Ianuenses, qui inter eos iustitiam faciant et rationem, et quod mercatores eorum ubique libere possint habere suum pondus et suam mensuram, quibus inter se res mercesque suas recognoscant." *Friderici I. Constitutiones*, ed. Ludwig Weiland, in *Monumenta Germaniae Historica: Constitutiones et acta publica imperatorum et regum*, vol. 1 (Hannover: 1893), 292–93.

6   Die Urkunden Friedrichs I. 1158–1167, 65. See also Giancarlo Andenna, "Il monastero e l'evoluzione urbanistica di Brescia tra XI e XII secolo," in *Santa Giulia di Brescia: Archeologia, arte, storia di un monastero regio dai Longobardi al Barbarossa*, ed. Clara Stella and Gerardo Brentegani (Brescia: Grafo, 1992), 93–118: 98.
7   Die Urkunden Friedrichs I. 1158–1167, 339.
8   Die Urkunden Friedrichs I. 1158–1167, 373.
9   Giovanni Tabacco, *Sperimentazioni del potere nell'alto medioevo* (Turin: Einaudi, 1993). Tabacco's original interpretation is discussed in Paolo Cammarosano, "Giovanni Tabacco, la signoria e il feudalesimo," in *Giovanni Tabacco e l'esegesi del passato*, ed. Giuseppe Sergi (Turin: Accademia delle scienze, 2006), 37–46.
10 Gianluca Raccagni, *The Lombard League, 1167–1225* (Oxford: British Academy, 2010), 55–56.
11 Ettore Falconi, "La documentazione della pace di Costanza," in *Studi sulla pace di Costanza* (Milan: Giuffrè, 1984), 44–104. Raccagni, *The Lombard League*, 81–103.
12 Raccagni, *The Lombard League*, 77.
13 Pietro Rocca, *Pesi e misure antiche di Genova e del Genovesato* (Genoa: Tipografia del Real istituto sordo-muti, 1871), 7 and 45. On the difference between the concessions of rights to the league members and to other cities, see Giuliano Milani, *I comuni italiani* (Rome and Bari: Laterza, 2009), 47. By way of comparison, I also recall the foot sculpted in the base of a column in Saint Paul's Cathedral, London, first documented in the 1180s. See Philip Grierson, *English Linear Measures: An Essay in Origins* (Reading: University of Reading, 1972), 18–19.
14 Charles M. Radding and Antonio Ciaralli, *The Corpus Iuris Civilis in the Middle Ages: Manuscripts and Transmission from the Sixth Century to the Juristic Revival* (Leiden: Brill, 2007), 67–68. On the preeminence of the Bolognese jurists, see Paolo Alvazzi del Frate et al., *Tempi del diritto: Età medievale, moderna, contemporanea* (Turin: Giappichelli, 2016), 62–63.
15 Justinian, *Digesta*, ed. Theodor Mommsen and Paul Krüger, vol. 1 of *Corpus iuris civilis* (Berlin: Weidmann, 1870), 202 (6.1.6, Paul), 347 (11.6.1, Ulpian) and 348 (11.6.7, Ulpian).
16 "Et has mensuras et pondera in sanctissima uniuscuiusque civitatis ecclesia servari." Justinian, *Novellae*, ed. Rudolf Schöll and Wilhelm Kroll, vol. 3 of *Corpus iuris civilis* (Berlin: Weidmann, 1895), 641 (128.15).
17 *Statuti pistoiesi del secolo XII*, ed. Natale Rauty (Pistoia: Società pistoiese di storia patria, 1996), 311.
18 "Modios aeneos atque lapideos cum sextariis atque ponderibus per mansiones singulasque civitates iussimus collocari, ut unusquisque tributarius sub oculis constitutis rerum omnium modis sciat, quid debeat susceptoribus dare: ita ut, si quis susceptorum conditorum modiorum sexagintorumque vel ponderum normam putaverit excedendam, poenam se sciat competentem esse subiturum." Justinian, *Codex*, ed. Paul Krüger, vol. 2 of *Corpus iuris civilis*

19. (Berlin: Weidmann, 1892), 426 (10.72.9). See also Antonio Pertile, *Storia del diritto italiano dalla caduta dell'impero romano alla codificazione*, 8 vols. (Turin: Unione tipografica-editrice, 1892–1902), 4:557.
  19. Natale Rauty, *Pistoia: Città e territorio nel medioevo* (Pistoia: Società pistoiese di storia patria, 2003), 185–87. The bishop had held the market since 998.
  20. "Colligere et ordinari mensuram et pillam blave." Cited in Rauty, *Pistoia: Città e territorio*, 235.
  21. *Statuti pistoiesi del secolo XII*, 311.
  22. Pilio da Medicina, *Quaestiones aureae* (Rome: Antonio Blado, 1560), 145–47. Pilio's *Quaestiones* are dated c. 1182, when he left Bologna for Modena. See Ugo Nicolini, *Pilii Medicinensis Quaestiones Sabbatinae: Introduzione all'edizione critica* (Modena: Università degli studi, 1933), 19–24. On the importance of Pilio's *Quaestiones*, see Kenneth Pennington and Wolfgang P. Müller, "The Decretists: The Italian School," in *The History of Medieval Canon Law in the Classical Period, 1140–1234: From Gratian to the Decretals of Pope Gregory IX*, ed. Wilfried Hartmann and Kenneth Pennington (Washington: Catholic University of America Press, 2008), 121–173: 168.
  23. Wolfgang Ernst, "The Glossators' Monetary Law," in *The Creation of the Ius Commune: From Casus to Regula*, ed. John W. Cairns and Paul J. du Plessis (Edinburgh: Edinburgh University Press, 2010), 219–75: 230–31.
  24. Caterina Santoro, *La politica finanziaria dei Visconti*, 3 vols. (Milan: Giuffrè, 1976), 1:215–16. I am grateful to Timothy McCall for introducing me to this document.

CHAPTER 12

  1. John Koenig, *Il "popolo" dell'Italia del nord nel XIII secolo* (Bologna: Il Mulino, 1986); Giovanni de Vergottini, "Arti e 'Popolo' nella prima metà del secolo XIII," in *Scritti di storia del diritto italiano*, ed. Guido Rossi, 3 vols. (Milan: Giuffrè, 1977), 1, 387–467; in particular 462.
  2. *Statuti inediti della città di Pisa dal XII al XIV secolo*, ed. Francesco Bonaini, 3 vols. (Florence: Vieusseux 1854–70), 1:268–9, note. See also Gioacchino Volpe, "Studi sulle istituzioni comunali a Pisa, città e contado, consoli e podestà. Secoli XII–XIII," *Annali della R. Scuola normale superiore di Pisa* 5 (1902): 250, no. 3; Cinzio Violante, *Economia, società, istituzioni a Pisa nel medioevo: Saggi e ricerche* (Bari: Dedalo, 1980), 104; Gioacchino Volpe, *Studi sulle istituzioni comunali a Pisa: Città e contado, consoli e podestà* (Pisa: Nistri, 1902), 4–5.
  3. *Statuti inediti Pisa*, 1:39.
  4. Lodovico Muratori, *Antiquitates Italicae Medii Aevi*, 5 vols. (Milan: Typographia societatis palatinae, 1738–41), 2:87–90. See also Ugo Tucci, "Pesi e misure nella storia della società," in *I documenti*, vol. 5.1 of *Storia d'Italia*, ed. Ruggiero Romano and Corrado Vivanti (Turin: Einaudi, 1973), 581–612: 595. The original diploma, dated January 31, 1178, was granted by

Frederick I Barbarossa. See *Die Urkunden Friedrichs I. 1168–1180*, ed. Heinrich Appelt, in *Monumenta Germaniae Historica: Diplomata regum et imperatorum Germaniae*, vol. 10, no. 3 (Hannover: Hahn, 1985), 268–69. See also Michele Luzzati, "Note di metrologia pisana," *Bollettino storico pisano* 31–32 (1962–63), 121–220: 196, note 13; Gioacchino Volpe, *Studi sulle istituzioni comunali*, 5. On the canons's rights, see also Gabriella Garzella, *Pisa com'era: Topografia e insediamento dall'impianto tardoantico alla città murata del secolo XII* (Naples: Liguori, 1990), 201.

5   Francesca Bocchi, "Le imposte dirette a Bologna," *Nuova rivista storica* 57 (1973), 273–312: 284.

6   *Il constituto del Comune di Siena dell'anno 1262*, ed. Lodovico Zdekauer (Milan: Hoepli, 1897), 178–79.

7   "Lo camerlengo et due ufficiali di penitentia . . . facciano fare una sceda di legno di tegole . . . bone et bene fatta, sì che per ciascuna tegola, fatta ad essa sceda o vero simile, sia buona et grossa convenevolmente at abbia buono orlo; et l'altra per li docci. Et tutti li tegolari e' quali fanno regole, sieno tenuti poscia fare tutte le tegole e docci a le dette scede." *Il Costituto del Comune di Siena volgarizzato nel 1309–1310*, ed. Mahmoud Salem Elsheikh, 4 vols. (Siena: Monte dei Paschi, 2002), 2:547–48. See also Duccio Balestracci, "Produzione ed uso del mattone a Siena nel medioevo," in *La brique antique et médiévale: Production et commercialisation d'un metériau*, ed. Patrick Boucheron, Henri Broise, and Yvon Thébert (Rome: École française de Rome, 2000), 417–28: 422.

8   *Statuta Communis Parmae*, ed. Amadio Ronchini, 4 vols. (Parma: Pietro Fiaccadori 1855–60), 2:71–72; *Statuti bonacolsiani*, ed. Ettore Dezza, Anna Maria Lorenzoni, and Mario Vaini (Mantua: Arcari, 2002), 263.

9   "Omnes mercatores regni . . . ista pondera, et cannas debent recipere a regia curia, scilicet, marcatas marco regiae curiae." *Constitutionum Regni Siciliarum libri III*, 2 vols. (Naples: Antonio Cervoni 1773), 1:408, rubric 50. We know that the emperor's instructions did not remain dead letter. In his *Liber Introductorius*, Frederick II's astrologer, Michael Scot, recalls that the standard of the ten-foot *pertica* was kept in the royal curia. See Oxford, Bodleian Library, ms 266, f. 24rA. See also Mario Rosario Zecchino, "Weights and Measures in the Norman-Swabian Kingdom of Sicily," in *People, Texts and Artefacts: Cultural Transmission in the Medieval Norman Worlds*, ed. David Bates, Edoardo d'Angelo, and Elisabeth van Houts (London: Institute of Historical Research, 2017), 253–66.

10   Andrea d'Isernia, "De adulterina moneta," in *Constitutionum Regni Siciliarum libri III*, 1:417.

11   "In vexillo suo sextarium defferebat." Giorgio Giulini, *Memorie spettanti alla storia, al governo, ed alla descrizione della città, e della campagna di Milano ne' secoli bassi*, 12 vols. (Milan: Bianchi, 1760–61), 8:128.

12   Lorenzo Tomasin, *Storia linguistica di Venezia* (Rome: Carocci, 2010), 15–18.

13   Guido di Nardo, "Misure medievali in Roma," *Capitolium* 2 (1926–27), 209–14. The *ruggitella* is also called *rubbiatella*.

14   *Corpus Inscriptionum Latinarum* vol. 6, no. 1 (Berlin: 1876), inscriptions no. 886–87. It may also be worth pointing out that the Capitoline Hill was full of granaries. See Dunia Filippi, "Il Campidoglio tra alto e basso medioevo: Continuità e modifiche dei tracciati romani," *Archeologia medievale* 27 (2000), 21–37: 33n. It is possible that the *congius vini*, the standard for wine and oil inscribed with Pope Boniface VIII's family crest, which today is in the Capitoline Museums, was also there.

15   "Iuxta erarium publicum, quod erat templum Saturni, ex alia parte fuit arcus miris lapidibus tabulatus, in quo fuit historia qualiter milites accipiebant a senatu donativa sua per saccellarium, qui amministrabat hoc; que omnia pensabat in statera antequam darentur militibus, ideo vocatur Salvator de Statera." *Codice topografico della città di Roma*, ed. Roberto Valentini and Giuseppe Zucchetti, 4 vols. (Rome: Istituto storico italiano per il medioevo, 1940–53), 3:17–65, ch. 24. I am grateful to Claudia Bolgia for this reference. See also Gabriella Maetzke, "La chiesa di S. Salvatore de' Stadera al Foro Romano," *Archeologia laziale* 10 (1990), 98–104: 98. See also Famiano Nardini, *Roma antica*, 4 vols. (Rome: De Romanis, 1818–20, 1st ed. 1666), 2:171.

16   See, for instance, Enrico Celani, "Lo statuto del Comune di Montelibretti," *Studi e documenti di storia e diritto* 13 (1892), 401–17: 414.

17   "Basis haec tertiam columnam porticus a dextro ianuae latere substinet." Vatican City, Biblioteca Apostolica Vaticana, Ms. Lat. 5253, f. 390. Aldo Manuzio the Younger (1547–97) was the grandson of the more famous typographer Aldo Manuzio the Elder (d. 1515). He traveled to Rome in 1561 to join his father Paolo and work on the press of Rome's commune, then based on the Capitoline Hill. See Francesco Barberi, *Paolo Manuzio e la stamperia del popolo romano (1561–1570)* (Rome: Gela, 1985).

18   Luca Peto, *De mensuris et ponderibus Romanis et Graecis* (Venice: Manuzio, 1573), 12–13.

19   "Lapides et cuppi, qui modo fiunt . . . ad modum designatum sub voltis comunis." *Statuti del Comune di Bologna dall'anno 1245 all'anno 1267*, ed. Luigi Frati, 3 vols. (Bologna: Tipografia regia, 1869–84), 2:153. Another mention is in volume 1, p. 189: "Staro et sadio lapideo sive marmoreo comunis, qui est solitus esse sub voltis palatii veteris comunis."

20   For the 1331 Statutes of the merchants in Monza, see Luciana Frangioni, *Milano e le sue misure: Appunti di metrologia lombarda fra Tre e Quattrocento* (Naples: Edizioni scientifiche italiane, 1992), 39. For Assisi, *Raccolte comunali di Assisi: Monete, gettoni, medaglie, sigilli, misure e armi*, ed. Maurizio Matteini Chiari (Perugia and Milan: Electa, 2000), 187.

21   "Item statuimus quod potestas et capitaneus et quilibet eorum teneantur et . . . faciunt mactones et tegulas secundum formam eis datam et designatam in pariete anteriori iuxta portam palatii veteris comunis Tuderti et si quidem invenerit ipsam formam non servare eum condempnent pro quolibem centenario mactonorum et tegularum in decem soldos cortoneses. Et quilibet contrafacientes accusare possit et habeat medietatem banni." *Statuta*

civitatis Tudertine (Todi, 1549), book 5, ch. 75. This section transcribes the folios 212v–213r of the 1337 Statutes today kept in Todi's Archivio Storico Comunale. See also Samuel Gruber, "Medieval Todi: Studies in Architecture and Urbanism" (PhD dissertation, Columbia University, 1998), 402. We know from the statutes of 1275 that the commune had a "pedem communis," but it is unclear where it was. See *Statuto di Todi del 1275*, ed. Getulio Ceci and Giulio Pensi (Todi: Trombetti, 1897), 47. On the statutes in general, see Maria Grazia Nico Ottaviani, "Todi e i suoi statuti (secoli XIII–XVI)," in *Todi nel Medioevo (secoli VI–XIV)*, 2 vols. (Spoleto: Centro italiano di studi sull'alto medioevo, 2010), 2:717–41. On the communal palaces, see Getulio Ceci and Umberto Bartolini, *Piazze e palazzi comunali di Todi* (Todi: Tipografia tiberina, 1979), 68–71. On the growing Consiglio Maggiore, whose members went from three hundred in 1292 to five hundred in 1337, see Getulio Ceci, *Todi nel medioevo* (Todi: Trombetti, 1897), 222.

22 In the Middle Ages, the term "palatium" referred to the seat of power. See Carla Uberti, "I palazzi pubblici," in *L'architettura civile in Toscana: Il medioevo*, ed. Amerigo Restucci (Siena: Monte dei Paschi, 1995), 151–223; Gigliola Soldi Rondinini, "Evoluzione politico-sociale e forme urbanistiche nella Padania dei secoli XII–XIII: I palazzi pubblici," in *La pace di Costanza 1183: Un difficile equilibrio di poteri fra società italiana ed impero* (Bologna: Cappelli, 1984), 85–98.

23 Giovanni Filippi, *L'Arte dei mercanti di Calimala in Firenze e il suo più antico statuto* (Turin: Bocca, 1889), 125. See also *Statuto del podestà del 1325*, ed. Romolo Caggese, vol. 2 of *Statuti della Repubblica Fiorentina* (Florence: Ariani, 1910–21), book 3, norm 27: "De pena qui vendiderit pannum nisi ad cannam de Kallismala." In a document of November 23, 1321, the *canna* is also designated as "ad mensuram Callismalae." In Domenico Maria Manni, *Osservazioni istoriche sopra i sigilli antichi de' secoli bassi*, 30 vols. (Florence: Albizzini, 1739–86), 21:49–50.

24 *Statuti del Comune di Padova dal secolo XII all'anno 1285*, ed. Andrea Gloria (Padua: Sacchetto, 1873), 286. The Statutes of 1277 speak of the bread shape as "designatam in lapidibus anguli palacii in capite scale ab avibus." For the dating, see Elisabetta Antoniazzi Rossi, *Palazzo della Ragione a Padova* (Milan: Skira, 2007), 25; repeated in Dennis Romano, *Markets and Marketplaces in Medieval Italy, c. 1100 to c. 1400* (New Haven and London: Yale University Press, 2015), 81–82. The *Palazzo della Ragione* was started in 1218, under the podestà of Giovanni Rusconi. See Sante Bortolami, "'Spaciosum, immo speciosum palacium': Alle origini del Palazzo della Ragione di Padova," in *Il Palazzo della Ragione di Padova: La storia, l'architettura, il restauro*, ed. Ettore Vio (Padua: Signum, 2008), 39–74: 39–43.

25 *Statuti del Comune di Padova*, 269 (norm 818), 271 (norm 821), and 281 (norm 842).

26 *Gli antichi statuti delle arti veronesi secondo la revisione scaligera del 1319*, ed. Luigi Simeoni (Venezia: Tipografia emiliana, 1914), xxv–xxvi and 248 (norm 48).

Verona also followed Venice's prescriptions as to the weighing of dyes, gold, and silver.

27   *Statuti del Comune di Padova*, 272 (norm 825.1) See also p. 269 (norms 817–18), as they identify the new Padua *passus* as the general linear standard. See Roberto Cessi, *Le corporazioni dei mercanti di panni e della lana in Padova fino a tutto il secolo XIV* (Venice: Real istituto di scienze, lettere e arti di Venezia, 1908), 39.

28   "Ad modum pertice de Verona consuete ab hinc retro." Maureen Fennell Mazzaoui, "The Emigration of Veronese Textile Artisans to Bologna in the Thirteenth Century," *Atti e memorie dell'Accademia di agricoltura, scienze e lettere di Verona* 19 (1967–68), 275–321; Andrea Castagnetti, *Mercanti, società e politica nella marca veronese-trevigiana* (Verona: Libreria universitaria editrice, 1990), 69–70.

29   The measurements are not my own, but those recorded in the mid nineteenth century, when the Bolognese *pietre* were less weathered. See Luigi Malavasi, *La metrologia italiana ne' suoi scambievoli rapporti desunti dal confronto col sistema metrico-decimale* (Modena: Malavasi, 1842–44), 190.

CHAPTER 13

1   Justinian, *Digesta*, ed. Theodor Mommsen and Paul Krüger, vol. 1 of *Corpus iuris civilis* (Berlin: Weidmann, 1870), 2 (1.1.10, Ulpian). See also Thomas Aquinas, *Summa theologiae* 2–2, q. 58, a. 1: "Iustitia est habitus secundum quem aliquis constanti et perpetua voluntate ius suum unicuique tribuit."

2   The concepts of *justitia distributiva* and *justitia commutativa* are Aristotelian. See Aristotle, *Nicomachean Ethics* 5.5. This text started circulating from around the year 1250 in a translation by Robert Grosseteste. See James McEvoy, "Robert Grosseteste's Greek Scholarship: A Survey of Present Knowledge," *Franciscan Studies* 56 (1998), 255–64: 261–62. It was further popularized by Thomas Aquinas, see his *Summa Theologiae*, 2.2.61. Nicolai Rubinstein, "Political Ideas in Sienese Art: The Frescoes by Ambrogio Lorenzetti and Taddeo di Bartolo in the Palazzo Pubblico," *Journal of the Warburg and Courtauld Institutes* 21 (1958), 179–207: 185. The idea that Lorenzetti's frescoes are based on Thomistic Aristotelianism is widespread. For an alternative view, see Quentin Skinner, "Ambrogio Lorenzetti: The Artist as Political Philosopher," *Proceedings of the British Academy* 72 (1986), 1–56: 2–3; Judith Resnik and Dennis Edward Curtis, *Representing Justice: Invention, Controversy, and Rights in City-States and Democratic Courtrooms* (New Haven and London: Yale University Press, 2011), 26–30.

3   Many art historians have misinterpreted the tools. See, for instance, Rubinstein, "Political Ideas," 182; Skinner, "Ambrogio Lorenzetti," 37–38; Chiara Frugoni, *Una lontana città: Sentimenti e immagini nel medioevo* (Turin: Einaudi, 1983), 183; Randolph Starn, *Ambrogio Lorenzetti: Palazzo Pubblico a Siena* (Turin: Società editrice internazionale, 1996), 54. To my knowledge,

Maria Monica Donato is the only scholar who suggested my interpretation, but she did so hesitantly, and her interpretation has not caught on. See Maria Monica Donato, "La 'bellissima inventiva': Immagini e idee nella sala della pace," in *Amborgio Lorenzetti: Il buon governo*, ed. Enrico Castelnuovo (Milan: Electa, 1995), 23–41: 35. On Siena's *staio, passetto*, and *canna*, the city's key standards, see *Il Costituto del Comune di Siena volgarizzato nel 1309–1310*, ed. Mahmoud Salem Elsheikh, 4 vols. (Siena: Monte dei Paschi, 2002), 1:110 (*staio*) and 205–06 (*canne* and *passetto*).

4 "Giustizia commutativa, che sta in non ingannare, e satisfare li debiti; ed in giustizia distributiva, che sta in distribuire il bene, ed il male, ed onore, e vergogna a ciascheduno, secondo che è degno." Domenico Cavalca, *Specchio di croce*, ed. Giovanni Gaetano Bottari (Rome: De' Rossi, 1738), 138. For the dating of Lorenzetti's fresco, Joseph Polzer, "Ambrogio Lorenzetti's 'War and Peace' Murals Revisited: Contributions to the Meaning of the 'Good Government Allegory,'" *Artibus et Historiae* 23, no. 45 (2002) 63–105: 63–64.

5 Frugoni, *Una lontana città*, 138; Skinner, "Ambrogio Lorenzetti," 38–39.

6 "Et li detti stai adrittati, tenere si debiano ne la città di Siena, di ferro et non di rame." Salem Elsheikh, *Il Costituto del Comune di Siena*, 1:110 (norm 118).

7 "El quale passetto da l'uno capo et da l'altro, sia ferrato." Elsheikh, *Il Costituto del Comune di Siena*, 1:206 (norm 242).

8 "Li Pisani fecero la pacie co' dicti comuni, [Florence and Lucca] e presero le mizure da' dicti comuni." Giovanni Sercambi, *Croniche*, ed. Salvatore Bongi, 3 vols. (Lucca: Giusti, 1892), 1:41.

9 Elsheikh, *Il Costituto del Comune di Siena*, 1:111 (norm 241).

10 The idea of rendering the angels specularly may derive from Giotto, who did the same in the Arena Chapel in Padua. Giotto also made use of the allegorical power of measurements in Padua's Communal Palace. See Eva Frojmovic, "Giotto's Allegories of Justice and the Commune in the Palazzo della Regione di Padua: A Reconstruction," *Journal of the Warburg and Courtauld Institutes* 59 (1996), 24–45: 35–38.

11 Elsheikh, *Il Costituto del Comune di Siena*, 1:205–6 (norms 240–1). On the history of the Mercanzia, see William M. Bowsky, *A Medieval Italian Commune: Siena under the Nine, 1287–1355* (Berkeley: University of California Press, 1981), 184–259; Mario Ascheri, *Siena nel Rinascimento: Istituzioni e sistema politico* (Siena: Il Leccio, 1985), 111–33.

12 Luigi Gambirasio, *Le corporazioni milanesi d'arti e mestieri nel medio evo* (Siena: San Bernardino, 1897), 29–30. Luciana Frangioni, *Milano e le sue misure: Appunti di metrologia lombarda fra Tre e Quattrocento* (Naples: Edizioni scientifiche italiane, 1992), 32.

13 In Venice, the *iusticerii* were instituted by doge Sebastiano Ziani in 1173, together with the *Statutum de edulis vendensis*. See Andrea da Mosto, *L' Archivio di stato di Venezia: Indice generale, storico, descrittivo ed analitico*, 2 vols. (Venice: Biblioteca dell'archivio di stato, 1937–40), 1:191. In Padua we also find *iusticierii*. See Sante Bortolami, "La città del santo e del tiranno: Padova nel primo

14 Duecento," in *Sant'Antonio, 1231–1981: Il suo tempo, il suo culto e la sua città*, ed. Giovanni Gorini (Padua: Signum, 1981), 244–61: 252–53. On Volterra, see *Statuti di Volterra I (1210/1224)*, ed. Enrico Fiumi (Siena: Tipografia nuova, 1951), 160. On Bologna, see Francesca Bocchi, *Il Duecento*, vol. 2 of *Atlante storico di Bologna* (Bologna: Grafis, 1995), 16 and 50.

14 Rossella Rinaldi, "La normativa bolognese del '200: Tra la città e il suo contado," in *Acque di frontiera: Principi, comunità e governo del territorio nelle terre basse tra Enza e Reno (secoli XIII–XVIII)*, ed. Franco Cazzola (Bologna: Clueb, 2000), 139–64: 147–48.

15 "Statuimus quod potestas Regii teneatur et debeat mercatum in die sabbati facere custodiri, ne pannus falsus pilis bovis asine canipe vel alterius generis falsitatis ibi portetur vel vendatur; et si alicui inventus fuerit pannus falsus in c soldi rexan. puniatur et pannum perdat." *Consuetudini e statuti reggiani*, ed. Aldo Cerlini, Corpus statutorum italicorum 16 (Milan: Hoepli, 1933), 9–10.

16 Giorgio Giulini, *Memorie spettanti alla storia, al governo, ed alla descrizione della città, e della campagna di Milano ne' secoli bassi*, 12 vols. (Milan: Bianchi, 1760–1761), 8:128.

17 "Qui autem falsitatem, aut fraudem aliam in mensuris, aut ponderibus, aut cannis commiserit, vel pannos tiraveris, per sententiam punietur poena unius librae auri, quam si solvere non poterit, per civitatem, vel terram publice fustigetur, et ad eius collum appendatur in sui poenam, et aliorum exemplum illud pondus, vel mensura, cum quo, vel qua, fraudem commisit, et si secundo in simili delicto fuerit apprehensus, amputetur ei manus, et si tertio iteraverit, suspendatur, ita ut eius anima separetur a corpore." *Constitutionum Regni Siciliarum libri III*, 2 vols. (Naples: Antonio Cervoni, 1773), 1:409, rubric 51.

18 Elsheikh, *Il Costituto del Comune di Siena*, 1:205–6. For the yearly fee, see Maurizio Tuliani, "Il Campo di Siena: Un mercato cittadino in epoca comunale," *Quaderni medievali* 46 (1998), 59–100: 100.

19 Luigi Simeoni, *L'antico mercato veronese e i suoi supposti capitelli* (Verona: Marchioni, 1899), 36–37. The *capitello* is inscribed with the year 1380. See also Francesca Bocchi, *Per antiche strade: Caratteri e aspetti delle città medievali* (Rome: Viella, 2013), 218.

20 "Datium minarum comunis et omnium mensuram in civitate et districtu." *Liber grossus antiquus Comunis Regii*, ed. Francesco Saverio Gatta, 6 vols. (Reggio Emilia: Costi/Goretti, 1944–62), 1:289–90, norm 158.

21 The act is dated June 1164. The same reoccured in 1187. See *I capitolari delle Arti veneziane sottoposte alla Giustizia e poi alla Giustizia Vecchia dalle origini al 1330*, ed. Giovanni Monticolo, 3 vols. (Rome: Tipografi del Senato, 1905), 2:liv. Venice's *Giustizia* was founded in 1173 to administer the food trade. Its duties soon expanded to include checks on prices and measurement standards. Modern historians refer to it as "Giustizia Vecchia" to separate it from the "Giustizia Nuova," which was created in 1265 to manage the wine business and the taverns. See James Shaw, *The Justice of Venice: Authorities and Liberties in the Urban Economy, 1550–1700* (London: British Academy, 2006), 22–23.

22 Gerhard Rösch, "Lo sviluppo mercantile," in *L'età del comune*, ed. Giorgio Cracco and Gherardo Ortalli, vol. 2 of *Storia di Venezia* (Rome: Treccani, 1995), 131–54: 146–48.

23 Natale Rauty, *Pistoia: Città e territorio nel medioevo* (Pistoia: Società pistoiese di storia patria, 2003), 227–46: 243. Pressing and beating may be a reference to Luke 6:38.

24 "Et quod quilibet emens granum vel bladum et alia legumina, salem vel salinam que mensurantur debeat implere et culmare mensuram absque aliqua percussione cum manibus facienda." *Statuto del podestà del 1325*, ed. Romolo Caggese, vol. 2 of *Statuti della Repubblica Fiorentina* (Florence: Ariani, 1910–21), 301, book 4, norm 32.

25 Charles-Marie de la Roncière, *Prix et salaires à Florence au XIVe siècle, 1280–1380* (Rome: Ecole française de Rome, 1982), 26–27.

26 "Et ad hoc ut habeatur istorum notitia potestas faciat hoc banniri publice in locis consuetis, et semel in mense." Caggese, *Statuto del podestà*, 301.

27 "In mensuratione pannorum debeat servari talis modus, videlicet quod ab uno capite canne super ponatur et revolvatur pannus usque ad quartum brachii signatum in ipso capite canne ut consuetum est, deinde mensuretur pannus sine aliqua tiratione usque ad signum octavi brachii quod est in alio capite canne et postea compleatur ipsa mensura usque ad summum et complementum canne." Giovanni Filippi, *L'Arte dei mercanti di Calimala in Firenze e il suo più antico statuto* (Turin: Bocca, 1889), 2:179, book 3, norm 27.

28 "Debeat ponere pannum in tabula vel storam vel in alio loco plano, et super illud ponere passum, non tirando pannum ex quo plane positum fuerit, et in capite passi policem, et post policem passum iterum, quousque mensuratum fuerit quod placuerit contrahentibus." Cerlini, *Consuetudini e statuti reggiani*, 254.

29 *Gli statuti del Comune di Bassano dell'anno 1259 e dell'anno 1295*, ed. Gina Fasoli (Venice: Real deputazione di storia patria per le Venezie, 1940), 95 (salt sellers) and 61 (wine makers).

30 This norm returns in the statutes of both 1279 and 1342. See *Statuto del Comune di Perugia del 1279*, ed. Severino Caprioli and Attilio Bartoli Langeli, 2 vols. (Perugia: Deputazione di storia patria per l'Umbria, 1996), 1:417, rubric 470; and *Statuti di Perugia dell'anno MCCCXLII*, ed. Giustiniano degli Azzi, 2 vols. (Rome: Loescher, 1913–16), 2:441.

31 Laura Ticciati, *L'ordine dei mercanti a Pisa nei secoli XII–XIII* (Pisa: ETS, 1992), 243–56.

32 Enrica Salvatori, "Ceti sociali e struttura urbana: La popolazione pisana delle cappelle di S. Michele in Borgo, S. Jacopo al Mercato, S. Cecilia e S. Lorenzo alla Rivolta nei secoli XI–XV," in *Pisa e la Toscana occidentale nel medioevo*, 2 vols. (Pisa: ETS, 1991), 1: 231–300; in particular 235–36.

33 This reflection is inspired by Giorgio Agamben, *Che cos'è un dispositivo?* (Rome: Nottetempo, 2006), 18–20.

34 *Statuti inediti della città di Pisa dal XII al XIV secolo*, ed. Francesco Bonaini, 3 vols. (Florence: Vieusseux 1854–70), 3:116–22.

35 | Daniel Philip Waley, *The Italian City-Republics*, (London: Weidenfeld and Nicolson, 1969), 59–64. See also John Hine Mundy, "In Praise of Italy: The Italian Republics," *Speculum* 64 (1989), 815–34: 826.

CHAPTER 14

1 | Pierluigi Castagneto, *L'Arte della lana a Pisa nel Duecento e nei primi decenni del Trecento: Commercio, industria e istituzioni* (Pisa: ETS, 1996), 124–25. The institutions of the *sensales* was not limited to Pisa. See, for instance, *Statuti della colonia genovese di Pera*, ed. Vincenzo Promis (Turin: Stamperia reale, 1871), 599–600.

2 | In the twelfth century, *estimo* was a form of value quantification. Starting in 1162, Pisa required its citizens to declare what they had. Such declarations were collected by *estimatores*, who also quantified the possessions of absentees and minors. Shortly afterwards, such officers are recorded in other Italian cities (Siena in 1168, Lucca 1182, and Florence 1202). See Cinzio Violante, *Economia, società, istituzioni a Pisa nel medioevo* (Bari: Dedalo, 1980), 107–15; Patrizia Mainoni, "Sperimentazioni fiscali e amministrative nell'Italia settentrionale," in *Pensiero e sperimentazioni istituzionali nella "Societas Christiana" (1046–1250)*, ed. Giancarlo Andenna (Milan: Vita e pensiero, 2007), 705–59: 718–19.

3 | "Et si minus inventum fuerit in una uncia, debeant consules vel potestas tollere inde baendum .V. soldorum pro uncia ei apud quem pondus minus inventum fuerit." *Statuti di Volterra I (1210/1224)*, ed. Enrico Fiumi (Siena: Tipografia nuova, 1951), 160.

4 | I have come across one other indication of medieval degrees of precision. It comes from Como, where the fine for deceitful bread makers increased of 12 *denari* for each *uncia* in excess ("et si invenerit aliquem pristinarium panem... minus pensa, puniatur in denariis duodecim pro qualibet uncia"). *Liber statutorum Comunis Novocomi*, ed. Antonio Ceruti, vol. 16, no. 2 of *Historiae patria monumenta* (Turin: Bocca, 1876), columns 181–82, rubric 221.

5 | *Statuti urbanistici medievali di Lucca*, ed. Domenico Corsi (Venice: Neri Pozza, 196), 69.

6 | *Statuti di Lucca*, 45.

7 | Aristotle, *Categories* 6. See also John E. Bell, *The Continuous and the Infinitesimal in Mathematics and Philosophy* (Monza: Polimetrica, 2006), 21–62.

8 | See, for instance, Pilio da Medicina, Tancredi da Bologna, and Graziano da Arezzo, *Libri de ordine iudiciorum*, ed. Frederick Bergmann (Göttingen: Vandenhoeck and Ruprecht, 1842), 8 and 169.

9 | Giles of Rome, *De mensura angelorum* (Venice: 1503), 35v–38r, quaestio 1. The treatise was originally composed in 1289. See Pasquale Porro, *Forme e modelli di durata del pensiero medievale* (Leuven: Leuven University Press, 1996), 122. For the dating, see Silvia Donati, "Studi per una cronologia delle opere

NOTES TO PAGES 96–100

10 di Egidio Romano I. Le opere prima del 1285: I commenti aristotelici," *Documenti e studi sulla tradizione filosofica medievale* 1 (1990), 57–66.
Aristotle, *Metaphysics*, trans. William David Ross (London: Methuen & Co., 1923) 10.1.1052b20.

11 Uguccione da Pisa, *Derivationes*, ed. Enzo Cecchini and Guido Arbizzoni, 2 vols. (Florence: Edizioni del Galluzzo, 2004), 2:766. Uguccione's list comes from Isidore, *Etymologiarum libri* 15.15.1–2.

12 Jan M. Ziolkowski, "Cultures of Authority in the Long Twelfth Century," *Journal of English and German Philology* 108 (2009), 421–48.

13 Franco Sacchetti, *Il trecentonovelle*, ed. Emilio Faccioli (Turin: Einaudi, 1970), XCII:249–51.

14 On such capes, Alessandro Tassoni, *Annotazioni sopra il Vocabolario degli accademici della crusca* (Venice: Marino Rossetti, 1698), 177.

15 On regional stereotyping in the Middle Ages, see Giorgio Cadorini, "Friulano, veneto e toscano nella storia del Friuli," in *Manuale di linguistica friulana*, ed. Sabine Heinemann and Luca Melchior (Berlin: De Gruyter, 2015), 316–37: 328.

16 "Il friolano metteva, e tirava il panno più su che la canna, quando uno sommesso, e quando più, e stavasi tanto attento che ad altro non guatava. Il fiorentino, che nel principio subito se ne fu avveduto, quando mettea il panno su la canna lasciava mezzo braccio della canna a drieto, e quando più sì che ogni quattro braccia tornavano al bun uomo forse tre e mezzo." Sacchetti, *Il trecentonovelle*, 249.

17 The word *saltamindosso* is northern, in keeping with Sacchetti's attention to dialects. See Vittorio Sant'Albino, *Gran dizionario piemontese-italiano* (Turin: Unione tipografico-editrice, 1859), 1155.

18 Enrico Rebuffat, "Per il significato di *cagnazzo* nella 'Commedia,'" *Studi di filologia italiana* 71 (2013), 123–64: 144–46.

19 Elisa Tosi Brandi, "Il sarto tra medioevo e prima età moderna a Bologna e in altre città dell'Emilia Romagna" (PhD dissertation, Università di Bologna, 2012), 39.

20 *Gli antichi statuti delle arti veronesi secondo la revisione scaligera del 1319*, ed. Luigi Simeoni (Venice: Tipografia emiliana, 1914), 13.

21 Giovanni Balbo, *Catholicon* (Strassburg: Adolf Rusch, 1475–77), entry "Mensura."

22 "Uno conperò il 100 della lana 18 lire e tolsen 490 libre e llavolla, e lavata tornò 184 libbre. Vo xapere che gli venne il 100 lavata." Paolo dell'Abbaco, *Trattato d'aritmetica*, ed. Gino Arrighi (Pisa: Domus galileana, 1964), 38, exercise 37.

23 Giordano da Pisa, *Quaresimale fiorentino: 1305–1306*, ed. Carlo Delcorno (Florence: Sansoni, 1974), 252.

24 "Non habebis in sacculo diversa pondera, maius et minus; nec erit in domo tua modius maior et minor. Pondus habebis iustum et verum, et modius equalis et verus erit tibi." Deuteronomy 25:13–15. This verse comes close to Proverbs 20:10.

25 | "Si tamen corruptio esset talis quod unam mensuram tenerent ad emendum et aliam ad vendendum; tunc alter cum quo mercatur hoc sciat, est semper mortale peccatum." Petrus Iohannis Olivi, *Quodlibeta quinque*, ed. Stefano Defraia (Grottaferrata: Editiones Collegii S. Bonaventurae ad Claras Aquas, 2002), 326, rubric 5.12. See also Pietro di Giovanni Olivi, *Usure, compere e vendite: La scienza economica del XIII secolo*, ed. Amleto Spicciani, Paolo Vian, and Giancarlo Andenna (Novara: Europía, 1990).

26 | "Tirano i panni e tragono loro le budella di corpo." Giordano da Pisa, sermon of January 1, 1304, in Oxford, Bodleian Library, Ms. Canon. Ital. 132, f. 75vb. See also Fra Giordano da Rivalto [da Pisa], *Prediche* (Florence: Viviani, 1739), 33, sermon 8. On the questionable reputation of merchants in the Middle Ages, see Régine Pernoud, *Storia della borghesia della Francia* (Milan: Jaca book, 1986), 102–22.

27 | *Constitutionum Regni Siciliarum libri III*, 2 vols. (Naples: Antonio Cervoni 1773), 1:408, norm 3.51 ("Conjungimus ut vendentes pannos in posterum ipsos ultra non distrahant nisi quantum canna protenditur, sed sine abstractionis alicuius violentia") and 409, norm 3.52 ("quicumque falsitatem, aut fraudem aliam in mensuris, seu ponderibus, vel cannis inventus fuerit commisisse, aut qui pannos extraverit . . . condamnatus").

28 | Rebuffat, "Per il significato di *cagnazzo*," 149.

CHAPTER 15

1 | See, for instance, a description of criminals as those who "non palesemente ne le loro boteghe, ma in cose oculte e riposte tengono le misure." *Statuti delle Arti degli oliandoli e pizzicagnoli e dei beccai di Firenze (1318–1346)*, ed. Francesca Morandini (Florence: Olschki, 1961), 55.

2 | *I capitolari delle arti veneziane sottoposte alla Giustizia e poi alla Giustizia vecchia dalle origini al 1330*, ed. Giovanni Monticolo, 3 vols. (Rome: Tipografi del senato, 1905), 2:557 and 568. On the shrinking of fabrics, see also vol. 2:467–68.

3 | "Qui autem incidere volent (sic) vestes de dicto novo panno teneatur et debeat ipsam talem petiam panni novi mensurare in presentia duorum officialium ad hoc ordinandorum per rectores et consiliarios et duodecim bonos viros de hac arte, deinde balneare et demum iterum mensurare." *Constitutum artis monetariorum Civitatis Florentie*, ed. Piero Ginori Conti (Florence: Olschki, 1938), ch. 57. The statutes of wool traders also has a similar norm: "Et nullus mensursator possit et debeat mensurare pannos, sine presentia alicuius sensalium pannorum." *Statuto dell'Arte della lana di Firenze (1317–19)*, ed. Anna Maria Agnoletti (Florence: Le Monnier, 1940), 66.

4 | Steven Runciman, "Byzantine Trade and Industry," in *Trade and Industry in the Middle Ages*, ed. Edward Miller, Cynthia Postan, and Michael Moissey Postan, vol. 2 of *The Cambridge Economic History* (Cambridge: Cambridge University Press, 1987), 132–67: 158–59.

5   "Quilibet possit accusare contrafacientes, et habeat medietatem banni." *Statuta Communis Parma*, ed. Amadio Ronchini, 4 vols. (Parma: Fiaccadori 1855–60), 2:71–72. The expression returns also at 2:73 and 75.

6   *Statuti di Bologna dell'anno 1288*, ed. Gina Fasoli and Pietro Sella, 2 vols. (Vatican City: Biblioteca apostolica vaticana, 1937–39), 2:121, norm 9.17; *Statuti di Perugia dell'anno MCCCXLII*, ed. Giustiniano degli Azzi, 2 vols. (Rome: Loescher, 1913–16), 2:437; *Statuto del Podestà dell'anno 1325*, ed. Giuliano Pinto, Francesco Salvestrini, Andrea Zorzi, vol. 2 of *Statuti della Repubblica Fiorentina*, ed. Romolo Caggese (Florence: Olschki, 1999), 301.

7   *Statuti del Comune di Padova dal secolo XII all'anno 1285*, ed. Andrea Gloria (Padua: Sacchetto, 1873), 269.

8   *Gli statuti del Comune di Bassano dell'anno 1259 e dell'anno 1295*, ed. Gina Fasoli (Venice: Real deputazione di storia patria per le Venezie, 1940), 75 and 318.

9   As a term of comparison, notice that in Justinian's Digest 11.6, measuring is framed as an activity in which only measurers take part.

10   The "quilibet possit accusare" formula was already employed in Bologna in the 1250s, but it seems to have been extended to measuring only in 1288. See *Statuti del Comune di Bologna dall'anno 1245 all'anno 1267*, ed. Luigi Frati, 3 vols. (Bologna: Tipografica regia, 1869–84), 2:211.

11   On the rise of the *popolo*, see Giovanni de Vergottini, *Arti e popolo nella prima metà del secolo XIII* (Milano: Giuffrè, 1943), 461. For the phenomenon in Parma and Bologna, see Roberto Greci, "Origini, sviluppi e crisi del Comune," in *Parma medievale*, vol. 3 of *Storia di Parma* (Parma: Monte università, 2010), 115–67: 153–67; Roberto Greci, "Bologna nel Duecento," in *Bologna nel medioevo*, ed. Ovidio Capitani, vol. 2 of *Storia di Bologna*, ed. Renato Zangheri (Bologna: Bononia University Press, 2007), 499–579: 537–51.

12   Massimo Vallerani, *La giustizia pubblica medievale* (Bologna: Il Mulino, 2005), 31.

13   Jean-Claude Maire Vigueur, "Justice et politique dans l'Italie communale de la seconde moitié du XIIIe siècle: L'exemple de Pérouse," *Comptes rendus de l'Académie des Inscriptions et Belles-Lettres* 130–32 (April-June 1986), 312–30.

14   Jean-Claude Maire Vigueur, "Giudici e testimoni a confronto," in *La parola all'accusato*, ed. Jean-Claude Maire Vigueur and Agostino Paravicini Bagliani (Palermo: Sellerio, 1991), 105–23: 107.

15   *Statuti di Padova*, ed. Gloria, 281, norms 841–43.

16   One *gonnella* costs 2 *soldi*, while "a mantle lined with squirrel fur five" (*pro mantello vario . . . solidi quinque*). *Statuti di Padova*, ed. Gloria, 282, norm 845.

17   "In qua [ecclesia] qualibet die missam audiebat et totum diurnum offitium atque nocturnum temporibus oportunis. Et, tempore quo in ecclesiastico offitio occupatus non erat, sedebat cum vicinis suis sub porticu communi iuxta palatium episcopi et loquebatur de Deo vel loquentem de Deo audiebat libenter. Non patiebatur quod aliquis puer prohiberet lapides contra baptisterium vel contra maiorem ecclesiam, ad destruendum celaturas et picturas. Quod cum videret . . . ac si pro custodia deputatus fuisset ibidem." Salimbene

18   de Adam, *Cronica*, ed. Giuseppe Scalia, 3 vols. (Rome and Bari: Laterza, 1966), 3:888.

18   These observations are inspired by Patrick Joyce, *The Rule of Freedom: Liberalism and the Modern City* (London and New York: Verso, 2003) and Michel Foucault, *Discipline and Punish: The Birth of the Prison* (London: Allen Lane, 1977), 200–204.

19   "Sed forte non peccat mortaliter quando est corruptio communis, quia satis ex hoc data est ratio quod huiusmodi corruptio omnibus innotescat. Si vero omnes, tam domestici quam forenses, communiter sciunt quantitatem ponderum et mensuram qua nunc post factam corruptionem, utuntur, tunc non cadit fraus vel falsitas in mensuris." Petrus Iohannis Olivi, *Quodlibeta quinque*, ed. Stefano Defraia (Grottaferrata: Editiones Collegii S. Bonaventurae ad Claras Aquas, 2002), 326, rubric 5.12.

20   Stefano Defraia, "Prefazione," in Olivi, *Quodlibeta quinque*, vii–xvii: xii. See Palémon Glorieux, *La littérature quodlibétique*, 2 vols. (Paris: Vrin, 1935), 2:205. See also Efrem Bettoni, *Le dottrine filosofiche di Pier Giovanni Olivi* (Milan: Vita e pensiero, 1959), 430.

21   Tiziana Suarez-Nani, "Notes pour l'histoire de la réception de Pierre de Jean Olivi," in *Pierre de Jean Olivi: Philosophe et théologien*, ed. Catherine König-Pralong, Olivier Ribordy, and Tiziana Suarez-Nani (Berlin: De Gruyter, 2010), 327–53: 350–52. Olivi's ideas spread also through his pupils in Florence, such as Petrus de Trabibus, whom Dante probably knew. See Sylvain Piron, "Le poète et le théologien: Une rencontre dans le studium de Santa Croce," in *'Ut philosophia poesis': Questions philosophiques dans l'oeuvre de Dante, Pétrarque et Boccace*, ed. Joël Biard and Fosca Mariani Zini (Paris: Vrin, 2008), 73–109.

22   Siena, Biblioteca Comunale degli Intronati, Ms U.V.7, ff. 244r–303v. The transcription of Olivi's *quaestio* on weights and measures is at f. 261v, under the title "Queritur an scienter utens." See also Dionisio Pacetti, *I codici autografi di S. Bernardino da Siena* (Florence: Quaracchi, 1937), 528–29.

23   For instance, Bernardino referred to Olivi's passage in Bernardino da Siena, *Quadragesimale de evangelio aeterno*, vol. 4 of *Opera Omnia*, ed. Pacifico M. Perantoni and Augustin Sépinski (Florence: Quaracchi, 1950–65), a. 2, c. 7, sermon 33.

24   "Colui . . . che tira il panno in su la canna . . . tiral bene. E l'altro che ha due canne, l'una da vendere, l'altra da comprare." Bernardino da Siena, *Prediche volgari sul campo di Siena 1427*, ed. Carlo Delcorno, 2 vols. (Milan: Rusconi, 1989), 2:1117. On Bernardino and merchants, see also Evelyn Welch, *Shopping in the Renaissance: Consumer Cultures in Italy, 1400–1600* (New Haven and London: Yale University Press, 2005), 70–73.

25   The Campo di Siena had been the site of the city's market since the 1170s. See Maurizio Tuliani, "Il Campo di Siena: Un mercato cittadino in epoca comunale," *Quaderni medievali* 46 (1998), 59–100. See also Fabio Gabbrielli, "Il palazzo del comune di Siena e il suo campo," in *Siena nello specchio del*

suo costituto in volgare del 1309–1310, ed. Nora Giordano and Gabriella Piccinni (Pisa: Pacini, 2014), 51–66: 57.

26   To my knowledge, the only other case was Florence, which for a period had two or three equivalent *pietre*: "et ne aliqua fraus vel deceptio fiat aut diminuito vel augmentum, stent et ponantur dicte mensure in duobus vel tribus loci ad minus in civitate Florentie." See *Statuto del podestà*, book 4, rubric 32.

27   "Accusatori segreti`. . . al cui detto et denuntiagione, si debia stare sença altre pruove; et coloro e' quali saranno accusatori sopra le predette cose, non debiano ad alcuno manifestare." *Il Costituto del Comune di Siena volgarizzato nel 1309–1310*, ed. Mahmoud Salem Elsheikh, 4 vols. (Siena: Monte dei Paschi, 2002), 2:320, distinction 5, rubric 170. See also Tuliani, "Il Campo di Siena: Un mercato cittadino," 68–69.

28   Nirit Ben-Aryeh Debby, "War and Peace: The Description of Ambrogio Lorenzetti's Frescoes in Saint Bernardino's 1425 Siena Sermons," *Renaissance Studies* 15, no. 3 (2001): 272–86; Stefan Visnjevac, "Cooperative Intervention: Sermons Supporting the Governing Authority in Fifteenth-Century Italy," in *Churchmen and Urban Government in Late Medieval Italy, c. 1200–c. 1450*, ed. Frances Andrews and Maria Agata Pincelli (Cambridge: Cambridge University Press, 2013), 288–303: 285–87.

CHAPTER 16

1   *Statuta Communis Parmae*, ed. Amadio Ronchini, 4 vols. (Parma: Pietro Fiaccadori 1855–60), 1:157.

2   Caterina Bruschi, "Aspetti di vita religiosa a Parma nel medioevo: Confraternite, ordini religiosi, ospedali, dissenso e pietà dei laici," in *Vivere il medioevo: Parma al tempo della cattedrale* (Cinisello Balsamo: Silvana, 2006), 46–51.

3   Alfonso Maierù, "Formazione culturale e tecniche d'insegnamento nelle scuole degli ordini mendicanti," in *Studio e studia: Le scuole degli ordini mendicanti tra XIII e XIV secolo* (Spoleto: Centro italiano di studi sull'alto medioevo, 2002), 3–31; Mariano d'Alatri, "Italia," in *Le scuole degli ordini mendicanti Secoli XIII–XIV)* (Todi: Accademia tudertina, 1978), 49–72: 62–63.

4   For a summary of the questions that scholars have asked themselves as to why religious orders worked for the commune, see Frances Andrews, epilogue to *Churchmen and Urban Government in Late Medieval Italy, c.1200–c.1450* (Cambridge: Cambridge University Press, 2013), 348–57. I want to take the opportunity to thank Frances Andrews for her encouragement during the last stages of the writing of this book.

5   I overlooked it, too. Emanuele Lugli, *Unità di misura: Breve storia del metro in Italia* (Bologna: Il Mulino, 2014), 34.

6   "Ut autem nulla fraudis vel laesionis provinciarum nascatur occasio, iubemus in illis mensuris vel ponderibus species vel pecunias dari vel suscipi, quae

beatissimo papae vel amplissimo senatui nostra pietas in praesenti contradidit." Justinian, *Novellae*, ed. Rudolf Schöll and Wilhelm Kroll, vol. 3 of *Corpus iuris civilis* (Berlin: Wedimann, 1895), 801, appendix VII, ch. 19. This norm was part of the *Pragmatica sanctio pro petitione Virgilii*, issued by Justinian on August 14, 554, at the request of Pope Vigilius.

7   See chapter 11. Cassiodorus, the sixth-century statesman retired to conventual life, wrote in his *Historia ecclesiastica* that Emperor Constantine proscribed to move the Egyptian cubit from the temple of Serapis to a church: "Et cubitum, quo mensuratur Nilus, non iam in templum Serapidis annis singulis, sed in Ecclesiam portari sancivit." See Cassiodorus, *Historia ecclesiastica* 2.18, in *Opera omnia*, 2 vols. (Paris: Dezallier, 1679), 1:232b.

8   Alain Guerreau, "*Mensura* et *metiri* dans la vulgate," *Micrologus* 21 (2011), 3–19. In the Bible, the term *mensura* returns 120 times and the verb *metiri* 60. The term *caritas*, in comparison, appears 104 times.

9   "Quis mensus est pugillo aquas et caelos palmo ponderavit?" Isaiah 40:12.

10   "Sed omnia in mensura, et numero et pondere disposuisti." Wisdom 11:20.

11   Augustine, *De Genesi ad litteram* 4.3.7; Augustine, *De Genesi adversus Manicheos* 1.16.26. On Augustine's interpretations, see William James Roche, "Measure, Number, and Weight in Saint Augustine," *New Scholasticism* 15 (1941), 350–76; and Alain Besançon, *The Forbidden Image: An Intellectual History of Iconoclasm* (Chicago: University of Chicago Press, 2000), 155–56.

12   Evgeny A. Zaitsev, "The Meaning of Early Medieval Geometry: From Euclid and Surveyors' Manuals to Christian Philosophy," *Isis* 90, no. 3 (1999), 522–553: 530–31.

13   "Corpus vero omnino locale et circumscriptibile; deus vero omnino inlocalis et incircumscriptibilis." Peter Lombard, *Sententiae* 1.37.6. See, in particular, *Magistri Petri Lombardi Parisiensis episcopi Sententiae in IV libris distinctae*, 3rd ed. (Grottaferrata: Editiones Collegii S. Bonaventurae ad Claras Aquas, 1971), 270.

14   "Mensuras rerum venalium necesse est in diversis locis esse diversas, propter diversitatem copiae et inopiae rerum, quia ubi res magis abundant, consueverunt esse maiores mensurae. In unoquoque tamen loco ad rectores civitatis pertinet determinare quae sunt iustae mensurae rerum venalium, pensatis conditionibus locorum et rerum. Et ideo has mensuras publica auctoritate vel consuetudine institutas praeterire non licet." Thomas Aquinas, *Summa theologica* 2.2, q. 77, a. 2. The point returns in Bernardino da Siena, *Quadragesimale de evangelio aeterno*, vol. 4 of *Opera omnia*, ed. Pacifico M. Perantoni and Augustin Sépinski (Florence: Quaracchi, 1950–65), 155–56.

15   "Omnia creavit in numero, pondere, et mensura, et isti est mensurator primus et certissimus." Venice, Biblioteca Marciana: Ms Lat. VI, 222. See also Anneliese Maier, *Zwischen Philosophie und Mechanik: Studien zur Naturphilosophie der Spätscholastik* (Rome: Edizioni di storia e letteratura, 1958), 24–25n; James McEvoy, *The Philosophy of Robert Grosseteste* (Oxford: Clarendon Press, 1982), 175–77.

16   Saint Bonaventure, *Collationes in Hexaemeron* 9.3; Saint Bonaventure, *Sermo de Trinitate* 9. See also Amy Neff, "'Palma debit Palmam': Franciscan Themes in a Devotional Manuscript," *Journal of the Warburg and Courtauld Institutes* 65 (2002), 22–66: 46 and 51.

17   "In qua mensura mensi fueritis, in ea remetietur vobis." Matthew 7:1–2. The verse was also commented upon in Saint Augustine, *De diversis quaestionibus octoginta tribus* 71.6.

18   Maria Giuseppina Muzzarelli, *Penitenze nel medioevo: Uomini e modelli a confronto* (Bologna: Patron, 1994), 41–43.

19   "Ostendens quod gravitas peccati majorem exigeret poenitentiam, sed commensurare vult eam juxta humanae fragilitatis impotentiam." Alain de Lille, *Liber poenitentialis*, ed. Jean Longère, 2 vols. (Leuven, Belgium: Nauwelaerts, and Lille, France: Giard, 1965), 2:48.

20   "Fecisti falsitatem vel fraudem aliquam in mensuris aut ponderibus? Ita dico ut falso modio cum ponderibus iniustis tua bona vendere aliis chrixtianis? Si fecisti aut consensisti viginti dies peniteas." Siena, Biblioteca Comunale, Ms F.VI.8, ch. 10. See Mino Marchetti, *Peccati, peccatori e penitenze nella chiesa medievale di Siena* (Siena: Cantagalli, 2000), 62–63. One observation: twenty days was also the penance for those who did not pay tithes in full, thus placing measuring and counting on the same level. On the dating, see pp. 10–13.

21   "In his que pondere, numero et mensura constant, veluti frumento, vino, oleo, argento, modo ea servanture que in ceteris, ut simul atque de pretio convenerit videatur perfecta venditio modo non etiam ut si de pretio convenerit, non tantum aliter videatur perfecta venditio quantum ad periculum quam si admensa vel adpensa vel adnumerata sint. Nam si omne vinum vel oleum vel frumentum uno pretio venierit, ideam iuris est quod in ceteris rebus." *I costituti della legge e dell'uso di Pisa (sec. XII)*, ed. Paola Vignoli (Rome: Istituto storico italiano per il medio evo, 2003), 269. In the twelfth century, the city of Pisa received the so-called *Littera Florentina*, a sixth-century manuscript containing the whole of Justinian's *Digest*, so the influence of the *Corpus iuris civilis* on the *Constitutum* is direct. After Pisa was conquered by the Florentines in 1406, the manuscript moved to Florence, hence its current name and location. Today it is in the Biblioteca Medicea Laurenziana. See Michael H. Hoeflich and Jasonne M. Grabher, "The Establishment of Normative Legal Texts: the Beginnings of the 'Ius commune,'" in *The History of Medieval Canon Law in the Classical Period, 1140–1234*, ed. Wilfried Hartmann and Kenneth Pennington (Washington: Catholic University of America Press, 2008), 1–21: 5; Davide Baldi, "Il 'Codex Florentinus del Digesto' e il 'Fondo Pandette' della Biblioteca Laurenziana, *Segno e testo* 8 (2010), 99–186.

22   "In his rebus quae pondere, numero mensura constant." Justinian, *Digesta*, ed. Theodor Mommsen and Paul Krüger, vol. 1 of *Corpus iuris civilis* (Berlin: Wedimann, 1870), 202 (18.1.5). This rubric also defines quantification as the key aspect of sales: "Quia venditio quasi sub hac conditione videtur fieri, ut

in singulos modios, quos, quasve admensus eris, aut in singulas libras, quas adpenseris, aut in singula corpora, quae adnumeraveris." See also Cornelia Cogrossi, "Il vino nel 'Corpus Iuris' e nei glossatori," in *La civiltà del vino: Fonti, temi e produzioni vitivinicole dal medioevo al novecento*, ed. Gabriele Archetti (Brescia: Centro culturale artistico di Franciacorta e del Sebino, 2003) 499–531: 508–10.

23   "Pondere, numero, mensurave constant." Gaius, *Institutes* 2.196.

24   Natalie B. Dohrmann, "Manumission and Transformation in Jewish and Roman Law," in *Jewish Biblical Interpretation and Cultural Exchange*, ed. Natalie B. Dohrmann and David Stern (Philadelphia: University of Pennsylvania Press, 2008), 51–65: 52.

25   Jacqueline Rambaud-Buhot, "Le 'Corpus Juris Civilis' dans le Décret de Gratien, d'après le manuscrit lat. nouv. acq. 1761 de la Bibliothèque nationale," *Bibliothèque de l'ecole des chartes* 111, no. 1 (1953), 54–64; Ennio Cortese, *Il diritto nella storia medievale*, 2 vols. (Rome: Il cigno Galileo Galilei, 1995), 2:197–213.

26   Michael Baxandall, *Painting and Experience in Fifteenth-Century Italy: A Primer in the Social History of Pictorial Style* (Oxford: Oxford University Press, 1988), 86–108: 101.

27   Anthony Randolph Bridbury, "Markets and Freedom in the Middle Ages," in *The Market in History*, ed. Bruce L. Anderson and Anthony John Heaton Latham (London: Croom Helm, 1986), 79–119: 108.

28   "E questo potremo provare e vedere [the greatness of heavenly rewards], se misurremo quelli [secular goods] con questi [heavenly rewards]. E possonsi misurare quelli beni con questi per quattro vie, e qui vedremo l'arte come si misurano." Giordano da Pisa, *Quaresimale fiorentino: 1305–1306*, ed. Carlo Delcorno (Florence: Sansoni, 1974), 183–91: 186.

29   "Considera, che farebbe lo coltello lungo uno braccio che fusse ficto in corpo d'uno homo? Certo, sommo dolore sarebbe. Or quanto è maggiore lo dolore del coltello di Dio che sempre si ficca et è infinito!" Giordano da Pisa, *Prediche inedite*, ed. Cecilia Iannella (Pisa: ETS, 1997), 250.

30   "Mensura est res aliqua modo suo vel tempore circumscripta." Isidore of Seville, *Etymologiarum libri* 16.25.

31   Inspired by Antonio Gramsci, *Quaderni del carcere*, ed. Valentino Gerratana, 4 vols. (Turin: Einaudi, 2007), 2:1384.

CHAPTER 17

1   "Et ordinamus quod potestas bon[oniae] infra iii menses ab ingressu sui regiminis teneatur facere inquiri modum lapidum et cupporum per fratres penitentie et videri diligenter et interrogari si lapides et cupi, qui modo fiunt, sunt crosi et ampli et lungi ad modum designatum sub voltis comunis." *Statuti del Comune di Bologna dall'anno 1245 all'anno 1267*, ed. Luigi Frati, 3

NOTES TO PAGES 115–16   261

    vols. (Bologna: Tipografia regia, 1869–84), 2:153. This norm comes from the statutes drawn in 1252–53. By 1259, the period was extended to four months.
2   *Statuta Communis Parmae*, ed. Amadio Ronchini, 4 vols. (Parma: Fiaccadori 1855–60), 1: 66 and 157; Augustine Thompson, *Cities of God: The Religion of the Italian Communes, 1125–1325* (University Park: Pennsylvania State University Press, 2005). 94; Frances Andrews, "The Commune of Parma and Its Religious Personnel," in *Churchmen and Urban Government in Late Medieval Italy, c. 1200–c. 1450* (Cambridge: Cambridge University Press, 2013), 45–66: 49–50.
3   *Statuta Communis Parmae*, 1:66–67. In 1317, the friars of Reggio were described as "qui stant ad portas pro conductore." Andrea Balletti, "Ordinamenti finanziari nell'età dei comuni," *Giornale degli economisti* 29 (1904), 172–89: 179.
4   "Integraliter sicuti faciebant layci." *Statuta Communis Parmae*, 1:157. See also Caterina Bruschi, "Gli umiliati a Parma (XIII–XIV sec.): Instaurazione e sviluppo di rapporti molteplici," *Rivista di storia e letteratura religiosa* 36 (2000), 209–38: 225–31.
5   Giuseppe Sassatelli, *Da Felsina a Bononia: Dalle origini al XII secolo*, vol. 1 of *Atlante storico di Bologna* (Bologna: Grafis, 1995), 50.
6   Julian of Speyer recalls that Saint Francis of Assisi founded three orders, among which the last was made of penitents: "tres ordines ordinavit . . . tertius . . . ordo penitentium dicitur." Julian of Speyer, "Vita Sancti Francisci," *Analecta franciscana* 10 (1926–1941), 333–71: 346. On the institutionalization of Penitents, Raffaele Pazzelli, "L'orientamento francescano del movimento penitenziale lungo il secolo XIII," in *Angela da Foligno terziaria francescana*, ed. Enrico Menestò (Spoleto: Centro italiano di studi sull'alto medioevo, 1992), 3–16.
7   *Statuti Bologna*, 1, p. 132.
8   Vincenzo Petriccione, "I penitenti francescani di Bologna nel secolo XIII," in *I frati penitenti di San Francesco nella società del Due e Trecento*, ed. Mariano d'Alatri (Rome: Istituto storico dei cappuccini, 1977), 259–69.
9   The lists are reproduced in *Analecta franciscana* 9 (Florence: Quaracchi, 1927), index III (Tertius ordo).
10   Petriccione, "I penitenti di Bologna," 259–69.
11   Francesca Pucci Donati and Rossella Rinaldi, "Il commercio al dettaglio a Bologna tra Due e Trecento: La piazza, l'osteria, la bottega," in *Il commercio al minuto: Domanda e offerta tra economia formale e informale. Secc. XIII–XVIII*, ed. Giampiero Nigro (Florence: Firenze University Press, 2015), 241–58: 247n.
12   Luigi Zanoni, *Gli umiliati nei loro rapporti con l'eresia, l'industria della lana ed i comuni nei secoli XII e XIII* (Rome: Multigrafica, 1970, 1st ed. 1911), 217; Raoul Manselli, "Gli umiliati, lavoratori di lana," in *Produzione, commercio e consumo dei panni di lana*, ed. Marco Spallanzani (Florence: Olschki, 1976), 231–36.
13   Zanoni, *Gli umiliati*, 208.
14   Christ's prohibition is in Matthew 5:34–36.

15   Maria Teresa Brolis, "'Quibus fuit remissum sacramentum': Il rifiuto di giurare presso gli umiliati," in *Sulle tracce degli umiliati*, ed. Maria Pia Alberzoni, Annamaria Ambrosioni, and Alfredo Lucioni (Milan: Vita e pensiero, 1997), 251–65: 257–59.

16   Frances Andrews, *The Early Humiliati* (Cambridge: Cambridge University Press, 2004): 129–32.

17   Zanoni, *Gli umiliati*, 223. Zanoni does not mention Modena and Ferrara, where Humiliates are documented as *massarii* and *fratres de boleta* from 1260 and 1288 respectively. See Ludovico Antonio Muratori, *Dissertazioni sopra le antichità italiane*, 5 vols. (Milan: Società tipografica de' classici italiani, 1836–37) 5:255–56.

18   On Bergamo, Maria Teresa Brolis, *Gli umiliati a Bergamo nei secoli XIII e XIV* (Milan: Vita e pensiero, 1991), 179–81; on Cremona, Zanoni, *Gli umiliati*, 233.

19   "Et quod una recta et iusta statera pro communi Brixie constituantur et teneatur apud portam Turlungae apud suprascriptos morantes ibidem. Item quod cum ipsa statera possit et debeat ponderari totum ferrum exportandum per ipsam portam, quod dictis fratribus videbitur ponderandum, ne comune Brizie in ipso ferro valeat fraudari." *Statuti bresciani*, ed. Federico Odorici, vol. 16, no. 2 of *Historiae patriae monumenta* (Turin: Deputazione subalpina di storia patria, 1876), column 1584 (109).

20   *Liber statutorum Comunis Novocomi*, ed. Antonio Ceruti, vol. 16, no. 2 of *Historiae patria monumenta* (Turin: Bocca, 1876), columns 181–87, rubrics 221–36. The Humiliates had a copy of the 1218 statutes and read it every other week. See columns 181–82: "et in totidem si non signaverit secundum formam statuti facti MCCXVIII die lune tercio mensis octubris; qui fratres teneantur habere exemplum hiuius statuti et legere omni quindena."

21   Bonvesin de la Riva, *De magnalibus Mediolani*, ed. Paolo Chiesa (Milan: Scheiwiller, 1997), 4.14.

22   Bonvesin, *De magnalibus Mediolani* 3.4 (chapels), ovens (3.27), heralds (3.20), tombstones (3.35), and cherries (4.3).

23   Pio Pecchiai, "I documenti sulla biografia di Buonvicino della Riva," *Giornale di storia della letteratura italiana* 78 (1921), 96–127: 108–11; Zanoni, *Gli umiliati*, 236–38. Zanoni's remarks at p. 217 on the Humiliates' relations with the commune are useful.

24   On the numerous manuscripts of Bonvesin's text, see Paolo Chiesa, "Introduzione," in Bonvesin de la Riva, *De magnalibus mediolani*, xi–lxxxiv: xiv.

25   Bonvesin, *De magnalibus Mediolani* 2.3. In Milan, a *pertica* was a unit for measuring area, not length.

26   "F[ossati autem latit]udinem ipsemet XXXVIII cubitis m[ensuravi] . . . et nota quod cubitus, de quo fit presens locucio, duorum pedum in longum et totidem digitorum in latum hominis magne statute longitudinem contigit." Bonvesin, *De magnalibus Mediolani* 2.5.

27   "Muris circuitus diligentissime [mensuratus XMCLI] cubitorum veraciter est inventus." Bonvesin, *De magnalibus Mediolani* 2.5.

28 | Brolis, *Gli umiliati a Bergamo*, 181.

29 | "Sane ad nostram noveris audientiam pervenisse, quod potestas et commune mediolani fratres primi et secundi ordinis humiliatorum officia publica seu communitatis gerere, ac exigere pedagia, et ad portam civitatis stare ut blada et farinas ponderent seu mensurent, per captionem bonorum suorum, per inhibitionem quod homines mercimonia cum ipsis non exerceant, neque molant in molendinis eorum pro sua voluntate compellunt." Girolamo Tiraboschi, *Vetera Humiliatorum Monumenta*, 3 vols. (Milan: Giuseppe Galeazzi, 1766–68), 2:247–49. The two other papal letters are at p. 216 (1247) and 258 (1253).

30 | The Humiliates' rule is in the *Propositum* section of Innocent III's letter of June 7, 1201. See Gilles Gérard Meersseman, *Dossier de l'ordre de la pénitence au XIIIe siècle* (Fribourg: Ed. Universitaires, 1961), 282.

31 | The pope's clarification may have reflected the fact that Tertiaries occupied an ambiguous position within the Franciscan order, with some communities solidly under Franciscan oversight, and others with no tie other than a general allegiance to the Tertiary rule. Ambiguity often meant tensions and attacks. As Penitents had possessions and families, hard-core Franciscans thought them more likely to fall into temptation, and accused them of weak participation. See, for example, Saint Bonaventure, *Opuscula varia ad theologiam mysticam et res ordinis fratrum minorum spectantia*, vol. 8 of *Opera omnia* (Florence: Quaracchi, 1898), 369, pars II, q. 16.

32 | Reggio Emilia, Archivio di Stato: Archivio del Comune, part I, Statuti, 1, cc. 19v–20v. See also Balletti, "Ordinamenti finanziari," 176.

33 | The norm defines "village" as any community with at least eight "hearths" (*focis*). *Statuta Communis Parmae*, 2:71–72.

34 | "Lo camerlengo et due ufficiali di penitentia . . . facciano fare una sceda di legno di tegole . . . bone et bene fatta." *Il Costituto del Comune di Siena volgarizzato nel 1309–1310*, ed. Mahmoud Salem Elsheikh, 4 vols. (Siena: Monte dei Paschi, 2002), 2:547–48.

35 | "Et tutte le finestre . . . reducere farò a la misura del passetto. Et queste cose farò fare per tutto el mese di gennaio. Et per queste cose fare, elegerò tre buoni huomini et leali di penetentia, e' quali queste cose facciano fare. Et neuno possa tenere desco, predella over banca fuore de la casa ovver bottiga ne le dette strade, oltre el detto passetto." Elsheikh, *Il Costituto del Comune di Siena*, 2, 40.

36 | Ugo Sorbi, "Catasti senese e fiorentino del XIV e XV secolo," *Rivista di storia dell'agricoltura* 2 (December 1986), 161–213: 167.

37 | The inscription reads: "HIC EST LIBER RELIGIOSI VIRI DOMPNI GUIDONIS MONACHI SANCTI GALGANI, CAMERARII COMUNIS SENARUM IN PRIMIS SEX MENSIBUS REGIMINIS VIRI NOBILIS DOMINI ALBERCI SIMONIS PICCIOLI DE BONONIA, DEI GRATIA SENENSIS POTESTATIS." See Alessandro Tomei, *Le biccherne di Siena: Arte e finanza all'alba dell'economia moderna* (Rome: Retablo, 2002).

38   *Statutum Potestatis Comunis Pistorii anni 1296*, ed. Lodovico Zdekauer (Milan: Hoepli, 1888), 118.

39   *Statutum Pistorii*, 175 and 190. Canals were called *gore* in Pistoia. The Gora d'Ombroncello referred to in the passage was the canal flowing near the Franciscan church.

40   *Statutum Pistorii*, 161–62.

41   This information comes from an inventory. Pistoia, Archivio di Stato: S. Iacopo 1, ff. 103r–v. See also Natale Rauty, *Pistoia: Città e territorio nel medioevo* (Pistoia: Società pistoiese di storia patria, 2003), 239. On the *Opera*, see Lucia Gai, "Opera di S. Iacopo di Pisoia," in *Opera: Carattere e ruolo delle fabbriche cittadine fino all'inizio dell'età moderna*, ed. Margaret Haines and Lucio Riccetti (Florence: Olschki, 1996), 310–13: 312. Measurements were placed in San Jacopo as early as 1265.

42   *L'Opera di S. Jacopo in Pistoia e il suo primo statuto in volgare (1313)*, ed. Lucia Gai and Giancarlo Savino (Ospedaletto: Pacini, 1994), 200; Carlo Vivoli, "L'archivio dell'Opera di San Jacopo alla metà del secolo XVIII," *Bullettino storico pistoiese* 109 (2007), 175–89: 188.

43   "Ut melius et diligentius et sine defectu fiat illud officium." Sarah Tiboni, "Pistoia: A Case Study," in *Churchmen and Urban Government in Late Medieval Italy, c. 1200–c. 1450*, ed. Frances Andrews (Cambridge: Cambridge University Press, 2013), 166–80: 174.

CHAPTER 18

1   The two bulls, which are identical, were issued on June 5, 1274, and August 15, 1274. See Venturino Alce, "Il convento di S. Domenico in Bologna nel secolo XIII," *Culta Bononia* 4, no. 2 (1972), 127–74: 159–61.

2   Enrico Guidoni, *La città dal medioevo al rinascimento* (Rome and Bari: Laterza, 1984), 138–52.

3   Luigi Pellegrini, *Insediamenti francescani nell'Italia del Duecento* (Rome: Laurentianum, 1984), 123–53; Sandra Farina, "I conventi mendicanti nel tessuto urbanistico di Bologna," *Storia della città* 9 (1978), 56–61. The bibliography on mendicant orders' urban settlements is vast. The debate, however, largely builds on Jacques Le Goff, "Apostolat mendiant et fait urbain dans la France médiévale: L'implantation des ordres mendiants: Programme-questionnaire pour une enquête," *Annales: Economies sociétés civilisations* 23, no. 2 (1968), 335–52. The conclusions of that essay also reappeared in Jacques Le Goff, "Ordres mendiants et urbanisation dans la France médiévale: Etat de l'enquête," *Annales: Economies sociétés civilisations* 25, no. 4 (1970), 924–46.

4   David Flood and Thadee Matura, *The Birth of a Movement: A Study of the First Rule of St. Francis* (Chicago: Franciscan Herald, 1975); William A. Hinnebusch, *I Domenicani: Breve storia dell'ordine* (Milan: Edizioni paoline, 1992), 18.

5   The pope's advisors were two Dominicans, Paolo di Segni and Pietro da

Tarantasia, and two Franciscans, Eudes Rigaud and Bonaventure of Bagno Regio. See Pietro Silanos, "'... Quos evidens ex eis utilitas ecclesiae universalis ...': Gregorio X e l'ordine dei frati minori," in *Gregorio X Pontefice tra occidente e oriente*, ed. Massimiliano Bassetti and Enrico Menestò (Spoleto: Centro italiano di studi sull'alto medioevo, 2015), 65–96: 93–94.

6 Roberto Rusconi, "'Clerici secundum alios clericos': Francesco d'Assisi e l'istituzione ecclesiastica," in *Frate Francesco d'Assisi* (Spoleto: Centro italiano di studi sull'alto medioevo, 1994), 71–100.

7 Micheline de Fontette, "'Religionum diversitatem' et la suppression des ordres mendiants," in *1274. Année charnière: Mutations et continuités* (Paris: Editions du CNRS, 1977), 223–29.

8 Jacques Le Goff, "La perception de l'espace de la chrétienté par la curie romaine et l'organisation d'un concile oecuménique en 1274," in *L'imaginaire médiéval* (Paris: Gallimard, 1985), 76–83. Le Goff saluted the council as a landmark since it marked the Church's acceptance of rational organization. When speaking of reason, though, Le Goff might have fallen for the ideology of the communes, as there is nothing "rational" in choosing an arbitrary distance.

9 Caroline Levine, *Forms: Whole, Rhythm, Hierarchy, Network* (Princeton, NJ: Princeton University Press, 2015), 82–111.

10 Guidoni, *La città*, 134. The norm, titled "De vicinitate," is attributed to Abbot Guido (1133–34). It first appeared in the *Instituta generalis capituli apud Cistercium*, the statutes approved by the Cistercian chapter in the late twelfth century, and was reworked, over and over, up to the beginning of the thirteenth. Some manuscripts do not even mention it. It reads: "Si cui locus ad abbatiam construendam oblatus fuerit non presumat accipere, nisi prius eum distare a ceteris abbatiis nostri ordinis decem leugis Burgundie pro certo cognoverit.... Grandie autem diversarum abbatiarum distent inter se ad minus duabus leugis." Cited in Joseph Turk, *Cistercii statuta antiquissima* (Rome: Tipografia Pio X, 1949), 21.

11 Claudio Stercal and Milvia Fioroni, *Le origini cistercensi: Documenti* (Milan: Jaca book, 2004), 157–58. See also Chrysogonus Waddell, *Narrative and Legislative Texts from Early Citeaux* (Citeaux: Comentarii Cistercienses, 1999).

12 "Statuimus etiam, ut a predicta urbe infra sex miliaria castella non hedificentur, et si aliquis munire presumpserit, nostro imperio et auxilio destruantur." *Die Urkunden Heinrichs IV. 1077–1106*, ed. Dietrich von Gladiss and Alfred Gawlik, in *Monumenta Germaniae Historica, Diplomata regum et imperatorum Germaniae*, vol. 6, no. 2 (Weimar: Böhlau Verlag, 1952 [1959]), 438. See also Michael E. Bratchel, *Medieval Lucca and the Evolution of the Renaissance State* (Oxford: Oxford University Press, 2008), 31–33.

13 *Statuti di Pistoia del secolo XII*, ed. Francesco Berlan (Bologna: Romagnoli, 1882), 1 (rubric 1, for protection of churches), 74 (rubric 93, for murders), 75 (rubric 95, for public streets). As the first statute of a commune, the norm echoes imperial directives. See Francesca Bocchi, *Attraverso le città italiane nel medioevo* (Bologna: Grafis, 1987), 31–32.

14 | Justinian, *Novellae*, ed. Rudolf Schöll and Wilhelm Kroll, vol. 3 of *Corpus iuris civilis* (Berlin: Weidmann, 1895), 334–35. See also Catherine Saliou, *Les lois des bâtiments: Voisinage et habitat urbain dans l'Empire romain. Recherches sur les rapports entre le droit et la construction privée du siècle d'August au siècle de Justinien* (Beirut: Institut français d'archéologie du Proche-Orient, 1994), 238.

15 | Giorgio Zacchello, "Il cardinale Simone Paltanieri. Breve profilo biografico," in *Monselice. Storia, cultura e arte di un centro minore del Veneto* (Monselice: Rigon, 1994), 625–31.

16 | Pope Alexander IV set up a distance of three hundred *canne* on August 25, 1260. See Giovanni Giacinto Sbaraglia, *Bullarium franciscanum* 4 vols. (Rome: De propaganda fide, 1759–68), 2:405–6. For Ascoli, see Raniero Giorgi, *Le clarisse in Ascoli* (Fermo: La Rapida, 1968), 25. More generally, on the application of the norm for the Franciscans, Giuseppe Fratini, *Storia della basilica e del convento di San Francesco in Assisi* (Prato: Ranieri Guasi, 1882), 86–87; Guidoni, *La città*, 136–37. For the Dominicans, see Alce, "Il convento di S. Domenico," 159–61; Alfonso d'Amato, *I domenicani a Bologna*, 2 vols. (Bologna: Edizioni Studio domenicano 1988), 1:119–23.

17 | Bologna, Archivio di Stato: Archivio del Convento di San Domenico, C I a, B. n. 10.

18 | The bull that Pope Clement IV issued on November 30, 1265, the *Ad consequendam gloriam*, still speaks of 300 *canne*. See Sbaraglia, *Bullarium franciscanum*, 3:59. Yet with the bull of June 5, 1268, the *Quia plerumque in futurorum eventibus*, Clement IV reduced the distance to 140. "Ad scandalum cessisse comperimus, diversis propterea litigiorum materiis suscitatis: nos super iis cum fratribus nostris deliberatione praehabita... ipsum ad centum et quadraginta cannas duximus reducendum." Sbaraglia, *Bullarium franciscanum*, 3:158.

19 | Sbaraglia, *Bullarium franciscanum*, 3:219 (August 13, 1274).

20 | Sbaraglia, *Bullarium franciscanum*, 3:586 (November 27, 1286). Pope Honorius IV defended the 140-*canne* rule by pointing at Clement IV's change of mind.

21 | Alfredo Pacini, *La chiesa pistoiese e la sua cattedrale nel tempo*, 12 vols. (Pistoia: CRT, 1994), 1:60–61; Lucia Gai, "Insediamento e prima diffusione degli ordini mendicanti a Pistoia," in *Gli ordini mendicanti a Pistoia (secc. XIII–XV)*, ed. Renzo Nelli (Pistoia: Società pistoiese di storia patria, 2001), 69–113: 101. The Carmelites moved out in the seventeenth century.

22 | Gentile da Montefiore and Niccolò Boccasini were appointed on July 8, 1300. Sbaraglia, *Bullarium franciscanum*, 5: 4–6. Pope Benedict XI's resolution came on December 5, 1303. *Le registre de Benoît XI*, ed. Charles A. Grandjean, 3 vols. (Paris: Thorin, 1883–1905), 2:707–11, no. 1165. See also Julian Gardner, "'Aedificia iam in regales surgunt altitudines': The Mendicant Great Church in the Trecento," in *Communes and Despots in Medieval and Renaissance Italy*, ed. John E. Law and Bernadette Paton (Farnham: Ashgate, 2010), 307–27: 322, esp. note 43.

23 | For a recent case, Giulia Barone, "L'età medievale (XIII–XIV secolo)," in *L'ordine dei predicatori. I Domenicani: Storia, figure e istituzioni (1216–2016)*, ed. Gianni Festa and Marco Rainini (Rome and Bari: Laterza, 2017), 5–29: 19–22.

24 | It took the Franciscans of Verona a long time to move to the church of San Fermo. The pope gave his permission in 1248, but they managed to move only in 1260, after which they undertook a major renovation of the church. See Carla Pompei-Perez, *La chiesa di San Fermo Maggiore* (Verona: Edizioni di vita veronese, 1954), 10–12 and 119–25; Gianpaolo Trevisan, "L'architettura (secoli XI–XIV)," in *I Santi Fermo e Rustico: Un culto e una chiesa in Verona*, ed. Paolo Golinelli and Caterina Gemma Brenzoni (Milan: Motta, 2004), 169–83.

25 | "Per aerem, etiam ubi alias recte mensurari loci dispositio non permittit." Sbaraglia, *Bullarium franciscanum*, 3:60.

26 | The start and end points of the measuring are described as "angulo extremo ecclesie Sancte Marie de la Scala de Verona propinquiore ecclesie Sancti Firmi Maioris de Verona ponito et iacentem versus oriens et meridiem . . . angulo propinquiorem dicte ecclesie Sancti Firmi." Verona, Archivio di Stato: Santa Maria della Scala, parchment b1, n. 25d (March 16, 1327). See also Gianpaolo Trevisan, "'Cum squadra et cordula et aliis edificiis ingeniosi::' La facciata della chiesa di San Fermo Maggiore a Verona e la misurazione della distanza da Santa Maria della Scala nel 1327," in *Arredi liturgici e architettura*, ed. Arturo Carlo Quintavalle (Milan: Electa, 2007), 143–51.

27 | "Adequata longitudine fili canapi transmissi per domini Episcope de Asiso pertinentis longitudo quantitatem cum canne cum bracchia 8." Verona, Archivio di Stato: Santa Maria della Scala, parchment b1, n. 25a (December 10, 1326).

28 | The definition of the *canna* can be found in Clement IV's bull of 1265, the one imposing 300 *canne*—not the one of 1268, when he set the length at 140: "Et quamlibet cannarum ipsarum octo palmarum longitudinem continere, non obstantibus varia locorum consuetudine." See Sbaraglia, *Bullarium Franciscanum*, 3:60. The letter by the bishop of Assisi is in Verona, Archivio di Stato: Santa Maria della Scala, parchment b1, n. 25b (March 2, 1327).

29 | *Tavole di ragguaglio dei pesi e delle misure già in uso nelle varie provincie del Regno col peso metrico decimale approvate con decreto reale 20 maggio 1877, n. 3836* (Rome: Stamperia reale, 1877), 745.

30 | The Roman architectural *canna*, of ten palms, is around 223 centimeters. Rome's mercantile *canna*, of eight palms, is 199 centimeters. See *Tavole di ragguaglio* (1877), 624. For the measurements of Assisi, see Manuela Bernardi, "Misure lineari e per laterizi," in *Raccolte comunali di Assisi: Monete, gettoni, medaglie, sigilli, misure e armi*, ed. Maurizio Matteini Chiari (Perugia and Milan: Electa, 2000), 183–86. The Assisi *passetto*, called "canna" in the city's fifteenth-century statutes, measured about 104 centimeters.

CHAPTER 19

1 | Bologna, Archivio di Stato: Archivio del Convento di San Domenico, C I a, B. n. 10.

2 | Franco Bergonzoni, "Un rapido profilo storico," in *Le mura perdute: Storia e immagini dell'ultima cerchia fortificata di Bologna*, ed. Giancarlo Roversi (Bologna:

Grafis, 1985), 11–48; Francesca Bocchi, *Il Duecento*, vol. 2 of *Atlante storico di Bologna* (Bologna: Grafis, 1995), 33–36. We know nothing about the dates of the *circla*. A 1250 mention of the internal side of the palisade has been taken as indicating that the ring was somewhat complete by then. See Giovanni Gozzadini, "Studi archeologici-topografici sulla città di Bologna," *Atti e memorie della deputazione di storia patria per le province di Romagna* 7 (1868), 1–104: 86. It is possible that the southern sector of the *circla* was not completed, hence Bishop Ottaviano's precise perimeter.

3   The stakes were made of "sallice," "lapidus de maxegna," or antique marble. See Jacques Heers, *Espaces publics, espaces privés dans la ville: Le liber terminorum de Bologne (1294)* (Paris: Editions du Centre national de la recherche scientifique, 1984), 125.

4   The *libri terminorum* are in Bologna's Archivio di Stato. The 1245 survey of the *circla* can be found both in the *Liber iuramentorum diversarum civitatum et memorabilium communis Bononie*, ff. 239r–246v, and in the *Registro Nuovo*, ff. 332v–341r. The *Liber terminorum* of 1286 is in *Registro grosso*, II, ff. 123r–126v. It is followed by the 1294 *Liber* (at ff. 127r–158r). On the *Liber grossus*, see Gianfranco Orlandelli, *Il sindacato del podestà: La scrittura da cartulario di Ranieri da Perugia e la tradizione tabellionale bolognese del secolo XII* (Bologna: Patron, 1963). On the *libri*, see also Maria Venticelli, "I Libri terminorum bolognesi," in *Medieval Metropolises*, ed. Francesca Bocchi (Bologna: Grafis, 1999), 223–34.

5   Bocchi, *Il Duecento*, 49–52; Giorgio Tamba, "I documenti del governo del Comune bolognese," *Quaderni culturali bolognesi* 2, no. 6 (1978), 47–66. On the *Ars notaria*, see James Jerome Murphy, *Rhetoric in the Middle Ages: A History of Rhetorical Theory from Saint Augustine to the Renaissance* (Berkeley: University of California Press, 1974), 264–68.

6   Bocchi, *Il Duecento*, 50.

7   In Bologna, removal of a stake was fined fifty *lire*. See *Statuti di Bologna dell'anno 1288*, ed. Gina Fasoli and Pietro Sella, 2 vols. (Vatican City: Biblioteca apostolica vaticana, 1937–39), 1:230, rubric 270. Punishments for unauthorized repositioning of boundary stones are documented also for Roman times. In *De condicionibus agrorum* Siculus Flaccus wrote, "Numa Pompilius decreed that whoever ploughed up a boundary stone would be outlawed, he and his oxen." *Die Schriften der römischen Feldmesser*, ed. Friedrich Blume and Karl Lachmann, 2 vols. (Berlin: Reimer, 1848–52), 1:302. In 59 BCE, the *Leges Mamilia, Roscia, Peducea, Alliena*, and *Fabia* set up hefty fines for similar infractions. See Stefano Del Lungo, *La pratica agrimensoria nella tarda antichità e nell'alto medioevo* (Spoleto: Centro italiano di studi sull'alto medioevo, 2004), 65.

8   The *libri* never say that the boundaries are invisible. Nonetheless, in the 1294 *Liber* we read that the stakes need to be found ("fuit unus terminus repertus," "item repertus fuit alius terminus olim positus in angulo domus"). The practice of burying stakes underground dates back to Roman times. The ancient surveyors Gaius and Theodosius wrote, "We buried squared boundary stones

underground" (*terminos quadratus sub terra conlocavimus*). Brian Campbell, *The Writings of the Roman Land Surveyors* (London: Society for the Promotion of Roman Studies, 2000), 250–52.

9 The *terminatio* is described in the *Lex Mamilia*, set by Caligula to deal with agricultural properties, and which proscribed a fine of fifty *aurei* for moving a stake. The norm had been included in the *Corpus iuris civilis*: "Qui terminos statutos extra suum gradum finesve moverint dolo malo, pecuniaria poena constituta est: nam in terminos singulos, quos eiecerint locove moverint, quinquaginta aureos in publico dari iubet." Justinian, *Digesta*, ed. Theodor Mommsen and Paul Krüger, vol. 2 of *Corpus iuris civilis* (Berlin: Weidmann, 1870), 792 (47.21.3). See also Adolf August Friedrich Rudorff, "Uber die sogenannte lex Mamilia de coloniis," *Zeitschrift für geschichtliche Rechtswissenschaft* 9 (1938), 379–420.

10 *Liber iurium Reipublicae Genuensis*, Historiae patriae monumenta 7, 2 vols. (Turin: Officina regia, 1854), 1: cols. 328–29. See also Luciano Grossi Bianchi and Ennio Poleggi, *Una città portuale del medioevo: Genova nei secoli X–XVI* (Genoa: Sagep, 1980), 96–97.

11 "Quia sepissime magne lites et contentiones inter cappellanos Padue pro discordia confiniorum fiebant idcirco domino Gerardo paduano episcopo placuit elligere tre sacerdotes... qui bona fide confinia dividerent et certibus finibus terminarent. Prefati ergo tres sacerdotes civitatem circumeuntes et dicta veterum hominum super hoc audientes tamdiu laboraverunt, quousque certos fines predictis confiniis imposuerunt." Andrea Gloria, *Codice diplomatico padovano*, 2 vols. (Venice: Real deputazione veneta di storia patria, 1879–81), 2:378–80, doc. 1307. See also Lionello Puppi and Mario Universo, *Padova* (Rome and Bari: Laterza, 1982), 34. The chronicle is *Rolandini Patavini cronica in factis et circa facta Marchie Trivixane*, ed. Antonio Bonardi (Città di Castello: Lapi, 1905–8), 183, 199, and 221.

12 In 1203, notary Martino records houses in Savona that poach on public land for half a foot. See *Il cartulario del notaio Martino, 1203–06*, ed. Dino Puncuch (Genoa: Società ligure di storia patria, 1974), 29. The cities of Vicenza, Novara, and Volterra carried out surveys between 1208 and 1209. Siena charted its streets in 1218, and in 1222 undertook the *terminatio* of the Terzo di San Martino, a suburban vineyard meant to accommodate an extension of the city. See Francesca Bocchi, Manuela Ghizzoni, and Rosa Smurra, *Storia delle città italiane: Dal tardoantico al primo rinascimento* (Turin: Utet, 2002), 248–49; Enrico Guidoni, *Il Duecento* (Rome and Bari: Laterza, 1989), 337–48 and 378–79. The report of Siena's "Terminazione del Terzo di S. Martino" is in Lodovico Zdekauer, *La vita pubblica dei Senesi nel Duegento* (Siena: Lazzeri, 1897), 108–10. On Siena, see also Mario Ascheri, "Le più antiche norme urbanistiche del Comune di Siena," in *La bellezza della città: Stadtrecht und Stadtgestaltung im Italien des Mittelalters und der Renaissance*, ed. Michael Stolleis and Ruth Wolff (Tübingen: Max Niemeyer, 2004), 241–67. Dimensions and ownerships are also at the core of the enlargement of the square in Assisi. See

Alberto Grohmann, *Assisi* (Rome and Bari: Laterza, 1989), 55. Stakes were first used in Parma in 1217. After acquiring the old imperial palace and surveying its plan for restoration, the commune placed "stone markers" (*terminos lapidis*) around it. See *Statuta Communis Parmae*, ed. Amadio Ronchini, 4 vols. (Parma: Fiaccadori, 1855–60): 1, 41–42. As we do not know anything more about this *terminatio*, I focus my attention to the Bolognese *terminatio* of 1245.

13   "Assiduo studio invenit rationem mensurandi non solum iacentes longitudines sed etiam tantes, et hoc non solum accessibiles sed etiam inaccessibiles." Robert Kilwardby, *De ortu scientiarum*, ed. Albert G. Judy (London: British Academy, 1976), 29, chapter 11, par. 61.

14   On the sizes and positions of market stalls, see Dennis Romano, *Markets and Marketplaces in Medieval Italy, c. 1100 to c. 1440* (New Haven and London: Yale University Press, 2015), 83 (Florence) and 95–96 (Piacenza and Pistoia). In Padua, anyone could accuse sellers if they operated outside of designated areas and if their shops exceeded prescribed boundaries. Paduan butchers, for instance, could only exercise under the porch of the *beccarie*, and the poles from which their meat hung could not stick out more than two feet. *Statuti del Comune di Padova dal secolo XII all'anno 1285*, ed. Andrea Gloria (Padua: Sacchetto, 1873), 278, norm 836.

15   Puppi and Universo, *Padova*, 34.

16   *Liber potheris Communis civitatis Brixiae*, ed. Francesco Bettoni Cazzago and Luigi Francesco Fè d'Ostiani (Turin: Bocca, 1899); Guidoni, *Il Duecento*, 354–67.

17   The first *terminatio* of Brescia is recorded on February 21, 1173, when *mensuratores* determined the four sides of the then new market square (today's Piazzale Tebaldo Brusato). It is described as a rectangle of twenty-one and one-half *pertiche* (north side), sixty-six *pertiche* (west), nineteen *pertiche* "minus one foot and a half" (south), and sixty-eight *pertiche* (east). See *Liber potheris*, 3, col. 566. On the creation of the square in 1172, see *Annales Brixienses*, ed. Ludwig Conrad Bethmann, in *Monumenta Germaniae Historica*, Scriptores, vol. 18 (Hannover: Hahn, 1863), 811–20: 813. On Brescia's twelfth-century markets, see Giancarlo Andenna, "Il monastero e l'evoluzione urbanistica di Brescia tra XI e XII secolo," in *Santa Giulia di Brescia: Archeologia, arte, storia di un monastero regio dai Longobardi al Barbarossa*, ed. Clara Stella and Gerardo Brentegani (Brescia: Grafo, 1992), 93–118.

18   *Inventario del Real archivio di stato in Lucca*, ed. Salvatore Bongi, 4 vols. (Lucca: Giusti, 1872–88), 2:127–29. For the documents, see pp. 142–167. For the charting in Siena, see Ugo Sorbi, "Catasti senese e fiorentino del XIV e XV secolo," *Rivista di storia dell'agricoltura* 2 (December 1986), 161–213.

CHAPTER 20

1   "Per humum defluentia vestimenta ultra unum palmum dependentem ab humero chlamydis plicaturam ultra duos palmos deferre praesumat."

NOTES TO PAGES 134–35   271

Sacrorum conciliorum nova et amplissima collectio, ed. Giovanni Domenico Mansi, 31 vols. (Florence and Venice: Zatta, 159–98), 24, columns 252–53. For the history of the *De habitu mulierum* norm, see Lodovico Antonio Muratori, *Antiquitates italicae medii aevi*, 6 vols. (Milan: Società palatina, 1738–42), 2:315. More recently, Ronald Rainey, "Sumptuary Legislation in Renaissance Florence" (PhD dissertation, Columbia University, 1985), 42–43.

2 Catherine Kovesi Killerby, *Sumptuary Law in Italy 1200–1500* (Oxford: Clarendon Press, 2002), 23–29; Maria Giuseppina Muzzarelli, "Una società nello specchio della legislazione suntuaria: Il caso dell'Emilia-Romagna," in *Disciplinare il lusso: La legislazione suntuaria in Italia e in Europa tra medioevo ed età moderna*, ed. Maria Giuseppina Muzzarelli and Antonella Campanini (Rome: Carocci, 2003), 17–31. The first extant sumptuary law is in Genoa's legal code, the *Breve della campagna* (1157), and bans expensive fur hems. The prohibition, however, is treated as an isolated case by fashion historians, given that it was said to last for four years only, while sumptuary laws have no time limits.

3 Maria Giuseppina Muzzarelli, *Guardaroba medievale: Vesti e società dal XIII al XIV secolo* (Bologna: Il Mulino, 1999): 276.

4 Elisa Tosi Brandi, "Il sarto tra medioevo e prima età moderna a Bologna e in altre città dell'Emilia Romagna" (PhD dissertation, Università di Bologna, 2012), 99; Maria Giuseppina Muzzarelli, "Una società nello specchio," 25.

5 The plea concludes the *De habitu mulierum*. See *Sacrorum conciliorum*, 24, column 253.

6 Robert Davidsohn, *Storia di Firenze*, 5 vols. (Florence: Sansoni, 1956), 1:213; Kristina M. Olson, "Uncovering the Historical Body of Florence: Dante, Forese Donati, and Sumptuary Legislation," *Italian Culture* 33, no. 1 (March 2015), 1–15: 5.

7 Salimbene de Adam recalls the change twice. Salimbene de Adam, *Cronica*, ed. Giuseppe Scalia, 3 vols. (Rome and Bari: Laterza, 1966), 1:146 and 436.

8 *La legislazione suntuaria: Secoli XIII–XVI. Emilia-Romagna*, ed. Maria Giuseppina Muzzarelli (Rome: Direzione generale per gli archivi, 2002), 54 and 449. On fashion bans, see also Diane Owen Hughes, "Sumptuary Law and Social Relations in Renaissance Italy," *Disputes and Settlements: Law and Human Relations in the West*, ed. John Bossy (Cambridge, UK, and New York: Cambridge University Press, 1983), 69–99: 73. On Venice, Luca Molà, "Leggi suntuarie in Veneto," in *Disciplinare il lusso: La legislazione suntuaria in Italia e in Europa tra medioevo ed età moderna*, ed. Maria Giuseppina Muzzarelli and Antonella Campanini (Rome: Carocci, 2003), 47–57: 51.

9 Muzzarelli, *La legislazione suntuaria*, 49.

10 "Et quod dominus Capitaneus et dominus Potestas et Executor ordinamentorum iustitie teneatur de predictis ornamentis et vestibus inquirere et per suam familiam scruptari facere in ecclesiis et predicationibus et aliis locis decentibus, et [ad] apothecas et domos aurificum et sartorum et ipsos condempnare, ut supra dictum est, et ea banniri fatiat quolibet mense semel." *Statuti della*

	Repubblica Fiorentina, ed. Romolo Caggese, 2 vols. (Florence: Olschki, 1999), 1:206–7 (book 5, rubric 13).
11	*Draghi rossi e querce azzurre: Elenchi descrittivi di abiti di lusso (Firenze 1343–1345)*, ed. Laurence Gérard-Marchant (Florence: SISMEL, 2013), 421.
12	Giordano da Pisa, *Quaresimale fiorentino: 1305–1306*, ed. Carlo Delcorno (Florence: Sansoni, 1974), 13. See also Lina Bolzoni, *The Web of Images: Vernacular Preaching from Its Origins to St. Bernardino da Siena* (Aldershot: Ashgate, 2004), 19.
13	Gilles Gérard Meersseman, *Dossier de l'ordre de la pénitence au XIIIe siécle* (Fribourg: Editions universitaires, 1961), 7.
14	Giancarlo Angelozzi, *Le confraternite laicali: Un'esperienza cristiana tra medioevo e età moderna* (Brescia: Queriniana, 1978), 79–85.
15	The rule of the Cistercians, the *Instituta generalis capituli apud cistercium*, states that all monasteries should follow the same principle: "Ut autem inter abbatias unitas indissolubilis perpetuo perseveret, stabilitum est primo quidem ut ab omnibus regulat beati benedicti uno modo intelligatur, uno modo teneatur; de hinc ut hiidem libri quantum dumtaxat ad divinum officium pernitent, idem victus, idem vestitus, idem denique per omnia mores inveniantur." *Le origini cistercensi: Documenti*, ed. Claudio Stercal and Milvia Fioroni (Milan: Jaca Book, 2004), 157–251: 168–69. For the standardization of pillows, food, and clothing, see norms 39 and 58 at pp. 204–5 and 218–19. See also norm 12 at pp. 52–53.
16	Aristotle, *Ethica Nicomachea: Translationis antiquioris quae supersunt: A. liber I sive 'Ethica noua,'* tr. Burgundio da Pisa, ed. René Antoine Gauthier (Leiden: Brill, and Bruxelles: Desclée de Brouwer, 1972).
17	Nicoletta Sarti, *Inter vicinos praesumitur aemulatio: Le dinamiche dei rapporti di vicinato nell'esperienza del diritto comune* (Milan: Giuffrè, 2003), 113–20. Sarti explains that Bolognese glossators such as Pilio of Medicina referred to *aemulatio* to forbid activities whose primary goal was competition for competition's sake.
18	"Si enim et idem est uni et civitati, maiusque et perfeccius quod civitatis videtur et suscipere et salvare. Amabile quidem enim et uni soli, melius vero et divinius genti et civitatibus." Aristotle, *Ethica Nicomachea*, tr. Robert Grosseteste, ed. René Antoine Gauthier (Leiden: Brill, and Bruxelles: Desclée de Brouwer, 1973), 375–588: 376. This is the translation that Robert Grosseteste produced in 1246–47, which was then revised by William of Moerbeke in the 1260s. For an analysis of the idea, see Matthew S. Kempshall, *The Common Good in Late Medieval Political Thought* (Oxford: Clarendon Press, 1999).
19	Thomas Aquinas, *Libros Politicorum Aristotelis expositio*, ed. Raimondo M. Spiazzi (Turin: Marietti, 1951), 6, note 11. On Grosseteste, see note above.
20	For instance, friar Giovanni della Marca (1393–1476) directly contributed to the communal statutes of Terni, Recanati, and Fermo. See Alberto Ghinato, "Apostolato religioso e sociale di San Giacomo della Marca in Terni," *Archivum franciscanum historicum* 49 (1965), 106–42 and 352–90. Friar Matteo

of Agrigento worked on the sumptuary laws for Palermo and Agrigento. See Paolo Evangelisti, "Identità e appartenenza comunitaria, pubblica prosperità e mercato nel linguaggio e nell'azione politica di due esponenti minoriti del Mediterraneo occidentale: Francesc Eiximenis e Matteo d'Agrigento (XIV–XV s.)," in *I Francescani e la politica (secc. XIII– XVII)*, ed. Giuliana Musotto, 2 vols. (Palermo: Musco, 2007), 1:387–413, in particular 393–94.

21 Florence, Biblioteca Nazionale, Conv. soppr. G4.936, f. 171r a-b. See Emilio Panella, "Dal bene comune al bene del comune: I trattati politici di Remigio dei Girolami nella Firenze dei bianchi-neri," *Memorie domenicane* 16 (1985), 1–122: 31.

22 *Statuti del Comune di Bologna dall'anno 1245 all'anno 1267*, ed. Luigi Frati, 3 vols. (Bologna: Tipografia regia, 1869–84), 1:132.

23 Tosi Brandi, "Il sarto," 35.

24 The event occurred in 1285. Salimbene de Adam, *Cronica*, 2:881. See also Dennis Romano, *Markets and Marketplaces in Medieval Italy, c. 1100 to c. 1400* (New Haven and London: Yale University Press 2015), 85–86.

25 Salimbene records that in 1204 one *sextarius* of spelt was sold for four *soldi* but that in 1227, a year of famine, its price grew by 50 percent. Salimbene de Adam, *Cronica*, 1:36 and 52.

26 Odd Langholm, *Economics in Medieval Schools: Wealth, Exchange, Value and Usury according to the Paris Theological Tradition, 1200–1350* (Leiden: Brill, 1992); Roberto Rusconi, *L'ordine dei peccati: La confessione fra medioevo ed età moderna* (Bologna: Il Mulino, 2002), 9–55.

27 Giacomo Todeschini, *I mercanti e il tempio: La società cristiana e il circolo virtuoso della ricchezza fra medioevo ed età moderna* (Bologna: Il Mulino, 2002), 349–70.

CHAPTER 21

1 "Perché sono le città debili? Perocché le cose vi vanno disordinate: se le cose andassero ordinate per tutti i cittadini, troppo sarebbe forte cittade; ma per le cose disordinate si guastano le cittadi, e nascono le discordie, ed è fatta debile la cittade." Giordano da Rivalto [Giordano da Pisa], *Prediche recitate in Firenze dal 1303 al 1306*, ed. Domenico Moreni, 2 vols. (Florence: Magheri, 1831), 2:83–84.

2 "Una agevole cittade che fosse ordinata, e bene in concordia, sarebbe si forte che non si potrebbe giammai vincere, e vincerebbono ogni altra gente." Giordano da Rivalto, *Prediche*, 2:84.

3 "[Il] bene ha virtù di dirizzare in queste opere . . . quali sono le cose bene diritte? Tutte quelle che sono ordinate . . . però che l'arte tu fai diritta e non ad inganno: la gonnella tua e la borsa e l'altre cose, tutte le fai diritte." Giordano da Pisa, *Quaresimale fiorentino: 1305–1306*, ed. Carlo Delcorno (Florence: Sansoni, 1974), 76–77.

4 "Quia si imaginemur corpus mathematicum existens actu oportet quod imaginemur ipsum sub aliqua forma, quia nihil est actu nisi per suam for-

mam . . . figura quae termino vel terminis comprehenditur." Thomas Aquinas, *Summa theologica* I, q. 73, art. 1. On Aquinas's definition of beauty as order, see Rosario Assunto, *La critica d'arte nel pensiero medievale* (Milan: Il Saggiatore, 1961), 243–57.

5 "Insieme si riposero a tai norme." Dante, *Inferno* 25.103. "Et quel poco, ch'i' sono / mi fa di loro una perpetua norma." Petrarch, *Rerum vulgarium fragmenta* 20.4.

6 Enrico Guidoni, *Il Duecento* (Rome and Bari: Laterza, 1989), 337. *Aequalitas* was both a moral and an aesthetic value. Bonaventura of Bagnoregio, the general minister of the Franciscan order, wrote that that beauty is "balance" (*aequalitas*) between all components. See Assunto, *Critica*, 239.

7 Nicola Ottokar, "Criteri d'ordine, di regolarità e d'organizzazione nell'urbanistica ed in genere nella vita fiorentina dei secoli XIII–XIV," in *Studi comunali e fiorentini* (Florence: La Nuova Italia, 1948), 143 and 146.

8 On the *nettezza* of Florentine streets, see Finotto, *La città chiusa: Storia delle teorie urbanistiche dal medioevo al Settecento* (Venice: Marsilio, 1992), 110. For the 1262 cleaning norm in Siena, see *Il constituto del Comune di Siena dell'anno 1262*, ed. Lodovico Zdekauer (Milan: Hoepli, 1897), 3.56.

9 "Potestas vel consules teneantur observare modum .XXV. brachiorum at .XXX. brachiorum de turribus factis et faciendis tempore sui consulatus vel potestarie, ita videlicet in Prato." *Statuti di Volterra I (1210/1224)*, ed. Enrico Fiumi (Siena: Tipografia nuova, 1951), 153–54. On the towers of the other cities, see Antonio Ivan Pini, *Città, comuni e corporazioni nel medioevo italiano* (Bologna: Clueb, 1986), 146.

10 The dimensional norm is in the 1228 Constitutions of the order. See Richard A. Sundt, "'Mediocres domos et humiles habeant fratres nostri': Dominican Legislation on Architecture and Architectural Decoration in the 13th Century," *Journal of the Society of Architectural Historians* 46, no. 4 (1987), 394–407.

11 Both cases are discussed in Finotti, *La città chiusa*, 102–3.

12 "Er tutte le finestre . . . reducere farò a la misura del passetto [that is, 2 *braccia*]. Et queste cose farò fare per tutto el mese di gennaio. Et per queste cose fare, elegerò tre buoni huomini e leali di penetentia, e' quali queste cose facciano fare. Et neuno possa tenere desco, predella over banca fuore de la casa ovver bottiga ne le dette strade, oltre el detto passetto." *Il Costituto del Comune di Siena volgarizzato nel 1309–1310*, ed. Mahmoud Salem Elsheikh, 4 vols. (Siena: Monte dei Paschi, 2002), 2:40.

13 See, for instance, Caroline Bruzelius, *Preaching, Building and Burying: Friars and the Medieval City* (New Haven: Yale University Press, 2014), 89–105 and 138–42.

14 Wolfgang Braunfels, *Mittelalterliche Stadtbaukunst in der Toskana* (Berlin: Mann, 1953), 86–130; Enrico Guidoni, "*Pulchritudo civitatis*: Statuti e fonti non statutarie a confronto," in *La bellezza della città: Stadtrecht und Stadtgestaltung im Italien des Mittelalters und der Renaissance*, ed. Michael Stolleis and Ruth Wolff (Tübingen: Max Niemeyer, 2004), 71–81.

NOTES TO PAGES 141-45 275

15 Andrew E. Benjamin, *Writing Art and Architecture* (Melbourne: re.press, 2010), 37–38.
16 Helen E. Longino, *The Fate of Knowledge* (Princeton, NJ: Princeton University Press, 2002).
17 "Nec domum in predictis terminis elevari usque ad triginta sex brachia interdici permittemus." Gabriella Rossetti, "Pisa e l'impero tra XI e XII secolo: Per una nuova edizione del diploma di Enrico IV ai Pisani," in *Nobiltà e chiese nel medioevo ed altri saggi*, ed. Cinzio Violante (Rome: Juvence, 1993), 165–66. See also Chris Wickham, *Sleepwalking into a New World: The Emergence of Italian City Communes in the Twelfth Century* (Princeton, NJ: Princeton University Press, 2015), 78–80.
18 This is inspired by Giorgio Agamben, *Stasis: La guerra civile come paradigma politico* (Turin: Bollati Boringhieri, 2015), 11.
19 Inspired by Marshall Berman, *All That Is Solid Melts Into Air: The Experience of Modernity* (New York: Penguin Books, 1982): 99.
20 Braunfels, *Stadtbaukunst*, 86–130.
21 Francesca Bocchi, *Il Duecento*, vol. 2 of *Atlante Storico di Bologna* (Bologna: Grafis, 1995), 89. The number of points was increased over time. In 1250 the heralds stopped at 32 points, in 1252 at 45, and in 1259 at 60. By 1267 the points were 87.
22 This is inspired by the book of my teacher, Marvin Trachtenberg, *Dominion of the Eye: Urbanism, Art, and Power in Early Modern Florence* (New York: Cambridge University Press, 2008), 258.
23 Bonvesin explains his goal in the prologue: "Veritate rerum cum ingenti diligentia et multo labore deliberate investigata, stimulo mediocri ad intelligendum legentibus habili, nullius pretii interventu, nullius inductione, nullius expectationis temporalis premii cause, sed potius inspiratione divina composui; eoque pacto ut, lecta et intellecta commendationis Mediolani veritate purissima." Bonvesin de la Riva, *De magnalibus Mediolani*, ed. Paolo Chiesa (Milan: Scheiwiller, 1997), 8–10.
24 Bonvesin, *De magnalibus Mediolani*, 156. Bonvesin repeats this observation also at p. 164.

CHAPTER 22

1 On the Genoese provenance of the manuscript, see Laura Dal Prà, "Bernardo di Chiaravalle: Realtà e interpretazione nell'arte italiana," in *Bernardo di Chiaravalle nell'arte italiana: Dal XIV al XVIII secolo* (Milan: Electa, 1990), 29–88: 77; Fabrizio Cigni, "Manuscrits en français, italien et latin entre la Toscane et la Ligurie à la fin du XIIIème siècle: Implications codicologiques, linguistiques, et évolution des genres narratifs," in *Medieval Multilingualism: The Francophone World and Its Neighbours*, ed. Christopher Kleinhenz and Keith Busby (Turnhoult, Belgium: Brepols, 2011), 187–217: 196. On its Franciscan

origins, see Amy Neff, "Byzantium Westernized, Byzantium Marginalized: Two Icons in the *Supplicationes variae*," *Gesta* 38, no. 1 (1999), 81–102: 96–97. The manuscript may have been in Florence by 1411, as parts of it were copied in the commonplace book of a Florentine merchant, Zanobi di Paolo Perini (Florence, Biblioteca Nazionale, Magl. VII 375, f. 12r). One of its prayers was also copied by Marsilio Ficino in the 1470s. See Paul Oskar Kristeller, "Marsilio Ficino as a Man of Letters and the Glosses Attributed to Him in the Caetani Codex of Dante," *Renaissance Quarterly* 36 (1983), 1–34: 21.

2   "Haec linea bis sexties ducta mensuram dominici corporis monstrat." Florence: Biblioteca Laurenziana, Plut. 25.3, 15v.

3   Caroline Bynum, *Christian Materiality: An Essay on Religion in Late Medieval Europe* (New York: Zone Books, 2011); Aden Kumler, "Manufacturing the Sacred in the Middle Ages: The Eucharist and Other Medieval Works of Ars," *English Language Notes* 53, no. 2 (2015), 9–44.

4   Michele Bacci, *The Many Faces of Christ: Portraying the Holy in the East and West, 300 to 1300* (London: Reaktion, 2014), 36.

5   "Sumpta est autem de Constantinopoli ex aurea cruce f[a]c[t]a ad formam corpor[is] Christi." Florence: Biblioteca Laurenziana, Plut. 25.3, 15v. See also Amy Neff, "'Palma dabit palmam': Franciscan Themes in a Devotional Manuscript," *Journal of the Warburg and Courtauld Institutes* 65 (2002), 22–66: 58–61. Because of this quote, Neff reads the *mensura Christi* as an abbreviation of the cross.

6   George P. Majeska, "Notes on the Skeuophylakion of St Sophia," *Vizantijskij vremennik* 55 (1998), 212–25. The *Diegesis* was translated into Latin as the "Historia et descriptionis templi Sanctae Sophiae," in *Patrologia Graeca* 122 (1889), cols. 1289–1316, from which I will be quoting.

7   "Veneranda autem illa crux, quae stat hodie in sceuophylacio, est mensura stature Jesu Christi Domini nostri, quam viri fideles et magni momenti ad ipsum Christum in Jerusalem accuratissime sunt dimensi. Eaque de causa argento illam et lapillis circumdederunt auroque obduxerunt: et ad hanc usque diem morbos curat, agritudines amovet et daemonia pellit." *Historia templi Sanctae Sophiae*, cols. 1305–6.

8   The Icelandic author is Níkulás Bergsson, abbot of the monastery of Þverá in Eyjafjörður, who mentioned the "silver cross of the same height of Christ" (*crux argentea aequalis staturae Christi*) in the chronicle of his pilgrimage to the Holy Land, 1151–54. See Xavier Barbier de Montault, "Les mesures, poids, fac-simile et empreintes de dévotion," in *Oeuvres completes*, 16 vols. (Paris: Vives, 1889–1902), 7:342; Michele Bacci, "Vera croce, vero ritratto e vera misura: Sugli archetipi bizantini dei culti cristologici del medioevo occidentale," in *Byzance et les reliques du Christ*, ed. Jannic Durand and Bernard Flusin (Paris: Association des amis du Centre d'histoire et civilisation de Byzance, 2004), 223–38.

9   Two examples should suffice. The first is Ælfwine's eleventh-century prayer-book of New Minster, Winchester, that is London: British Library, Ms Cott.

Titus D XXVI, fol. 3r. It links the height to the cross, describing it as made of precious wood: "De mensiu [mensura] saluatoris. haec figura dedecies multiplicata perficit mensuram domini nostri ihesu xpi corporis et assumpta a ligno pretioso dominice." The second is a twelfth-century version of Paterius's commentary on the scriptures. Cambrai, Bibliothèque municipale: Ms 508 (467), f. 1r. There the cross is described in terms that are very similar to those employed in the Genoese manuscript: "Linea a Constantinopoli ex aurea cruce sumpta que si sequieties fuerit ducta monstrat longitudinem corporis Christi." I am grateful to Hanna Vorholt for signaling this manuscript to me.

10 "Extra sanctuarium minus erecta est crux mensuralis, quae scilicet staturam Christi secundum carnem indicat." Barbier de Montault, "Les mesures," 343. See also George P. Majeska, "Russian Pilgrims in Constantinople," *Dumbarton Oaks Paper* 56 (2002), 93–108.

11 Cambridge, Trinity College, Ms B 1.40 (453). After copying the text, the scribe forgot to reproduce the segment.

12 Besides the manuscripts that I have already mentioned, other examples are Rouen: Bibliothèque municipale: Ms U 52 (1127), 157v; and Dijon, Bibliothèque municipale: Ms 448, 82v. The latter dates to the fourteenth century. See *Catalogue général des manuscrits des bibliothèques publiques de France*, vol. 5: Dijon (Paris: Plon, 1889), 106–9. A chapel of Valencia's cathedral is also dedicated to the "Santa Longitut." See Gabriel Llompart Moragues, "Longitudo Christi Salvatoris: Una aportación al conocimiento de la piedad popular catalana medieval," *Analecta Sacra Tarraconensia* 40 (1967), 93–115: 94.

13 "Questa e la misura del glorioso corpo del nostro Signore Geso Cristo cioe dele sedici parti luna dela lunghezza del suo santissimo corpo." Florence, Biblioteca Riccardiana: Ms 1294, c. 103. See Salomone Morpurgo, *I manoscritti della R. Biblioteca Riccardiana*, 2 vols. (Rome: Ministero dell'educazione nazionale, 1900), 1:356–63; Gustavo Uzielli, *Le misure lineari medioevali e l'effige di Cristo* (Florence: Seeber, 1899), 9.

14 Munich: Bayerische Staatsbibliothek, Clm. 536, 85r. See Elenora Cianci, "De Furtu. Il più antico incantesimo di area tedesca per riconoscere il ladro: Eredità e contesto culturale," *Itinerari* 4 (2014), 207–18.

15 "Questa e la misura e la lunghezza del chorpo del nostro Signiore Giesu Cristo che fu XVI volte lungho quanto e questa misura qui disopra facto come si vede cioe sedici volte. Al nome diddio amen Innomine patri et filio et spirito santo questa misura fu rechata in Chostantinopoli innuna croce doro." Florence, Biblioteca Riccardiana: Ms 1763, c. 56v.

16 "Chi questa misura adosso portara o in chasa la terra o che ogni di laveggha non potea in quel di morire di morte subitana ne in fuocho ne innacqua ne di demonio, tempesta non gli potea nuocere et se niuna femmina che fosse gravida e portasse questa misura adosso non potea morire in quello parto confidandosi senpre nel nome del nostro Signiore Yhu Xpo." Florence, Biblioteca Riccardiana: Ms 1763, c. 56v–57r.

17 | Saint Augustine, *De diversis quaestionibus LXXXIII* 6.
18 | Florence, Biblioteca Riccardiana: Ms 1763, c. 57r.
19 | I give three examples. Hamburg, Staatsbibliothek: Fragm. Lat. 361; London, British Library, Harley roll T.11; and London, British Library: Harley Charter 43 A.14. The London rolls represent Christ's height in the shape of a cross, a reference to the holy relic in Constantinople.
20 | "Itinerarium of Pseudo-Antoninus Placentinus (the Pilgrim of Piacenza)," ed. Paul Geyer, in *Itineraria et alia geographica*, Corpus Christianorum series latina 175 (Turnhout: Brepols, 1965), 129–53: 140–41. Gregory of Tours also mentions the practice. See John Wilkinson, *Jerusalem Pilgrims: Before the Crusades* (Warminster: Aris & Phillips, 1977), 84, n. 33, and 175. See also William J. Diebold, "Medium as Message in Carolingian Writing about Art," *Word & Image* 22, no. 3 (2006), 196–201. On sacred measurements in the Middle Ages, see Xavier Barbier de Montault, "Les mesures de dévotion," *Revue de l'art chrétien* 15 (1881), 360–416; Adolf Jacoby, "Heilige Längenmasse: Eine Untersuchung zur Geschichte der Amulette," *Schweizerisches Archiv für Volkskunde* 29 (1929), 1–17 and 181–216; Elisée Legros, "La mesure de Jésus et autres saintes mesures," *Enquêtes du Musée de la vie wallonne* 60 (1962), 313–17.
21 | Bacci, *The Many Faces of Christ*, 61–63 and 141–43.
22 | Belardo of Ascoli (1112–65) thought the chapel of the Ascension enshrined the impression of Christ's left foot, while others claimed they could discern both. See Denys Pringle, *The Churches of the Crusader Kingdom of Jerusalem: A Corpus*, 4 vols. (Cambridge: Cambridge University Press, 1993–2009), 3:73.
23 | As a property that can be shared by many things, a measurement is less a thing than a difference between things. Maurice Merleau-Ponty, "Le chiasme," in *Le visibile et l'invisible* (Paris: Gallimard, 1964), 176–82: 175.
24 | "Viri fideles et magni momenti ad ipsum Christum in Jerusalem accuratissime sunt dimensi." *Historia templi Sanctae Sophiae*, cols. 1305–6.
25 | On relics, see Georges Didi-Huberman, *La ressemblance par contact: Archéologie, anachronisme et modernité de l'empreinte* (Paris: Editions de Minuit, 2008), 76–91; Jean-Claude Schmitt, "Les reliques et les images," in *Les reliques: Objects, cultes, symboles*, ed. Edina Bozoky and Anne-Marie Helvetius (Turnhout, Belgium: Brepols, 1999), 145–59: 150.
26 | "Nota quod ubi non mutatur forma rei, non dicitur mutari res." In Ernst H. Kantorowicz, *The King's Two Bodies: A Study in Mediaeval Political Theology* (Princeton, NJ: Princeton University Press, 1957), 295.
27 | Accursio's titanic commentary on the *Corpus iris civilis* (more than 100,000 glosses!) became the reference text for any legal official. See Mario Ascheri, *I diritti del medioevo italiano: Secoli XI–XV* (Rome: Carocci, 2000): 27–51 and 212–15.
28 | "Tertium est in navi, quia licet particularatim fuit refecta, licet omnis tabula nova fuerit, nihilominus est eadem navis." In Kantorowicz, *King's Two Bodies*, 295. The gloss refers to *Digest* 5.1.76: "Cuius rei species eadem consisteret, rem quoque eandem esse existimari."

29 | The metaphor has attracted a lot of interest among structuralists and art historians. See Roland Barthes, *Roland Barthes*, trans. Richard Howard (New York: Hill and Wang, 1977), 46; Rosalind E. Krauss, *The Originality of the Avant-Garde and Other Modernist Myths* (Boston: MIT Press, 1985), 2–4.

30 | "Agens autem agit per suam formam, et non per suam materiam: omne igitur patiens recipit formam sine materia . . . et per hoc modum, sensus recipit formam sine materia, quia alterius modi esse habet forma in sensu, et in re sensibili. Nam in re sensibili habet esse naturale, in sensu autem habet esse intentionale et spirituale." Thomas Aquinas, *Sentencia libri de anima*, 2.24. See also John F. Wippel, *Metaphysical Themes in Thomas Aquinas* (Washington: Catholic University of America Press, 2007), 114.

31 | "Et ponitur conveniens exemplum de sigillo et cera. Non enim eadem dispositio est cerae ad imaginem, quae erat in ferro et auro. Et ideo subiungit, quod cera accipt signum idest imaginem sive figuram auream aut aeneam, sed non inquantum est aurum aut aes." Aquinas, *De anima* 2.24 n. 4.

32 | The convent printed them around January 1477. See Melissa Conway, *The Diario of the Printing Press of San Jacopo di Ripoli, 1476–1484* (Florence: Olschki, 1999), 96, entry 1476, n. 6. The dating of the diary reflects the Florentine calendar, with the new year beginning March 25. It is unclear if the late-fifteenth-century *Orazione della misura di Cristo* in New York's Pierpont Morgan Library (PML 16529) was a product of the San Jacopo printing press. See Curt F. Bühler, "An Orazione della misura di Cristo," *La Bibliofilía*, 39, no. 10–11 (1937), 430–33; Don C. Skemer, *Binding Words: Textual Amulets in the Middle Ages* (University Park: Pennsylvania State University Press, 2006), 228–30. On a similar devotion, for the print of Christ's wound, see David S. Areford, *The Viewer and the Printed Image in Late Medieval Europe* (Burlington, VT: Ashgate, 2010), 253–58.

33 | Giuseppe Baldacchini, Francesco Baldacchini, Lucia Casarosa, and Giovanna Falcone, "Crux mensuralis of Grotteferrata and Shroud of Turin," *Proceedings of the International Workshop on the Scientific Approach to the Acheiropoetos Images* (2010), http://www.acheiropoietos.info/proceedings/BaldacchiniWeb.pdf. Many memorial stones and relics were constructed in the seventeenth century. See, for instance, Johanna Heideman, "The Roman Footprints of the Archangel Michael: The Lost Shrine of S. Maria in Aracoeli and the Petition of Fioravante Martinelli," *Mededelingen van het Nederlands Instituut te Rome* 47 (1987), 147–56.

34 | Andrea Palladio, *Descritione de le chiese, stationi, indulgenze & reliquie de Corpi Sancti, che sonno in la Città de Roma* (Rome: Vincenzo Lucrino, 1554), [10].

35 | Andrea Fulvio, *Antiquitates urbis* (Rome, 1527), f. 23v. See also Carlo Ginzburg, *Indagini su Pietro* (Turin: Einaudi, 1981), 93–94, n. 35.

36 | "In Cappella . . . est lapis quadratus quatuor columnis marmoreis, sub cujus . . altitudo Domini Nostri Jesu Christi, antequam crucifigeretur, staturam corporisque magnitudinem denotat, supra quem numerati fuerunt triginta denari a Judaeis Judae traditori, ac etiam a Judaeis super Christi vestem jactae

sortes." *Tabula magna continens elenchum reliquiarum et indulgentiarum sacrosanctae ecclesiae Lateranensis* in Philippe Laurer, *Le Palais de Latran: Étude historique et archéologique* (Paris: Leroux, 1911), 296–301: 298. The lacunae are Laurier's.

37   Manfredo Tafuri, *Interpreting the Renaissance: Princes, Cities, Architects* (New Haven: Yale University Press, 2006), 24–27.

38   Sible de Blaauw, "Translations of the Sacred City between Jerusalem and Rome," in *The Imagined and Real Jerusalem in Art and Architecture*, ed. Jeroen Goudeau, Mariëtte Verhoeven, and Wouter Weijers (Leiden: Brill, 2014), 136–65.

39   For the demolition of the council hall, see Carlo Pietrangeli, *Il Palazzo apostolico lateranense* (Fiesole: Convivio, 1991), 195. It is unclear what happened to the canopy following the destruction. In 1644, Fioravante Martinelli described it standing near the Lateran's oratory of Saint Thomas. See Fioravante Martinelli, *Roma ricercata nel suo sito* (Rome: Bernardino Tani, 1644), 162–63. In 1688, Maximilien Misson saw it in the Lateran courtyard: "On y montre une mesure de la hauteur de Jesus Christ à la quelle dit-on, personne ne s'est jamais trouvé tout-à-fait égal." Maximilien Misson, *Nouveau Voyage d'Italie*, 3 vols. (The Hague: Van Bulderen, 1698, 3rd ed.), 2:177–78. The fact that Misson mentions the thickness ("six pieds d'Angleterre et l'epaisseur d'un Ecu") indicates that he thought that Christ's height was reproduced by the top slab, not the columns. Misson's account invalidates Barbier de Montault's hypothesis that the ciborium was moved to the cloister under Benedict XIV (1740–56). Barbier de Montault, "Les mesures," 340.

40   A map in Giovanni Severano's *Memorie sacre delle sette chiese di Roma* (Rome: Mascardi, 1630) also shows the four columns in the apse of the council hall.

41   Herbert L. Kessler and Johanna Zacharias, *Rome 1300: On the Path of the Pilgrim* (New Haven: Yale University Press, 2000), 34; Erik Thunø, *The Apse Mosaic in Early Medieval Rome* (Cambridge: Cambridge University Press, 2015), 52.

42   "Benedictione per pontificem, ut supra, data, ascendit per basilicam predictam ad palatium lateranense et quum pervenisset ad primam aulam magnam, que aula concilii nuncupatur, positum fuit faldistorium ante gradus lapidis, super quattuor columnas positi, qui mensura Christi appellatur." Johann Burchard, *Liber notarum ab anno MCCCCLXXXIII usque ad annum MDVI*, ed. Enrico Celani, 3 vols. (Città di Castello: Lapi, 1906–42), 1:83 (September 12, 1484).

43   "Quum D. N. fuerit apud Sanctum Joannem lateranensem in aula concilii, prior presbyteorum cardinalium, ferulam in manu habens, ordinat versus lapidem qui mensura Christi dicitur, subdiaconos, auditores, advocatos et secretarios hinc et inde pariter, ita ut digniores videlicet subdiaconi stent versus dictum lapidem, uibus sic ordinatis, idemprior stans sine mitra in medio inter primos subdiaconos, alta voce, quasi legendo, dicit: Exaudi Christe." Burchard, *Liber notarum*, 1:66 (August 26, 1484).

44   The similarities must have appeared even stronger as some people identified the slab at the top of the canopy, rather than the columns, as the object

that reproduced Christ's height. In his *De caerimoniis curiae romanae* (1488), for instance, Agostino Patrizi Piccolomini writes: "Et cum perventu[m] fuerit in prima aula, quem aula concilii appellatur, sedet Pontifex in sede sibi parata apud tabulam lapideam in capite aulae, que dicitur mensura Christi." *Rituum ecclesiasticorum sacrarum cerimoniarum SS. Romanae Ecclesiae* (Venice, 1516), xviii. For the success of this printed edition, which was reissued numerous times, see *Habemus papam: Le elezioni pontificie da San Pietro a Benedetto XVI*, ed. Francesco Buranelli (Rome: De Luca, 2006), 13–16. The confusion about what element reproduces Christ's height continues today. See Kessler and Zacharias, *Rome 1300*, 37. The ambiguity, as it is, was also an effect of abbreviations, as both "mensura" and "mensa" (table) are rendered as "mens" in many plans of the council hall.

45 | "Une salle en laquelle y a ung grand marbre, sur IIII colonnes, quy sont du marbre, et samble assez estre ung autel; mais on dict que c'est la haulteur de Nostre Seigneur Jesucrist, et va-on dessoubz, en allant à la procession." Georges Lengherent, "Voyage archéologique au XVe siècle," *Annales archéologiques* 22 (1862), 91.

CHAPTER 23

1 | Alice-Mary Talbot, "Pilgrimage in the Eastern Mediterranean between the 7th and 15th centuries," in *Egeria: Monuments of Faith in the Medieval Mediterranean*, ed. Ersi Brouskari, Vasileios Skoulas, and Maria Kazakou (Athens: Hellenic Ministry of Culture, 2008), 37–46: 44.

2 | "Quidam Floretinae diocesis, tibiis contractus, ita quod nullo modo ambulare poterat, audito B. Ambrosij obitu & fama miraculorum, Senas se deferri fecit, & ad sepulcrum sancti viri veniens, maxima cum deuotione & lacrymis orans, vovit imaginem ad sui corporis mensuram ad eius tumulum apponere, si eius meritis sanitatem obtinere posset." *Miracula Beati Ambrosii Senesis*, in *Acta sanctorum* 9 (Paris and Rome: Palmé, 1865), 165.

3 | Hippolito of Florence, "Miracula intra Triennium ab obitu patrata," *Acta sanctorum* 5 (Antwerp: 1685), 403–7.

4 | "Virga quae orantis continet longitudinem." Quoted in Louis Gougaud, "La prière dite de Charlemagne et les pièces apocryphes apparentées," *Revue d'histoire ecclésiastique* 20 (1924), 211–38: 225–27. See also Anton E. Schönbach, "Zeugnisse zur deutschen Volkskunde des Mittelalters," *Zeitschrift des Vereins für Volkskunde* 12 (1902), 1–14. For the dating of Ebendorfer's treatise, see Alphons Lhotsky, *Thomas Ebendorfer: Ein österreichischer Geschichtsschreiber, Theologe und Diplomat des 15. Jahrhunderts*, in *Monumenta Germaniae Historica: Scriptores*, vol. 15 (Stuttgart: Hiersemann Verlag, 1957), 81. On hs fame, see the exhibition catalogue *Thomas Ebendorfer von Haselbach (1388–1464): Gelehrter, Diplomat, Pfarrer von Perchtoldsdorf*, ed. Johannes Seidl (Perchtoldsdorf: Marktgemeinde, 1988).

5   Ernesto Piacentini, "Santa Rosa da Viterbo: Culto liturgico e popolare," in *Santa Rosa: Tradizione e culto*, ed. Silvio Cappelli (Rome: Vecchiarelli, 1999), 71–98; Giuseppe Abate, "S. Rosa da Viterbo, terziaria francescana, 1233–1251: Fonti storiche della vita e loro revisione critica," *Miscellanea francescana* 52 (1952), 113–278.

6   The surviving ribbons are listed in Eleonora Rava and Filippo Sedda, "Schede," in *Dalla reliquia alle reliquie: La santità di Rosa visibile e tangibile* (Viterbo: Centro studi Santa Rosa da Viterbo, 2012), 22–26. I would like to thank Eleonora Rava for granting me access to the archives of the Monastery of Santa Rosa in Viterbo.

7   "Chaintures, qui sont touchiés à lad[ette] sainte, pour les raporter et donner aux femmes enchaintes." Georges Lengherent, "Voyage archéologique au XVe siècle," *Annales archéologiques* 22 (1862), 90. Lengherent visited Viterbo in 1485. Saint Rose's popularity was high in the fifteenth century. See Simonetta Valtieri and Enzo Bentivoglio, *Il giubileo del 1450 e il culto di Santa Rosa* (Rome: Ginevra Bentivoglio, 2016), 31–45.

8   "Deuota quædam & Catholica mulier, Margarita nomine, filiam habuit quatuor annorum a natiuitate contractam, quod numquam ambulare potuit: hæc magna cum deuotione & fide, ad sepulcrum B. Ambrosij filiam detulit, eamque dictum sepulcrum tangere fecit: quo facto, domum regressa, sequenti die infantula per se surrexit, & per domum ambulare cœpit: sicque breui temporis spatio sancti viri Ambrosij meritis incolumis perfecte euasit." *Miracula Ambrosii Senesis*, col. 196b.

9   Marc Bloch, *Les rois thaumaturges* (Strasbourg and Paris: Istra, 1924), 115–16.

10   The pilgrim is often identified with an anonymous traveler from Piacenza. See *Itineraria et alia geographica*, Corpus Christianorum series latina 175 (Turnhout: Brepols, 1965), 140–41. Christ's footprints were in Pilate's *pretorium*. See John Wilkinson, Joyce Hill, and William F. Ryan, *Jerusalem Pilgrimage 1099–1185* (London: Hakluyt Society, 1988), 46–47.

11   Saint Colombanus, the famed founder of the monastery at Bobbio, left an imprint of his hand on a rock in an Apennine cave and many pilgrims flocked to gauge it with their own hands. See Janet Bord, *Footprints in Stone* (Loughborough: Heart of Albion, 2004), 66.

12   Measuring is a form of touch, but one that does not reproduce sensations such as cold and heat. See Thomas Aquinas, *Sentencia libri de anima* 2.11.

13   "Apud Bononiam scholaris quidam nomine Nicolaus gravi renum ac genuum dolore adeo est vexatus, quod de lecto surgere non valebat . . . Deo igitur et beato Dominico se devovens cum se filo, de quo fienda erat candela, totum mensus est in longum, coepit etiam corpus, collum et pectus cingere. Cum tandem genu filo ambiente cinxisset, invocato ad quamlibet mensurationem nomine Jesu et beati Dominici continuo se alleviatum sentiens exclamavit: ego sum liberatus." Jacobus de Voragine, *Legenda Aurea*, ed. Johann Georg Theodor Grässe (Leipzig: Arnold, 1850), 480–81.

14 | Christopher S. Wood, *Forgery, Replica, Fiction: Temporalities of German Renaissance Art* (Chicago: University of Chicago Press, 2008), 40.
15 | Richard Krautheimer, "Introduction to an 'Iconography of Mediaeval Architecture,'" *Journal of the Warburg and Courtauld Institutes* 5 (1942), 1–33: 5–7 and 13–16.
16 | The distance is around 42 meters, which is close to the interval of 28 *dexteri* (41.6 meters) between the corresponding sites in Jerusalem as reported by the *Commemoratorium de casis Dei*, a description of the Holy Land commissioned by Charlemagne. See Robert G. Ousterhout, "The Church of Santo Stefano: A 'Jersualem' in Bologna," *Gesta* 20 (1981), 311–21.
17 | Bologna: Biblioteca universitaria, Ms. 1473. The *Vita* is transcribed in *San Petronio vescovo di Bologna nella storia e nella leggenda*, ed. Francisco Lanzoni (Rome: Libreria pontificia di Federico Pustet, 1907), 219–40.
18 | "Ibi [in Jerusalem] enim per aliqua dierum curricula sibi fuerat per totam Iudeam colligere vectigalia." Lanzoni, ed., *San Petronio*, 231.
19 | For all three passages: "Ligneam crucem, quae in longitudine at latitudine undique per totum facta fuerat . . . et mensus est cum esset Ierusalem quot calamis Golgotha distaret a monte Oliveti: eodem modo ibi et mensus est . . . Fecit quoque tipice ingenti cura piscinam, secundum quod calamo mensus fuerat in Ierusalem, instar natatoriae Siloe." Lanzoni, ed., *San Petronio*, 232–33.
20 | "Ortus. . . quidem est locus, qui quamquam parvus esse videatur, tamen ab agricolis gratiosior, ac carios habetur omnibus arvis. Est quidem consitus arboribus omnium generum. Illic sunt poma ad commendum acceptabilia; ibi est olera per totius anni circulum; illic nascuntur erbe salutifere, spirantes odores suavissimos, croceos quoque et purpureos flores producentes specie diversi coloris." Lanzoni, ed., *San Petronio*, 233.
21 | "Huiusmodi vero lucus magis omnibus diligitur et custoditur ab agricolis." Lanzoni, ed., *San Petronio*, 233.

CHAPTER 24

1 | "Et auras cautus humum longo signavit limite mensor." Ovid, *Metamorphoses* 1.135–36. Ovid returns to the theme of the boundary as an archaic achievement in Ovid, *Fasti* 2.639–42.
2 | Leonard Barkan, *The Gods Made Flesh: Metamorphosis and the Pursuit of Paganism* (New Haven: Yale Univestiy Press, 1986), 103–36; Robert Black, "Ovid in Medieval Italy," in *Ovid in the Middle Ages*, ed. James G. Clark, Frank T. Coulson, and Kathryn L. McKinley (Cambridge: Cambridge University Press, 2011), 123–42.
3 | William S. Anderson, "On the *Tegernsee* Ms of Ovid's Metamorphoses (Munich Clm 29007)," *California Studies in Classical Antiquity* 11 (1978) 1–19.
4 | Genesis 4:2–12.

5 | Isidore of Seville, *Etymologiarum libri* 17.1.
6 | Gina Fasoli, "Prestazioni in natura nell'ordinamento economico feudale," in *Economia naturale, economia monetaria*, ed. Ruggiero Romano and Ugo Tucci, vol. 6 of *Storia d'Italia* (Turin: Einaudi, 1983), 65–89; Liubov A. Kotelnikova, "Rendita in natura e rendita in denaro nell'Italia medievale," in *Economia naturale, economia monetaria*, 94–111: 101.
7 | In Roman times, the word "modus" meant both "measured quantity" and "defined area." See the glossary of Brian Campbell, *The Writings of the Roman Land Surveyors: Introduction, Text, Translation and Commentary* (London: Society for the Promotion of Roman Studies, 2000), "modus."
8 | Pier Silverio Leich, *Studi sulla proprietà fondiaria nel Medioevo*, 2 vols. (Verona: Forni, 1903), 1:70.
9 | "Et per unoquoquoe manso iugera duodecim legittimos." Cited in Bruno Andreolli, "Misurare la terra: Metrologie altomedievali," in *Uomo e spazio nell'alto medioevo*, 2 vols. (Spoleto: Centro italiano di studi sull'alto medioevo, 2003), 1:151–87, at 173.
10 | Philip James Jones, "An Italian Estate, 900–1200," *Economic History Review* 7, no. 1 (1954), 18–32: 26.
11 | Mario Lupo, *Codex diplomaticus Civitatis et Ecclesiae Bergomatis*, 2 vols. (Bergamo: Antoine, 1784–99), 1, column 1078.
12 | Florence: Archivio di Stato, Vescovato di Pistoia, September 29, 1282. This act concerns a field measured in Liutprand feet. The act is also in Pistoia, Archivio di Stato: San Iacopo 31, f. 39r. See Natale Rauty, "Appunti di metrologia pistoiese," in *Pistoia: Città e territorio nel medioevo* (Pistoia: Società pistoiese di storia patria, 2003), 177–226: 183.
13 | "Prima pecia de terra jacet ad locum ubi Ponticello dicitur, est ipsam terciam portionem, exinde per mensura justa perticas jugealis novem et tabulas decesim. Coheret tam ad ipsam porcionem, quamque ad super totum de mane et mons vias." Giuseppe Cappelletti, *Le chiese suffraganee dell'arcidiocesi di Milano*, vol. 12 of *Le chiese d'Italia* (Venice: Antonelli, 1857), 304.
14 | On Liutprand's reform, see Ninto Tamassia, "Pesi e misure dell'Italia medioevale," in *Studi in Onore di Biagio Brugi nel XXX anno del suo insegnamento* (Palermo: Gaipa, 1910), 423–27.
15 | "Qui cum equitaret per districtum Mediolani, homines illarum partium ad eius presentiam accedentes, conquerebantur ei, quod non habebant mensuram, et quod unus alterum in mensura concupiebat. Qui posuit pedem suum super quendam lapidem magnum et spaciosum, volens in eo signum fieri ad modum pedis sui; ad quod emere et vendere deberent. Sed Dei potentia signum pedis eius in ipso lapide fuit scultum et signatum, quod usque in presentem diem ibi aparet, et dicitur pes Liprandus; ad cuius mensuram usque in hodiernum diem vendunt et emunt." *Epitomae ex Pauli historia factae*, in *Scriptores rerum Langobardicarum et Italicarum secc. VI–IX*, ed. Georg Waitz, Monumenta Germaniae Historica (Hannover: Hahn, 1878), 193–97: 197.
16 | Apocalypse 10:2–11.

17 | Leich, *Studi sulla proprietà fondiaria*, 1:15–16.
18 | "Idest nominative terras arabiles iuris ipse domne Beatricis in loco qui vocabatur prato longo, a pertica *legitima* de pedibus duodecim a pede Liprandi mensurata." Cited in Carlo D'Arco, *Dell'economia politica del municipio di Mantova a' tempi in cui si reggeva a repubblica* (Mantua: Negretti, 1842), 344.
19 | Vito Fumagalli, *Le origini di una grande dinastia feudale: Adalberto-Atto di Canossa* (Tübingen: Max Niemeyer, 1971), 48–49; Giuseppe Sergi, "The Kingdom of Italy," in *The New Cambridge Medieval History*, ed. Timothy Reuter (Cambridge: Cambridge University Press, 2000), 346–71: 356–59; Tiziana Lazzari, "Aziende fortificate, castelli e pievi: Le basi patrimoniali dei poteri dei Canossa," in *Matilde e il tesoro dei Canossa, tra castelli e città*, ed. Arturo Calzona (Cinisello Balsamo: Silvana, 2008), 96–115: 97–99.

CHAPTER 25

1 | On Carolingian measurements, Harry A. Miskimin, "Two Reforms of Charlemagne? Weights and Measures in the Middle Ages," *Economic History Review* 20, no. 1 (1967), 35–52; Michael McCormick, *Charlemagne's Survey of the Holy Land: Wealth, Personnel, and Buildings of a Mediterranean Church between Antiquity and the Middle Ages* (Washington: Dumbarton Oaks Library Research and Collection, 2011), 95–116. Witold Kula's commentary is brief, but spot-on as usual. Witold Kula, *Measures and Men* (Princeton, NJ: Princeton University Press, 1986), 161–63.
2 | Giovanni Tabacco, *Egemonie sociali e strutture del potere nel medioevo italiano* (Turin: Einaudi, 1974), 189.
3 | Catherine Fischer Drew, "The Immunity in Carolingian Italy," *Speculum* 37, no. 2 (1962), 182–97: 192. See Renata Salvarani, "La struttura territoriale delle diocesi in Italia settentrionale in età carolingia: Il caso di Mantova," in *Alle origini del Romanico: Monasteri, edifici religiosi, committenza tra storia e archeologia. Italia settentrionale, secoli IX–X*, ed. Renata Salvarani, Giancarlo Andenna, and Gian Pietro Brogiolo (Brescia: CESIMB, 2005), 31–57, at 46–47. The situation continued up to the twelfth century; see Chris Wickham, *Courts and Conflict in Twelfth-Century Tuscany* (Oxford: Oxford University Press, 2003), 6–9.
4 | Tabacco, *Egemonie sociali*, 193.
5 | *Regesta chartarum pistoriensium: Alto medioevo* (Pistoia: Società pistoiese di storia patria, 1973), p. 87 n. 105. The *omina* standard is documented in the canonry of San Zeno in 1083, see Natale Rauty, "Appunti di metrologia pistoiese," in *Pistoia: Città e territorio nel medioevo* (Pistoia: Società pistoiese di storia patria, 2003), 177–226: 182 and 185.
6 | François Menant, *Campagnes lombardes au Moyen Age: L'économie et la société rurales dans la région de Bergame, de Crémone et de Brescia du Xe au XIIIe siècle* (Rome: Ecole française de Rome, 1993), 798; Bruno Andreolli, "Misurare la terra: Metrologie altomedievali," in *Uomo e spazio nell'alto medioevo*, 2 vols.

(Spoleto: Centro italiano di studi sull'alto medioevo, 2003), 1: 151–87, at 152–53.

7 Natale Rauty, "Intervento del Comune nel controllo delle misure a Pistoia (secoli XII–XV)," in *Pistoia: Città e territorio nel medioevo* (Pistoia: Società pistoiese di storia patria, 2003), 227–46: 232–33.

8 *Die Urkunden Otto des II*, ed. Theodor Sickel, in *Monumenta Germaniae Historica, Diplomata regum et imperatorum*, vol. 2, no. 1 (Hannover: 1888), 223.

9 *Die Urkunden Konrad I, Heinrich I und Otto I*, ed. Theodor Sickel, in *Monumenta Germaniae Historica, Diplomata regum et imperatorum*, vol. 1 (Hannover: 1879–84), 145 (February 15, 952) and 242 (April 20, 962).

10 The distinction is made clear by two diplomas (from 883 and 925, respectively): "In quibuscumque patriis ac provintiis regni nostri . . . esset" and "in quibuscumque patriis ac provinciis regni nostri." See *Mediae Latinitatis Lexicon Minus*, ed. Jan Frederick Niermeyer and Co van de Kieft, 2 vols. (Darmstadt: Wissenschaftliche Buchgesellschaft, 2002), 2:1007–8. See also Ernst H. Kantorowicz, *The King's Two Bodies: A Study in Medieval Political Theology* (Princeton, NJ: Princeton University Press, 1957), 232–26.

11 Carlo D'Arco, *Della economia politica ai tempi in cui si reggeva a republica* (Mantua: Negretti, 1842), 344.

12 For Treviso, see *Codice diplomatico padovano dal secolo sesto a tutto l'undecimo*, ed. Andrea Gloria (Venice: Deputazione veneta di storia patria, 1877), 259, doc. 232 (December 11, 1076). For Bergamo, see Angelo Mazzi, *Il sextarius Pergami: Saggio di ricerche metrologiche* (Bergamo: Pagnoncelli, 1877), 18.

13 Angelo Mazzi, "Nota metrologica: Il patronus misura milanese del sale," *Archivio storico lombardo* 28 (1901), 1–19. See also Angelo Mazzi, "Questioni metrologiche lombarde," *Archivio storico lombardo* 38 (1911), 5–64. Mainoni thinks that the creation of a new standard was a response to demographic growth. See Patrizia Mainoni, "Sperimentazioni fiscali e amministrative nell'Italia settentrionale," in *Pensiero e sperimentazioni istituzionali nella "Societa Christiana" (1046–1250)*, ed. Giancarlo Andenna (Milan: Vita e pensiero, 2007), 705–59: 749.

14 The civil war ended in 1044. See Tabacco, *Egemonie sociali*, 226–31.

15 "Unusquisque mansus est quadraginta et octo bubulcarum." Andreolli, "Misurare la terra," 173.

16 "Et perticarum sex in multis petiis infra coherenthis." Mario Lupo, *Codex diplomaticus Civitatis et Ecclesiae Bergomatis*, 2 vols. (Bergamo: Antoine, 1784–99), 2, cols. 1025–26. See also Pier Silverio Leich, *Studi sulla proprietà fondiaria nel medioevo*, 2 vols. (Verona: Forni, 1903), 71.

17 "Ad passum qui signatum est in columna marmorea atrii Sancti Mathei apostoli huius Salernitani archiepiscopii." *Codex diplomaticus Cavensis*, vol. 10 (1073–80), ed. Giovanni Vitolo and Simeone Leone (Cava dei Tirreni: Badia di Cava, 1990), doc. 7, 22–23.

18 "Mensurata cum pede, qui designatus est in petra juxta Portam S. Pancratii posita." Cited in Vincenzo Borghini, *Discorsi*, ed. Domenico Maria Manni,

NOTES TO PAGES 169–70   287

4 vols. (Milan: Società tipografica de' classici italiani, 1809) 1:123–24, note. The earliest evidence of the existence of the San Pancrazio foot dates to 1108: "ad iusta mensura ad pede[m] qui dicitur de porta sancti pancratji." *Le carte del monastero di S. Miniato al Monte (Secoli IX–XII)*, ed. Luciana Mosiici (Florence: Olschki, 1990), 205. Other mentions return in 1166 and 1192. See, respectively, *Le carte del monastero di S. Miniato al Monte*, ed. Mosiici, 299; and *Le carte del monastero di S. Maria in Firenze (Badia)*, ed. Luigi Schiaparelli and Anna Maria Enriques, 2 vols. (Rome: Istituto storico italiano per il medioevo, 1990), 2:150.

19 On these walls, Florence's fourth ring, see Giovanni Fanelli, *Firenze: Architettura e città*, 2 vols. (Florence: Vallecchi, 1973), 1:11–12; Franek Sznura, *L'espansione urbana di Firenze nel Dugento* (Florence: La Nuova Italia, 1975), 70–72.

20 Costantino Caciagli, "Considerazioni sul braccio volterrano," *Volterra* 17, no. 12 (1978), 4–7.

21 Florence, Archivio di Stato: Diplomatico Normali 83 (Luco di Mugello, June 4, 1094), f. 33r. The Florentine line is 51.80 centimeters, which does not match Liutprand's foot (around 43.65 centimeters), as is often repeated. See, for instance, *Santa Maria del Fiore: Teorie e storie dell'archeologia e del restauro nelle città delle fabbriche arnolfiane*, ed. Giuseppe Rocchi Coopmans de Yoldi (Florence: Alinea, 2006), 365–66. Also, I do not agree with Rocchi that the incision on the column of the baptistery is a *pietra di paragone*. Rocchi notices that its size (17 centimeters) is a third of the Florentine line (51.80 centimeters), and that its height (39 centimeters) is a third of the *doppio braccio a panno* (116.73 centimeters). I have never encountered one *pietra* that reproduces a multiplicand of a standard, that merges two lengths into one shape, or that is incised on a round surface.

22 The document that certifies the existence of the foot in 1200 is transcribed in Schiaparelli and Enriques eds, *Le carte di Badia*, 2:197. On Florence's fifth ring of walls, see Fanelli, *Florence*, 1:24–25. It is unclear what happened to the Florentine standard after 1200 but, as it is well known, Florence's thirteenth-century communal documentation was lost in a fire. The statutes of 1325 affirm that the standards were then kept in the *camera* of the commune and ought to be incised in at least two or three public locations, as their superintendents saw fit: "Et ne aliqua fraus vel deceptio fiat aut diminuito vel augmentum, stent et ponantur dicte mensure in duobus vel tribus loci ad minus in civitate Florentie, silicet in camera Communis Florentine." *Statuto del podestà del 1325*, ed. Romolo Caggese, vol. 2 of *Statuti della Repubblica Fiorentina* (Florence: Ariani, 1910–21), book 4, rubr. 31. On measurements in 1325, see Richard A. Goldthwaite, *The Building of Renaissance Florence: An Economic and Social History* (Baltimore: Johns Hopkins University Press, 1980), 208–9.

23 All the passages can be found in chapter 10, but for the readers' convenience I repeat those of Faenza ("ante ianuas maioris ecclesiae") and Bergamo ("super rezzios selciati eccl. S. Mariae Maioris"). Also, the *braccio* of Pavia was above

the door of the basilica of San Giovanni in Borgo in 1265. It was destroyed in 1818. See Cesare Prelini, "La torre maggiore della città di Pavia detta il campanile del duomo," *Almanacco della provincia di Pavia*, 1879, ix—xl: xii.

24 The *Historia Langobardorum* survives in more than one hundred manuscripts. See Laura Pani, "Aspetti della tradizione manoscritta dell'Historia Langobardorum," in *Paolo Diacono: Uno scrittore fra tradizione longobarda e rinnovamento carolingio*, ed. Paolo Chiesa (Udine: Forum, 2000), 367—412.

25 Jörg Busch, "Die Lombarden und die Langobarden," *Frühmittelalterliche Studien* 29 (1995), 289—311.

26 Sicard of Cremona, *Chronicon*, Patrologia Latina 213 (Paris: Migne, 1855), 488b.

27 Alberto Milioli, *Liber de temporibus et etatibus*, ed. Oswald Holder-Egger, in *Monumenta Germaniae Historica: Scriptores*, vol. 31 (Hannover: Hahn, 1903), 572—668: 619—20.

28 Massimo Giansante, "Milioli, Alberto," in *Dizionario biografico degli Italiani* 74 (Rome: Istituto dell'enciclopedia italiana: 2010), 497—99.

CHAPTER 26

1 Philip J. Jones, "An Italian Estate, 900—1200," *Economic History Review* 7 (1954), 18—32: 30. Jones mentions cases from the Lucca territory, among the best-documented in Italy. The monasteries of San Frediano di Tolli and San Quirico became insolvent in the late twelfth century. San Sesto went bankrupt in 1335.

2 Georges Duby, *Rural Economy and Country Life in the Medieval West*, tr. Cynthia Postan (Philadelphia: University of Pennsylvania Press, 1998), 33—42.

3 Boethius, *Liber de geometria*, Patrologia Latina 63 (Paris: Migne, 1847), cols. 1352—64; Columella, *De re rustica* 5.1—3; Cassiodorus, *Institutiones* 2, preface, 3. On early medieval treatises about measuring, see Menso Folkerts, "The Importance of the Pseudo-Boethian Geometria During the Middle Ages," in *Boethius and the Liberal Arts: A Collection of Essays*, ed. Michael Masi (Bern: Peter Lang, 1981), 187—209.

4 Evgeny A. Zaitsev, "The Meaning of Early Medieval Geometry," *Isis* 90, vol. 3 (1999): 522—53.

5 Cassiodorus, *Institutiones* 6.1—5.

6 "Geometriae disciplina primum ab Aegyptiis reperta dicitur, quod, inundante Nilo et omnium possessionibus limo obductis, initium terrae dividendae per lineas et mensuras nomen arti dedit. Quae deinde longius acumine sapientium profecta et maris et caeli et aeris spatia metiuntur. Nam provocati studio sic coeperunt post terrae dimensionem et caeli spatia quaerere: quanto intervallo luna a terris, a luna sol ipse distaret, et usque ad verticem caeli quanta se mensura distenderet, sicque intervalla ipsa caeli orbisque ambitum per numerum stadiorum ratione probabili distinxerunt." Isidore of Seville, *Etymologiarum libri* 3.10.1—3.

NOTES TO PAGES 173–76 289

7 Roman land surveyors's goals were "restituere finem" and "dirigerem limitem." See Stefano del Lungo, *La pratica agrimensoria nella tarda antichità e nell'alto medioevo* (Spoleto: Centro italiano di studi sull'alto medioevo, 2004), 24–25.

8 Anita F. Moskowitz, *The Sculpture of Andrea and Nino Pisano* (Cambridge, UK, and New York: Cambridge University Press, 1986), 31–48.

9 The best edition of the *Corpus* is Brian Campbell, *The Writings of the Roman Land Surveyors* (London: Society for the Promotion of Roman Studies, 2000). For a recension of the surviving manuscripts, see Lucio Toneatto, *Codices artis mensoriae: I manoscritti degli antichi opuscoli latini d'agrimensura*, 3 vols. (Spoleto: Centro italiano di studi sull'alto medioevo, 1994–95), 1:4–21.

10 For instance, Thulin divided *agrimensores* manuscripts into three groups, arguing that they each stemmed from one archetype. See Carl Thulin, *Die Handschriften des Corpus Agrimensorum Romanorum* (Berlin: Verlag der Königlichen Akademie der Wissenschaften, 1911), 3–5. By analyzing a larger number of manuscripts, however, Toneatto has revealed that numerous texts do not fit into Thulin's taxonomy. See Toneatto, *Codices artis mensoriae*, 1:51–53.

11 Brussels, Bibliothèque Royale Albert I, Ms 4499–503. Toneatto, *Codices artis mensoriae*, 1:315–24.

12 The Brussels manuscript contains Balbus's letter in full, whereas most medieval copies omit its first section. See Serafina Cuomo, "Sources," in *The Oxford Handbook of Engineering and Technology in the Classical World*, ed. John Peter Oleson (Oxford: Oxford University Press, 2008), 15–34: 26–27.

13 Toneatto, *Codices artis mensoriae* 1, 25–26.

14 Isidore of Seville, *Etymologiarum libri* 15.13–15.

15 Toneatto, *Codices artis mensoriae* 1, 20.

16 Michael D. Reeve, "Agrimensores," in *Texts and Transmission: A Survey of Latin Classics*, ed. Leighton Durham Reynolds (Oxford: Clarendon Press, 1983), 1–6. On the marginalization of the *Corpus*, Bethold L. Ullman, "Geometry in the Mediaeval Quadrivium," in *Studi di bibliografia e di storia in onore di Tammaro de Marinis*, ed. Giovanni Mardersteig, 4 vols. (Verona: Stamperia Valdonega, 1964), 4:263–85, in particular 266; Lon R. Shelby, "The Geometrical Knowledge of Mediaeval Master Masons," *Speculum* 47, no. 3 (1972), 395–421: 400 and 404; *Ad Quadratum: The Practical Application of Geometry in Medieval Architecture*, ed. Nancy Y. Wu (Aldershot: Ashgate, 2002), 32 and 220–22. For an alternative view, see Tadeusz Zagrodski, "L'influence de la tradition antique de la distribution de l'étendue sur le tracé des plans des villes créés au moyen âge," in *Mélanges offerts à René Crozet*, ed. Pierre Gallais and Yves François Riou, 2 vols. (Poitiers: Société d'études médiévales, 1966), 1:451–60; Marie-Thérèse Zenner, "Imagining a Building: Latin Euclid and Practical Geometry," in *Word, Image, Number: Communication in the Middle Ages*, ed. John J. Contreni and Santa Casciani (Florence: SISMEL, 2002), 219–46: 224.

17 Jens Høyrup, "'Oxford' and 'Cremona': On the Relation between Two Versions of al-Khwarizmi's Algebra," in *Actes du 3me Colloque Maghrébin sur*

18   *l'Histoire des Mathématiques Arabes* (Algiers: Association algérienne d'histoire des mathématiques, 1998), 159–78.
The idea was first put forward in Folkerts, "The Importance of Pseudo-Boethian Geometria," 190. See also Ullman, "Geometry in the Medieval Quadrivium," 263–85. My position, however, is closer to that in David S. Bachrach, *Warfare in Tenth-Century Germany* (Woodbridge: Boydell, 2012), 108 n. 26.

19   Jon N. Sutherland, *Liudprand of Cremona: Bishop, Diplomat, Historian. Studies of the Man and His Age* (Spoleto: Centro italiano di studi sull'alto medioevo, 1988), 77–100. The section on land charters is at pp. 104–6. The return of the same surveyors in numerous documents hints at an appointment of considerable length. See, for instance, *Documenti dei fondi cremonesi 759–1069*, ed. Ettore Falconi, vol. 1 of *Le carte cremonesi dei secoli VIII–XII* (Cremona: Biblioteca statale di Cremona, 1979), docs. 69–72.

20   Barcelona, Archivo de la Corona de Aragón: Ripoll 106, ff. 76r-89r. The Ripoll manuscript re-elaborates many of the passages of the *Corpus* by an author about whom we know little more than his name, Gismundus. The most recent study on the manuscript is Ricard Andreu Expósito, *Edició crítica, traducció i estudi de l'ars gromatica sive geometria gisemundi* (PhD dissertation, Universitat Autònoma de Barcelona, 2012). At p. 12, Expósito argues that Gismundus's text has Visigothic origins.

21   Both letters are in Fritz Weigle, *Die Briefsammlung Gerberts von Reims* (Weimar: Hermann Böhlaus Nachfolger, 1966), 31 and 158. On the former letter, see also Toneatto, *Codices artis mensoriae*, 1, 198. On the latter, Gerbert d'Aurillac, *Opera mathematica*, ed. Nikolaj Michajlovič Bubnov (Berlin: Friedländer und Sohn, 1899), 43–45.

22   Menso Folkerts and Alphonse J. E. M. Smeur, "A Treatise on the Squaring of the Circle by Franco of Liège, of about 1050," *Archives Internationales d'histoire des sciences* 26 (1976): 59–105 and 225–53, esp. 232 and 247. See also Paul Tannery, "La Géometrie au XIe siècle," in *Sciences exactes au Moyen Age*, vol. 5 of *Mémoires scientifiques* (Toulouse: Privat, 1922), 79–102.

23   On the network of cathedral schools in northern Europe, including information on the scholars that I have mentioned, see C. Stephen Jaeger, *The Envy of Angels: Cathedral Schools and Social Ideals in Medieval Europe, 950–1200* (Philadelphia: University of Pennsylvania Press, 1994), 53–75.

24   Boethius, *Liber de geometria*, cols. 1355–57.

25   Marco Petoletti, "Un nuovo manoscritto della biblioteca di Petrarca: Il più antico codice degli *agrimensores* (Wolfenbüttel, Aug. fol. 3623)," *Studi petrarcheschi* 24 (2011): 1–28. I am thankful to Giulia Perucchi for telling me about this article.

26   About this manuscript, produced around 820–40 and "rediscovered" by Poggio Bracciolini in the library of Fulda in 1411, see Mechthild Palm, "Zum karolingischen Agrimensorencodex Vat. Pal. Lat. 1564," *Miscellanea Bibliothecae Apostolicae Vaticanae* 5 (1997), 157–73.

NOTES TO PAGES 178-81

27 On the identification of those structures as boundary markers, see James N. Carder, "Art Historical Problems of a Roman Land Surveying Manuscript: The Codex Arcerianus A, Wolfenbüttel" (PhD dissertation, University of Pittsburgh, 1976), 83-84.

28 Umberto Tocchetti Pollini, "Le città in età romana: L'inizio del fenomeno urbano e le sue trasformazioni," in *Archeologia urbana in Lombardia: Valutazione dei depositi archeologici e inventario dei vincoli*, ed. Gian Pietro Brogiolo (Modena: Panini, 1985), 34-47. The most important study of the centuriation in northern Italy is: *Misurare la terra: Centuriazione e comuni nel mondo romano*, ed. Salvatore Settis (Modena: Panini, 1983).

29 Varro, *De re rustica* 1.10. See also Columella, *De re rustica* 5.1, which draws from Varro's passage. On the centuriation system, Campbell, *The Writings of the Roman Land Surveyors*, 2-5 and 134-63. Less extensive is Oswald A. W. Dilke, *The Roman Land Surveyors: An Introduction to the* Agrimensores (Newton Abbot: David and Charles Press, 1971).

30 Only in exceptional conditions could the Roman Senate authorize a different division of the fields. For instance, in 173 BCE the lots of Cisalpine Gaul were increased to ten *iugera* to attract settlers. See Livy, *Ab urbe condita* 42.4.3. In northern Italy, few areas offer variations to the traditional centuriation reticulum. In Verona there are *centuriae* of 21 × 20 *actus*, near Cividale of 12 × 12 *actus*, and in Treviso of 21 × 21 *actus*. Yet none of them offers an alternative, as they are still based on the *iugerum*. See Settis ed., *Misurare la terra*, 85-87. On the assignment of the lots, see Ferdinando Castagnoli, "Le *formae* delle colonie romane e le miniature dei gromatici," *Atti della Reale Accademia d'Italia: Memorie della classe di scienze morali e storiche* 4, no. 4 (1943), 83-118.

31 On centuriation technology, see Pier Luigi dall'Aglio, "La bassa pianura parmense tra età romana e primo medioevo," *Archivio storico per le province parmensi*, 4, no. 55 (2003), 491-507, esp. 492; Joseph Rykwert, *The Idea of a Town: The Anthropology of Urban Form in Rome, Italy, and the Ancient World* (Princeton, NJ: Princeton University Press, 1976), 58; Michael J. T. Lewis, *Surveying Instruments of Greece and Rome* (Cambridge: Cambridge University Press, 2001). The *groma* was still used in the Middle Ages. The German canoness Hrotsvitha (c. 935-c. 1002) alludes to its use when describing the foundation of a church in Gandersheim, where she lived. See Hrotsvitha, *Primordia cenobii Gandeshemensis*, ed. Paul von Winterfeld, in *Monumenta Germaniae Historica: Scriptores*, vol. 34 (Berlin: Weidmann, 1902), 236.

32 Dilke, *Land Surveyors*, 59.

33 Gianfranco Tibiletti, "Problemi gromatici e storici," *Rivista storica dell'antichità* 2 (1972), 87-96.

34 Aristotle, *Metaphysics* 1.983a, trans. by Hippocrates G. Apostle (Bloomington: Indiana University Press, 1966), 16.

35 Peter Kidson, "A Metrological Investigation," *Journal of the Warburg and Courtauld Institutes* 53 (1990): 71-97; and Pierre Gros, "Nombres irrationels et nombres parfaits chez Vitruve," *Mélanges de l'Ecole française du Rome: Antiquité*

88 (1976): 669—704. For a summary of the various values of $\sqrt{2}$, see Stefaan van Liefferinge, "The Hemicycle of Notre-Dame of Paris: Gothic Design and Geometrical Knowledge in the Twelfth Century," *Journal of the Society of Architectural Historians* 69, no. 4 (2010), 491—507, esp. 499.

36 Giovanna Bonora Mazzoli, "Persistenze della divisione agraria romana nell'ager bononiensis," in *Insediamenti e visibilità nell'alto ferrarese dall'età romana al medioevo* (Ferrara: Accademia delle scienze, 1989), 87—101.

37 Fernando Lugli, *I numeri raccontano Carpi: Uno studio di archeologia della misura* (Carpi: Centro ricerche, 2005), 16—19. My father and I took the distances with a GPS system, which has a margin of error of around eighteen centimeters per terrestrial degree. Another example of Roman surveyors' precision comes from the twenty-nine-kilometer-long surviving section of the German frontier laid down under Domitian, which has an error of about two meters. See Dilke, *Land Surveyors*, 54.

38 Carlotta Franceschelli and Stefano Marabini, "Assetto paleoidrografico e centuriazione romana nella pianura faentina," *Agri Centuriati: International Journal of Landscape Archaeology* 1 (2004), 87—107.

39 In Pistoia, the *stioro* and the *pugnoro* corresponded to, respectively, the Roman *actus* (1,266 square meters) and the square *pertica* (around 8.8 square meters). See Natale Rauty, *Pistoia: Città e territorio nel medioevo* (Pistoia: Società pistoiese di storia patria, 2003), 202.

CHAPTER 27

1 On the use of "ropes" (*funiculi*) for measuring "boundaries" (*fines*), see Isidore, *Etymologiarum libri* 15.14.1.

2 Michele Bacci, *Lo spazio dell'anima: Vita di una chiesa medievale* (Rome and Bari: Laterza, 2005), 5—6.

3 Giovan Battista Visi, *Notizie storiche della città e dello stato di Mantova*, 2 vols. (Mantua: Pazzoni, 1732), 2:267.

4 See *Liber Pontificalis*, ed. Louis Marie Olivier Duchesne, 2 vols. (Paris: Thorin, 1886—92), 2:294. Urbanus II confirmed the ceremonial procedure at the Council of Piacenza of 1095.

5 Plans and elevations were also separated from a practical viewpoint. In his *Practica Geometriae*, Hugh of Saint Victor writes that elevation belongs to the field of *altimetria*, which relies on indirect measuring. Plans, on the other hand, are part of *planimetria* and can be measured directly and precisely. Hugh of Saint Victor, *Opera propaedeutica*, ed. Roger Baron (Notre Dame, IN: University of Notre Dame Press, 1966), 17.33—52. In general, Stephen K. Victor, *Practical Geometry in the High Middle Ages: "Artis Cuiuslibet Consummatio" and the "Pratike de Geometrie"* (Philadelphia: American Philosophical Society, 1979), 3—5 and 18.

6   Neda Bodner, "Earth from Jerusalem in the Pisan Camposanto," in *Between Jerusalem and Europe: Essays in Honour of Bianca Kühnel*, ed. Renana Bartal and Hanna Vorholt (Leiden: Brill, 2015), 74–93.

7   Paul Emmons, "Drawing Sites: Site Drawings," in *Architecture and Field/Work*, ed. Suzanne Ewing, Jérémie Michael McGowan, Christ Speed, and Victoria Clare Bernie (London: Routledge, 2011), 119–28.

8   Herbert L. Kessler and Johanna Zacharias, *Rome 1300: On the Path of the Pilgrim* (New Haven: Yale University Press, 2000), 127–30. For a review of the literature on the mosaic, see Marina Righetti Tosti-Croce, "La basilica tra Due e Trecento," in *La basilica romana di Santa Maria Maggiore*, ed. Carlo Pietrangeli (Florence: Nardini, 1987), 129–69: 159–61.

9   Marvin Trachtenberg, *Dominion of the Eye: Urbanism, Art, and Power in Early Modern Florence* (New York: Cambridge University Press, 1997), 224.

10 Gerbert d'Aurillac, *Opera mathematica*, ed. Nikolaj Michajlovič Bubnov (Berlin: R. Friedländer und Sohn, 1899), 91–92. See also pp. 60–61. For a commentary, Kurt Vogel, "L'aritmetica e la geometria di Gerberto," in *Gerberto: Scienza, storia e mito*, ed. Michele Tosi (Bobbio: Archivi storici bobiensi, 1985), 577–96: 594. Because of these geometrical schemes, philologists have suggested that Gerbert read and copied from the so-called "X" tradition of the *Corpus*, one example of which is Naples, Biblioteca Vittorio Emanuele III: V.A.13. See Lucio Toneatto, *Codices artis mensoriae: I manoscritti degli antichi opuscoli latini d'agrimensura*, 3 vols. (Spoleto: Centro italiano di studio sull'alto medioevo, 1994–95), 1:150–51.

11 Scholars have long identified Roman surveying techniques as a potential source for medieval architectural procedures, but either assume the possibility of diffuse oral traditions or stress the simplicity of these geometric structures (thus implying their atemporality: basic concepts that may spontaneously emerge at any time). Shelby marginalizes the role of the Roman *agrimensores* in shaping the medieval relationship to geometrical planning. See Lon R. Shelby, "The Geometrical Knowledge of Mediaeval Master Masons," *Speculum* 47, no. 3 (1972), 395–421: 400 and 404. See also Paul Frankl, "The Secret of Medieval Masons," *Art Bulletin* 27 (1945), 49–60; François Bucher, "Medieval Architectural Design Methods: 800–1560," *Gesta* 11, no. 2 (1972), 37–51: 37–41. The most updated treatment of the subject is *Ad Quadratum: The Practical Application of Geometry in Medieval Architecture*, ed. Nancy Y. Wu (Aldershot: Ashgate, 2002). And yet, despite the many contributions to this edited volume, the *agrimensores* are mentioned only twice, as transmitters of Euclidean principles (p. 32) and metric knowledge across the Mediterranean Sea (pp. 220–22).

12 Although a homogeneous grid, the centuriation adapted to the characteristics of the territory, thus producing not one unitary, potentially infinite grid, but a patchworklike arrangement of variously oriented schemes. The centuriation rotation changed as well. It is around thirty-four degrees in Mantua,

twenty in Cremona, and four in Ivrea, while the course of the Aemilian Way, the road that served as the reference line for all those area, oscillates between thirty and forty degrees. See Piero Ugolini, "Sistema territoriale e urbano della Valle Padana," in *Insediamenti e territorio*, ed. Cesare de Seta, vol. 8 of *Storia d'Italia: Annali* (Turin: Einaudi, 1985), 159–240: 164–66. The false idea that the centuriation conformed to the solar course emerges over and over in scholarship. See Joseph Rykwert, *The Idea of a Town: The Anthropology of Urban Form in Rome, Italy, and the Ancient World* (Princeton, NJ: Princeton University Press, 1976), 202. For a response, see Brian Campbell, *The Writings of the Roman Land Surveyors* (London: Society for the Promotion of Roman Studies, 2000), 134. The adaptability of the centuriation is articulated in Frontinus's *De ars mensoria*, part of the *Corpus agrimensorum romanorum*. See *Die Schriften der römischen Feldmesser*, ed. Friedrich Blume and Karl Lachmann, 2 vols. (Berlin: Reimer, 1848–52), 1:30.

13. Paolo Porta "La cattedrale paleocristiana nelle fonti storico-monumentali," in *Faenza: La basilica cattedrale*, ed. Antonio Savioli (Florence: Nardini, 1988), 13–25: 16.

14. Marco Rossi, *La rotonda di Brescia* (Milan: Jaca Book, 2004), 25.

15. Francesco Gandolfo, "Il cantiere dell'architetto Lanfranco e la cattedrale del vescovo Eriberto," *Archeologia medievale* 3, no. 1 (1989), 29–47.

16. For the beginning of the cathedral, usually associated to either the date inscribed on the facade, 1107, or following the devastating earthquake ten years later. See Paolo Galli, "I terremoti del gennaio 1117: Ipotesi di un epicentro nel Cremonese," *Il quaternario* 18, no. 2 (2005), 87–100. The church seems to have been completed within the first quarter of the twelfth century. See Arturo Calzona, *Il cantiere medievale della cattedrale di Cremona* (Cinisello Balsamo: Silvana, 2009), 26–40.

17. Hans Peter Autenrieth and Beate Autenrieth, "Der Bau des Domes in Cremona zur Zeit des Bischofs Oberto di Dovara (1117—1162)," in *Arte d'Occidente: Temi e metodi. Studi in onore di Angiola Maria Romanini*, ed. Antonio Cadei, 3 vols. (Rome: Sintesi d'informazione, 1999), 1:111–21. Vitruvius's passage: "ita enim erit duplex longitudo operis ad latitudinem." Vitruvius 3.4.3.

18. Paolo Piva, "Il battistero paleocristiano di Piacenza," *Antiquité tardive* 5 (1997), 265–74; Paolo Piva, "Architettura, 'complementi' figurativi, spazio liturgico (secoli IV/V–XIII)," in *Dall'alto medioevo all'età comunale*, ed. Giancarlo Andenna, vol. 2 of *Storia di Cremona* (Bergamo: Bolis, 2004), 364–445: 365–66.

19. Cristiano Zanetti, *La cattedrale di Cremona: Genesi, simbologia ed evoluzione di un edificio romanico* (Cremona: Biblioteca statale, 2008), 117–18. Other churches that follow the two-square ground plan are the cathedral of Salerno, Santa Maria in Gazzo, and Santa Maria in Castello in Carpi. See Emanuele Lugli, "Pietre di Paragone: The Production of Spatial Order in the Twelfth-Century Lombard City" (PhD dissertation, New York University, 2009), 178–79 and 191–95.

20   The eastern wall is 34.4 meters, which corresponds to 70.75 Pisan feet, with one Pisan foot being 48.6 centimeters. On the dimension of the Pisan foot, Piero Pierotti, "Pontedera: Un saggio di metrologia medievale," *Rivista geografica italiana* 94 (1987): 391–406.

21   Peleo Bacci, "Le fondazioni della facciata del secolo XI nel duomo di Pisa," *Il Marzocco* 22, no. 2 (1917), 2.

22   Ettore Casari, "Osservazioni sulla planimetria del duomo di Modena: Lanfranco, i quadrati, le diagonali," in *Lanfranco e Wiligelmo. Il duomo di Modena*, ed. Enrico Castelnuovo, Vito Fumagalli, Adriano Peroni, and Salvatore Settis (Modena: Panini, 1984), 223–27.

23   Menso Folkerts and Alphonse J. E. M. Smeur, "A Treatise on the Squaring of the Circle by Franco of Liège, of about 1050," *Archives internationales d'histoire des sciences*, 26 (1976): 59–105 and 225–53, esp. 232 and 247.

24   Since seventeen is a prime number, uncommon in mathematical series and traditional religious exegesis, its association to the Roman *centuriation* is quite secure. It is perhaps no chance that seventeen is the only number under twenty to which Isidore of Seville does not provide a symbolical interpretation in his *Liber numerorum*, a short treatise on the figural meanings of every number found in the Scriptures. See Isidore, *Liber numerorum qui in sanctis scripturis occurrunt*, Patrologia Latina 83 (Paris: Migne, 1850), cc. 179–200. The limited allegorical appeal of the number seventeen is discussed in Heinz Meyer and Rudolf Suntrup, *Lexicon der Mittelalterlichen Zahlenbedeutungen* (Munich: Fink, 1987), 661–64. I am grateful to Moritz Wedell for pointing me to this source.

25   Leonardo Pisano, *La pratica di geometria*, ed. Gino Arrighi (Pisa: Domus galileana, 1966), 26. There, Leonardo Pisano also recalls that Pisans used the six-foot *pertica* for building.

26   Keith D. Lilley: *City and Cosmos: The Medieval World in Urban Form* (London: Reaktion, 2009) 125–28; see also Eric C. Fernie, introduction to *Ad Quadratum: The Practical Application of Geometry in Medieval Architecture*, ed. Nancy Y. Wu (Aldershot, UK: Ashgate, 2002), 1–9: 6.

27   Boethius, *De arithmetica*, Patrologia Latina 63, 1079–1168: 1079; *The Metalogicon of John of Salisbury: A Twelfth-Century Defense of the Verbal and Logical Arts of the Trivium*, tr. Daniel D. McGarry (Gloucester, MA: Peter Smith, 1971), 105. There John of Salisbury writes, "What mathematics concludes, in regard to such things as numbers, proportions, and figures is indubitably true and cannot be otherwise."

28   Apocalypse 10:2–11; Ezekiel 40:1–43:17. See also Lilley, *City and Cosmos*, 127–28.

29   Ezekiel's passage was discussed in Saint Jerome, *Commentariorum in Hiezechielem* 13.43.10–122; and Gregorius Magnus, *Moralia in Iob*, 24.8. See Dominique Iogna-Prat, "Topographies of Penance in the Latin West (c. 800–c. 1200)," in *A New History of Penance*, ed. Abigail Firey (Leiden: Brill, 2008), 149–72: 149.

30   "Omnis autem mensurarum observatio et oritur et desinit signo. Signum est cuius pars nulla est." Balbus, *Ad Celsum*, in *Die Schriften der römischen Feldmesser*, ed. Friedrich Blume and Karl Lachmann, 2 vols. (Berlin: Reimer, 1848–52), 1:97.

CHAPTER 28

1   Isidore of Seville, *Etymologiarum libri* 15.15.1–2.
2   Bartholomeus Anglicus, *De proprietatibus rerum* 11.127–29.
3   "A metior hec meta -te, idest circuitus, vel alicuius rei finis vel terminus; unde hec methodus—di, idest remedium . . . hec methodica—ce, quedam species medicine que remedia sectatur et curam, unde methodici dicuntur hii qui nec elementorum rationem observant . . . item a metior hoc membrum, pars corporis." Uguccione da Pisa, *Derivationes*, ed. Enzo Cecchini and Guido Arbizzoni, 2 vols. (Florence: Edizioni del Galluzzo, 2004), 2:765–66. Isidore's passage is copied at p. 763.
4   "Omnis enim natura aut defraudatione aut enormitate rescinditur, proprietate mensurae conseruatur. Ita naturale erit statu, quod non naturale effici potest decessu uel excessu." Tertullian, *De anima* 53. Tertullian drew from Aristotle, but also from Roman physicians and the Scriptures. See Robert Brennan, *Describing the Hand of God: Divine Agency and Augustinian Obstacles to the Dialogue between Theology and Science* (Eugene, OR: Pickwick, 2015), 67.
5   Joel Kaye, *A History of Balance, 1250–1375: The Emergence of a New Model of Equilibrium and Its Impact on Medieval Thought* (Cambridge, UK, and New York: Cambridge University Press, 2014), 150–51.
6   "Utilitas geometria . . . est . . . ad sanitatem, ut medici." Boethius, *Liber de geometria*, Patrologia Latina 63 (Paris: Migne, 1847), col. 1353.
7   "Pedes proprie appellantur: quibus etiam mensuram sibi aliam homines attollentes pedem invenerunt." Pseudo Galen, "Introductio sive Medicus," in *Opera omnia*, ed. Karl Gottlob Kühn, 20 vols. (Leipzig: Knobloch, 1821–33), 14:708. This text was thought to be by Galen in the Middle Ages, during which it circulated widely because of its didactic format. It was downgraded to pseudo-Galenic works in 1525. Today, it is considered the work of a physician who was contemporary to Galen. See Galen, *Tome III. Le médecin: Introduction*, ed. and tr. Caroline Petit (Paris: Les belles lettres, 2009), cxix–cxx.
8   "Oportebat in pede proportione respondere manui, iuste natura omnia constituit." Galen, "De Usu Partium corporis humani," in *Opera omnia*, ed. Kühn, 3:234.
9   "Ad eundem modum et femoris ad tibiam et tibiae ad pedem proportionem considerans summam quandam naturae artem invenies; quemadmodum rursum et partium, quae tum pedi ipsi, tum summae manui insunt, nam hujus quoque partes mirabiliter inster se consentiunt, quo modo certe et brachii partibus ad cubitum, et cubiti ad summam anum, et huius etiam partibus

NOTES TO PAGES 195-96   297

10 inter se incredibilis quaedam est proportio." Galen, "De usu partium," in *Opera omnia*, ed. Kühn, 4:354-55.

"Proceritatem tironum ad incommam scio semper exactam, ita ut VI pedum uel certe V et X unciarum inter alares equites uel in primis legionum cohortibus probarentur." Vegetius, *De re militari* 1.5. The soldier's height is repeated at 2.19. In the fourteenth-century enrollment lists of Treviso, the soldier's height was the first feature to be considered. See Gian Maria Varanini, "Imperfezioni fisiche, esenzioni dagli obblighi militari, segnali di identità: Tipologie documentarie e popolazione maschile (Italia, sec. XIV-XV)," in *Deformità fisica e identità della persona tra medioevo ed età moderna*, ed. Gian Maria Varanini (Florence: Firenze University Press, 2015), 93-118: 117.

11 Paolo Ostinelli, "I chierici e il *defectus corporis*: Definizioni canonistiche, suppliche, dispense," in *Deformità fisica e identità della persona tra medioevo ed età moderna*, ed. Gian Maria Varanini (Florence: Firenze University Press, 2015), 3-30.

12 *Documentazione di vita assisana*, ed. Cesare Cenci, 3 vols. (Grottaferrata: Quaracchi, 1974), 1:43.

13 "Quidam enim mensurare consueverunt agros cum mensuris cubitalibus, vel ulnis, aut cum passibus. Alii cum perticis, sive cum alio quolibet mensurabili instrumento." Leonardo Pisano, *La practica geometriae*, vol. 2 of *Scritti*, ed. Baldassarre Boncompagni (Rome: Tipografia dell scienze matematiche, 1862), 3.

14 David C. Lindberg *Theories of Vision from al-Kindi to Kepler* (Chicago: University of Chicago Press, 1976), 1-17 and 87-90. See also Galen, *On the Usefulness of the Parts of the Body*, trans. Margaret T. May, 2 vols. (Ithaca, NY: Cornell University Press, 1968), 1:92-93.

15 In the opening of his treatise on optics, Euclid states that straight line-rays issued from the eye form a visual cone, which has its apex at the center of the eye and its base on the surface of the visual object: "Ponatur ab oculo educatas rectas lineas ferri spacio magnitudinum immensarum. Et sub visibus contentam figuram conum esse verticem quidem in oculo habentem basim vero ad terminos consepctorum." Euclid, *Optics*, propositions 1 and 2. For the medieval transmission of these passages, see Wilfred R. Thiesen, "*Liber de visu*: The Greco-Latin Translation of Euclid's *Optics*," *Medieval Studies* 41 (1979), 44-105: 62.

16 "Radii lineares exeunt de pupilla ad modum pyramidis, cuius caput est versus oculum et eius basis ad partem rei visae; propter quod autem fit verior visio est quod dicitur sagitta, quae est linea media radiis, aequidistans ab extremis, et aliquid videre est in illud hanc sagittam movere." Avicenna, *De anima* 3.5. In particular, see *Avicenna Latinus*, ed. Simone Van Riet (Leuven and Leiden: Peeters, 1972), 213.

17 Lindberg, *Theories of Vision*, 19-32.

18 An important medieval scholar who insisted that the laws of nature were congruent with the laws of optics was Robert Grosseteste (1175-1253). See his treatise "De lineis angulis et figuris" transcribed in *Die philosophischen Werke des Robert Grosseteste*, Bischofs von Lincoln, ed. Ludwig Baur (Münster:

Aschendorff, 1912), 59–65, esp. 60: "All causes of natural effects must be rendered as lines, angles, and figures. Otherwise it is impossible to know their reason." (*Omnes enim causae effectum naturalium habent dari per lineas, angulos et figuras. Aliter enim impossibile est sciri propter quid in illis*). See also Alistair Cameron Crombie, *Robert Grosseteste and the Origins of Experimental Science, 1100–1700* (Oxford, UK: Clarendon Press, 1953), 131.

19   Katherine H. Tachau, *Vision and Certitude in the Age of Ockham: Optics, Epistemology and the Foundations of Semantics, 1250–1345* (Leiden: Brill, 1988), xvi and 78–80; Gérard Simon, *Archéologie de la vision: L'optique, le corps, la peinture* (Paris: Editions de Seuil, 2003), 125–30.

20   Bacon mentions the date of 1267 in his writing. See Roger Bacon, *Opus maius*, ed. John Henry Bridges, 2 vols. (Oxford: Clarendon Press, 1897), 1:280–81. The "perspectiva" section was completed earlier, around 1263. See David C. Lindberg, *Roger Bacon and the Origins of "Perspectiva" in the Middle Ages* (Oxford: Clarendon Press, 1996), xxiii. The section on radiology is Bacon, *Opus maius*, ed. Bridges, 2:35–37 and 71–74. Relevant information can also be found in section number 4; see 1:113–24. On the popularity of the "perspectiva," see p. xv. On Bacon's theory of radiology, see also Robert Bartlett, *The Natural and the Supernatural in the Middle Ages* (Cambridge, UK, and New York: Cambridge University Press, 2008), 116–18.

21   "Sciendum ergo quod quando haec novem non egrediuntur temperamentum, scilicet nec excellunt nec diminuuntur, tunc fit visus certificatus." Bacon, *Opus maius*, 2:74.

22   "Conpressiorem autem vallem." *Die Schriften der römischen Feldmesser*, ed. Friedrich Blume and Karl Lachmann, 2 vols. (Berlin: Reimer, 1848–52), 1:34. See also David Paniagua, "Frontino, agrimensura ed esegesi tardoantica del testo tecnico nel commento dello pseudo-Agennio Urbico," *Incontri di filologia classica* 10 (2010/11), 29–79: 51–52.

23   "Est mons ab, accipe hastam duorum cubitorum longiorem te, et pone ante te in plano. Postea considera ipsam hastam, quae est c-d-e, et mitte visum tuum recte de f per d usqe a, dividens ipsam hastam super unum cubitum, et vide, quantum sit f-e ad e-d, tantum est f-g ad g-a. Ambula retro, quousque videas de h per c usque ad a, ubi est sumitas montis, et vide, quantum sit h-e ad e-c, tantum est h-g ad g-a." *Geometria incertis auctoris*, in *Opera mathematica*, ed. Nikolaj Michajlovič Bubnov (Berlin: R. Friedländer und Sohn, 1899), 332.

24   Catherine Jacquemard, "Recherches sur la composition et la transmission de la *Geometria incerti auctoris*," in *Science antique, science médiévale*, ed. Louis Callebat and Olivier Desbordes (Hildesheim and New York: Olms-Weidmann, 2000), 81–119.

25   The originality of the *Geometria incertis auctoris* is striking especially when compared to the so-called *Geometria II*, the most common geometrical treatise in the ninth and the tenth centuries, which also includes sections of the *Corpus*. See Menso Folkerts, "Boethius Geometrie II: Ein mathematisches Lehrbuch des Mittelalters" (PhD dissertation, Universität Göttingen, 1967), 109–11.

26 | Arianna Borrelli, *Aspects of the Astrolabe: "Architectonica Ratio" in Tenth and Eleventh-Century Europe* (Stuttgart: Steiner, 2008), 92–95.
27 | "Si cuiuslibet rei altitudinem investigare volueris, huius modi militari ingeniolo invenstigare poteris. Sume arcum cum sagitta et filo, et una fili summitate sagittae postremitati inhaerent, altera in manu remanente, sagitta arcu emissa altitudinis mensurandeae cacumen percutiat." *Geometria incertis auctoris*, ed. Bubnov, 334–35.
28 | "Quia visum humi adjungere difficile mensori." *Geometria incertis auctoris*, ed. Bubnov, 324.

CHAPTER 29

1 | Leonardo Pisano, *La practica geometriae*, vol. 2 of *Scritti*, ed. Baldassarre Boncompagni (Rome: Tipografia delle scienze matematiche, 1862), 42–43 (on surveyors) and 203–4 (example of the tree). Boncompagni's edition is today criticized for lacking philological rigor, a problem that has not been solved by the publication of Laurence Siegler's edition in 2002. See Giuseppe Germano, "New Editorial Perspectives on Fibonacci's *Liber Abaci*," *Reti medievali* 14, no. 2 (2013), 157–73. While waiting for a new critical edition of Fibonacci's texts, I refer to Boncompagni's publication.
2 | Jens Høyrup, "Leonardo Fibonacci and Abbaco Culture: A Proposal to Invert the Roles," *Revue d'histoire des mathématiques* 11 (2005), 23–56.
3 | Kurt Vogel, "A Surveying Problem Travels from China to Paris," in *From China to Paris: 2000 Years Transmission of Mathematical Ideas*, ed. Yvonne Dold-Samplonius and Benno Van Dalen (Stuttgart: Steiner, 2002), 1–8.
4 | "Quare amplectens strictius ipsum modum Indorum et attentius studens in eo, ex proprio sensu quedam addens et quedam etiam ex subtilitatibus Euclidis geometrice artis apponens, summam huius libri." Leonardo Pisano, *Il liber abaci*, vol. 1 of *Scritti*, ed. Baldassarre Boncompagni (Rome: Tipografia delle scienze matematiche, 1857), 1.
5 | Fibonacci added this autobiographical anecdote to the second (1228) edition of his *Liber abaci*, dedicated to Michael Scot, Frederick II's astrologer. See Raffaella Franci, "Il *Liber Abaci* di Leonardo Fibonacci," *Bollettino dell'Unione matematica italiana* 5 (2002), 293–328.
6 | For a new critical edition of the prologue to the *Liber abaci*, see Germano, "New Editorial Perspectives," 171–73.
7 | Charles Burnett, "Fibonacci's 'Method of the Indians,'" *Bollettino di storia delle scienze matematiche* 23, no. 2 (2003), 87–98. I am grateful to Michele Campopiano for showing me this article.
8 | The Florentine *Praticha d'arismetricha* (Vatican City, Biblioteca Apostolica Vaticana: Ms Ottoboniano Lat. 3307, from ca. 1450) states that the books by Fibonacci could be found in "Santo Spirito e in Santa Maria Novella, e anchora nella Badia di Firenze e in particularità l'ànno molti nostri citta-

9. dinj." And Benedetto da Firenze's *Pratica arismetrica* (1463), which is often regarded as the best mathematical compendium of the century, declares that its materials were taken from "Lionardo Pisano e d'altri auctorj.' See Siena, Biblioteca degli Intronati L IV. 21., fol. 1.

9. Leon Battista Alberti reworked some of Fibonacci's exercises in his *Ludi matematici*. See Stephen R. Wassell, "Commentary on *Ex ludis rerum mathematicarum*," in *The Mathematical Works of Leon Battista Alberti*, ed. Kim Williams, Lionel March, and Stephen R. Wassell (Basel: Birkhäuser, 2010), 75–140: 100–104. In 1442 the Pisan Cristofano di Gherardo di Dino translated Fibonacci's treatise into Italian. The result is contained in Florence, Biblioteca Riccardiana, Ms 2186 (R. III. 25), at ff. 83–129r. It has been printed as Leonardo Fibonacci, *La pratica di geometria volgarizzata da Cristofano di Gherardo di Dino*, ed. Gino Arrighi (Pisa: Domus Gaileana, 1966). As Eva Caianiello argues, Fibonacci's followers did not only replicate his text; they also improved it. See Eva Caianiello, "Les sources des textes d'abaque italiens du XIVe siècle: Les échos d'un débat en cours," *Reti medievali* 14, no. 2 (2013), 189–209.

10. We know of only one manuscript of the *Corpus* that was produced in the fourteenth century. See Lucio Toneatto, *Codices artis mensoriae: I manoscritti degli antichi opuscoli latini d'agrimensura*, 3 vols. (Spoleto: Centro italiano di studio sull'alto medioevo, 1994–95), 1:48–49.

11. "Incipit capitulum octavum de reperiendis preciis mercium per maiorem guisam." Leonardo Pisano, *Liber abaci*, 83–118.

12. Leonardo Pisano, *Liber abaci*, 84–90.

13. "Ex his superficialibus mensuis quidam in multiplicando colligunt quandam quantitatem, quam vocant iugerum, vel aripennium, sive carrucam, sive tornaturam, vel culturam, vel alias quantitates, que aliis censentur vocabulis." Leonardo Pisano, *Practica geometriae*, 3.

14. Leonardo Pisano, *Liber abaci*, 118.

15. "Canna pisana est palmorum 10, vel brachiorum 4; canna autem Ianue, ut dictum est, palmorum 9. Canna itaque provincie [Provence], et Sicilie, et surie [Syria] et Constantinopolis sunt unius mensure, scilicet palmorum 8." Leonardo Pisano, *Liber abaci*, 111.

16. "Opportet eos qui arte abbaci uti voluerint, ut subtiliores et igeniores appareant scire computum per figuram manuum, secundum mgistrorum abbaci usum antiquitus sapientissime inventam." Leonardo Pisano, *Liber abaci*, 5. On the illustration, see Bente K. Abbaddo, "Storia di un manoscritto duecentesco: L IV 20 della Biblioteca comunale degli Intronati di Siena (Fibonacci, *Liber abaci*)," in *Il codice miniato in Europa: Libri per la chiesa, per la città, per la corte*, ed. Giordana Mariani Canova and Alessandra Perriccioli Saggese (Padua: Il Poligrafo, 2014), 115–25.

17. Scholars such as Robert of Chester (c. 1145) and Gerard of Cremona (c. 1170) introduced Arabic numerals into Christian Europe by discussing them in their translations of the algebraic treatise by al-Khwarizmi. See Barnabas B. Hughes, "Gerard of Cremona's Translation of al-Khwarizmi's al-Jabr: A

NOTES TO PAGES 206–8

18 Critical Edition," *Mediaeval Studies* 48 (1986), 211–63; and *Robert of Chester's Latin Translation of al-Khwarizmi's Al Jabr*, ed. Barnabas B. Hughes (Stuttgart: Franz Steiner, 1989). "Gestus representabant ydoneos." Salimbene de Adam, *Cronica*, ed. Giuseppe Scalia, 3 vols. (Rome and Bari: Laterza, 1966), 1:61.

19 The document, dated between 1233 and 1241, is transcribed in Francesco Bonaini, "Memoria unica sincrona di Leonardo Fibonacci," *Giornale storico degli archivi toscani* 6, no. 2 (1858), 238–46.

20 The full passage reads: "Qui agrimensor sciat legere et scribere, et sit bene instructus in arte et ministerio colligendi et faciendi rationem ad ambacum; et hoc scribatur quando eligetur in apodixa: et etiam qui sciat de mensura et esmo." *Statuti inediti della Città di Pisa dal XII al XIV secolo*, ed. Francesco Bonaini, 3 vols. (Florence: Vieusseux 1854–70), 1:115.

21 Elisabetta Ulivi, "Scuole e maestri d'abaco in Italia tra Medioevo e Rinascimento," in *Un ponte sul Mediterraneo: Leonardo Pisano, la scienza araba e la rinascita della matematica in Occidente*, ed. Enrico Giusti (Florence: Polistampa, 2016), 121–59.

22 Leonardo Pisano, *Liber abaci*, 299–301.

23 Guido Zaccagnini, "L'insegnamento privato a Bologna e altrove nei secc. XIII e XIV," *Atti e memoria della Real deputazione di storia patria per le province di Romagna* 14 (1923–24), 254–301.

24 Ulivi, "Scuole e maestri d'abaco," 126. Magister Lotto was paid forty lire.

25 Nadia Ambrosetti, *L'eredita arabo-islamica nelle scienze e nelle arti del calcolo dell'Europa medievale* (Milan: LED 2008), 247–50. Enrico Giusti, "Matematica e commercio nel *Liber abaci*," in *Un ponte sul Mediterraneo: Leonardo Pisano, la scienza araba e la rinascita della matematica in Occidente*, ed. Enrico Giusti (Florence: Polistampa, 2016), 59–120.

26 "In Venexia la mexura a la qual se vende lo formento sì li vien dito stero e debis saver che le stera 3 de Venexia sì fa I cafesse [qafiz] de Tonisto [Tunis] e lle 300 stera avança in Tonisto cafessi 6." *Zibaldone de Canal, Manoscritto mercantile del sec. XIV*, ed. Alfredo Stussi (Venice: Comitato per la pubblicazione delle fonti relative alla storia di Venezia, 1967), 41–73: 44.

27 "In Ilaiaça se vende fromento e orço ad una mexura ch'è nome marçapan e per la volontade de li Armini nexun non può dire lo vero de un mexe a l'olltro como torna nexuna mexura cum questa, perché elli la crexe e menema a soa volluntade e perçiò li marchadanti de reçeve assè fiade dano." *Zibaldone de Canal, Manoscritto mercantile*, 61–62.

28 Roberto Sabatino López and Gabriella Airaldi, "Il più antico manuale di pratica della mercatura," *Miscellanea di studi storici* 2 (1983), 100–133.

29 Francesco Balducci Pegolotti, *La pratica della mercatura*, ed. Allan Evans (Cambridge, MA: Mediaeval Academy of America, 1936), 37.

30 Pegolotti, *La pratica della mercatura*, 46.

31 Tommaso Bertelè, *Misure di peso a Bisanzio* (Padua: Museo Bottacin, 2009), 28, note 29. Lopez and Airaldi, "Il più antico manuale," 114. For an alternative view, Eric Fernie, "Pegolotti's Cloth Lengths," in *The Vanishing Past: Studies*

of Medieval Art, Liturgy and Metrology Presented to Christopher Hohler, ed. Alan Borg and Andrew Martindale (Oxford: BAR International Series III, 1981), 13–28.

32 Pegolotti, *La pratica della mercatura*, 133–136. The measurements of Tunis are compared to those of Béjaïa, Annaba, Palermo, Gaeta, Naples, Pisa, Genoa, Champagne, Nîmes, Venice, Florence, Seville, Ancona, Tropea, Scalea, Ceuta, and Castel di Castro (now a neighborhood of modern Cagliari).

33 John S. Critchley, *Marco Polo's Book* (Aldershot and Brookfield, VT: Variorum, 1992), 48–50. See also Hans Ulrich Vogel, *Marco Polo Was in China: New Evidence from Currencies, Salts and Revenues* (Leiden and Boston: Brill, 2013), 24.

CONCLUSION

1 Cicero, *In Verrem* 2.1.130–136.

2 For an archaeological reading of Verres's rebuilding of the temple, see Frank Tenney and Gorham Phillips Stevens, "The First and Second Temples of Castor at Rome," *Memoirs of the American Academy in Rome* 5 (1925), 79–102: 80–81. Interestingly, after Cicero's death, the temple became the office for weights and measures.

3 "Itaque si loci natura permittit, rationem servare debemus: sin autem, proximum rationi." Brian Campbell, *The Writings of the Roman Land Surveyors* (London: Society for the Promotion of Roman Studies, 2000), 145.

4 Martin Heidegger, *What Is a Thing?* tr. W. B. Barton and V. Deutsch (Chicago: Henry Regnery, 1967), 33–34.

5 Bruno Latour, "Why Has Critique Run Out of Steam? From Matters of Fact to Matters of Concern," *Critical Inquiry* 30 (2004), 225–48.

6 The British Association for the Advancement of Science met in Liverpool between September 14–21, 1870. See *Report of the 40th Meeting of the British Association for the Advancement of Science, Held at Liverpool* (London: John Murray, 1871). Maxwell was president of the Section A for Mathematics and Physics. On the Parisian meeting, see Edvige Schettino, "L'Italia e la convenzione del metro: Negoziazioni tecniche, scientifiche e legislative," *Physis* 41, no. 2 (2004), 345–56: 347.

7 James Clerk Maxwell, "Sectional Proceedings," *Nature*, September 22, 1870, 419–22. I am grateful to Christie L. Harner for sending me this article.

8 Boethius, *De arithmetica*, Patrologia Latina 63 (Paris: Migne, 1860), 1079–1167a: 1079b.

Index

abbaco, 99, 125, 202–7
Abelard of Bath, 70
Académie des sciences. *See* Paris (France)
actus, 97, 178–82, 189–90
Adalbold of Utrecht, 177
Adam, Robert, 16
Adelman, 177, 188
Alain de Lille, 112
Alberti, Leon Battista, 52–53, 199, 300n9
Albert the Great, 104
Albino (Italy), 164
Aldo Manuzio the Younger, 83, 246n17
Alessandria (Italy), 117
Alexander IV (Rinaldo di Jenne), 123, 266n16
Al-Ma'mun, 42, 233n4
Amelia (Italy), 66
Anthony of Novgorod, 147
Aquinas, Thomas, 96, 112, 137, 140, 149–50, 154
Arbuthnot, Charles, 42
Aristotle, 70, 96, 137, 149, 181, 248n2, 296n4
Arrigoni, Bernardino, 73–74
arts (*societates artium*), 80–81, 83, 85, 90, 101, 103, 136, 138, 207
Ascoli (Italy), 123, 143
Assisi (Italy), 75–76, 85, 121, 123–25, 143, 196, 269n12; measures of, 126, 239n6

Asti (Italy), 76
Augustine, Saint, 104, 111, 148, 175–76
Austrian empire, measures of, 33–34, 39
Avicenna, 196

Babylonian standards, 45
Bacon, Roger, 197, 233n4
Badiou, Alain, 226n26, 240n25
Balbi, Giovanni, 99
Balbo, Prospero, 45
Baraldi, Pietro, 11
Barcelona (Catalonia), 124
Bartholomeus Anglicus, 194
Basile, Giovan Battista, 69
Bassano (Italy), 67, 93, 102
Beatrice of Canossa, 165–66, 167–68, 169
Beccaria, Cesare, 24, 28, 30, 222n5, 227n13
Benedict XI (Niccolò Boccasini), 124
Bergamo (Italy), 117; measures of, 31, 65, 77, 80, 169, 170
Bernardino of Siena, Saint, 104–5, 256nn22–23
Blondel, Jacques-François, 39
Bobbio (Italy), monastery of, 176–77, 198, 282n11
Boethius, 172, 177, 192, 195, 216
Boito, Camillo, 53

303

Bologna (Italy), xi–xii, 75, 87, 90, 102, 115–16, 123, 128, 134–35, 137, 138, 140–41, 143, 183, 184, 206; church of San Domenico, 121, 123–24, 135, 155; church of Santo Stefano, 156–58; *libri terminorum*, 128–33; measures of, 60, 72, 85, 87, 168–69; Saint Petronius, patron saint, 156–58; *studium*, 78, 128, 132, 177
Bonaventure of Bagnoregio, Saint, 104, 112, 122, 263n31
Boniface VIII (Benedetto Caetani), 124, 246n14
Bonvesin de la Riva, 72, 117–18, 143–44
Borso d'Este, 73
Boscovich, Roger Joseph, 27–28
bread, 112, 117; measures of, xiii, 87, 90, 95, 103, 120, 137, 252n4
Brescia (Italy), 41, 117, 177, 188; *Lber viis*, 132–33; measures of, 31, 65, 77
Burattini, Tito Livio, 11, 110
Burchard, Johann, 152
Burnet, Thomas, 46

Canina, Luigi, 11
Carafa, Giovanni, 22–23, 45
Carletti, Nicola, 22, 226n23
Carpi (Italy), 292n37, 294n19
Cassiodorus, 172, 258n7
cathedral: as economic institution, xiv, 67, 168, 237n9; as learning center, 172, 177, 188; planning of, 187–93; as support of *pietre di paragone*, 20, 59, 60, 65–66, 73, 77, 79, 85–87, 110, 169, 170
Cavalca, Domenico, 89
Celestine I, 156
centuriation, 178–84, 187–92, 196, 210
Charlemagne, 172, 184, 283n16; measurement reform, 285n1
Charles the Fat, 167
Chateaubriand, François-Auguste-René de, 49
Christ: footprints of, 148, 154; healing by touch, 154; height of, 145–52, 153; prohibition of measuring, 116
Cicero, 209–10
Cicognara, Leopoldo, 52, 53

Cinderella, 69
Cistercians, 122–23, 137
city walls, 81, 118, 122, 123, 144; the *circla* of Bologna, 128–30; of Florence, 169–70
Clement IV (Gui Foucois), 124, 125, 127
communal palace: of Reggio Emilia, 57–58; as seat of power, 143, 206, 247n22, 249n10; of Siena, 88–90; as support for *pietre di paragone*, 60, 72, 83–87, 115
commune: definition of, xiii, 63, 64, 80, 169; as economic institution, 57–58, 67, 72, 136, 137, 206; fighting for measuring rights, 63–64, 67, 75–79, 80–87, 90–92; official checkers of the, 128–29; relying on citizens for checks, 101–3; relying on *religiosi viri* for checks, 109, 115–20, 138–40; and spatial control, 128, 132, 136, 141–43
Como (Italy), 117; measures of, 252n4
Condillac, Etienne de, 32
Constance (Switzerland), Peace of, 77–78, 81, 85, 86
Constantinople, 92, 150, 156; Hagia Sophia, 147, 149; measures of, 101, 208
Corbie (France), monastery of, 172
Corpus agrimensorum romanorum, 173–78, 188, 192, 196–97, 202, 210
Corpus iuris civilis, 77–79, 88, 110, 112–13, 123, 149
counting, 96–97, 119, 141–42, 182, 204–6, 259n20. See also *abbaco*
Cremona (Italy), 117, 167, 171, 176–77, 293–94n12; cathedral of, 189–90; measures of, 31
Cristiani, Girolamo Francesco, 10, 22, 41–48, 198
Cumberland, Richard, 45

Dante, 140, 256n20
Decretals, 113, 195
Descartes, René, 46, 126
Dickens, Charles, 49
Disney, Walt, 69
distances, 6, 29, 37–38, 41, 42, 50, 57, 60, 70, 117, 121–32, 142–43, 156, 158, 178, 181, 197–98
Dominic, Saint, 121, 135, 155

INDEX

Dominicans, 121–24, 128, 140–41, 150, 206;
   preaching against false measures, 100,
   109, 137, 140, 153
earth, 9, 32, 42, 46–47, 162–63, 165, 181, 212;
   meridian, 11, 22, 27, 36–37; shape of, 37;
   size of, 27–28, 30–31, 42–44; standards
   extracted from, xv, 5, 7, 32, 41–42, 53;
   theory of the, 46 (*see also* geometry)
Ebendorfer, Thomas, 153
Egypt, 173, 178; measures of, 45, 258n7;
   Pyramid of Giza, 42
Elba (Italy), island of, 81
Erasmus of Rotterdam, 177
Etruscan measures, 45
Euclid, 70, 172, 175, 186, 201, 293n11, 297n15
Euripides, 46

Faenza (Italy), 65, 77, 170, 183, 187
fees: for checking standards, 81, 117, 168;
   import duties, 117; property taxes, 67,
   156, 178; tithes, 136, 163, 167, 172, 259n20
Ferdinand I, 21
Ferrara (Italy), 66–67, 76, 117, 241n35,
   262n17; measures of, 60–63
Fidenza (Italy), 60
Filarete (Antonio di Pietro Averlino), 66
fines, 102–3
Florence (Italy), 37, 87, 89, 100, 101, 102, 114,
   117, 134–35, 139, 140, 145, 207; cathedral
   of, 139, 173; church of Santa Croce, 100,
   153; church of Santa Maria Novella,
   114; convent of San Jacopo di Ripoli,
   150; Laurentian Library, 145, 259n21;
   measures of, 3, 24, 25, 28, 31, 85, 92–93,
   169–70, 257n25
Florian, Jean-Pierre Claris de, 39
France, measures of. *See* Paris (France)
Francesco della Marca, 104
Franciscans, 109, 112, 116, 119, 121–24, 137,
   145, 194
Francis of Assisi, Saint, 115–16, 121
frauds, 58–59, 69, 82, 102–3, 112, 118, 137
Frederick I Barbarossa, 66, 76–78, 170, 171,
   184, 242n3
Frederick II, 82, 91, 100, 206; measures of,
   245n9

Fréret, Nicolas, 42
Frontinus, 197, 293–94n12
Fulbert, 177, 188, 198
Fulvio, Andrea, 150

Galeazzo II Visconti, 78
Galen, 195, 296–97n9
Genoa (Italy), 76, 92, 129–30, 145, 271n2;
   measures of, 77, 208, 302n32
Genovesi, Antonio, 20, 22
Gentile da Montefiore, 124
geometry, 111, 176–77, 186, 190–93, 195,
   196–97, 214; *Geometria incertis auctoris*,
   197–99; as "measuring of the earth,"
   172–73, 194; *Practica geometriae*, 201–3,
   208. *See also* Euclid
Gerbert of Aurillac. *See* Silvester II (Gerbert
   d'Aurillac)
Giles of Rome, 96
Giordano da Pisa, 100, 104, 114, 136, 139–40,
   144
Giotto, 249n10
Giuliano da Maiano, 65
Goethe, Johann Wolfgang von, 17, 50, 52,
   235n6
Grado (Italy), 92
gravity, 36, 37
Greece: ancient measures of, 45–46;
   stadion, 42
Gregory VII (Hildebrand of Sovana), 185
Gregory IX (Ugolino di Conti), 195
Gregory X (Teobaldo Visconti), 121–23, 134
groma, 180–82, 196, 291n31
Grosseteste, Robert, 112, 137, 248n2, 297n18
Grottaferrata (Italy), 150
guilds. *See* arts (*societates artium*)

Halley, Edmond, 11
Heidegger, Martin, 214
Henry IV, 123, 142
Henry V, 66
Henry VI, 81
Homogirus, 162, 173
Hughes of Arles, 166, 167
Hugh of Saint Victor, 198, 292n5
Humiliates, 109, 116–18, 213
Husserl, Edmund, 221n4

Huygens, Christiaan, 10, 223n4, 231n30
Hyginus Gromaticus, 42, 210

industria, 138
Ingenhousz, Jan, 36
Innocent III (Lotario dei Conti di Segni), 116, 263n30
Innocent IV (Sinibaldo Fieschi), 118
Innocent VIII (Giovanni Battista Cybo), 152
Isidore of Seville, 162, 173, 176, 194, 295n24
Italy, national standards of, xiii, 6, 11, 29–30. *See also* metric system

Jacopo da Varagine, 155
Janety, Marc-Etienne, 38
Japan, introduction of the metric system, 6
Jerusalem, 147–50, 156–58, 186; chapel of the Ascension, 148–49, 154, 278n22; Holy Sepulcher, 155–56; measures of, 44, 283n16; Temple of Solomon, 165
John of Salisbury, 192
Juan, Jorge, 37
jurists, 57–58, 69, 74, 77, 78, 82, 83, 96–97, 103, 112–13, 128–29, 132, 137, 149, 164, 171. *See also* Pilio of Medicina
justice, 10, 64, 69, 75–76, 88–90, 111, 123, 138, 141, 162, 163, 210; symbols of, 67, 73, 88–90, 92, 165, 170, 249n13
Justinian, 77, 110, 112–13, 123, 147, 149–50, 255n9

Kilwardby, Robert, 130–31

La Condamine, Charles-Marie de, 37, 44, 231n25
Lalande, Joseph Jérôme Lefrançois de, 28, 38
Lanfranc, 188–92
Lengherent, Georges, 152, 281n45
Lenoir, Étienne, 33, 38, 230n9
Leonardo Fibonacci (Leonardo Pisano), 190, 196, 201–8, 284n, 287n
Lewis, William, 38
Liberius, 186–87
Liège (Belgium), 177
Liutprand, 45, 164–66, 167, 168–69, 171, 184
Lodi (Italy), 164

London (United Kingdom), 38, 45, 50, 208, 213; British Library, 22; Royal Society, 11, 223n4
Lorenzetti, Ambrogio, 88–90, 105
Lucca (Italy), 67, 95, 123, 132, 166, 252n2, 288n1; measures of, 89, 165, 227n11
Lyon (France), second council of, 121–24, 134

Mairano, Romano, 92
Malebranche, Nicolas, 9–12, 16, 44
manso, 44, 163–64, 169
Mantua (Italy), 165, 185, 293n12; measures of, 65, 66, 73–74, 77, 82, 184
Manuel I Komnenos, 92
Marco Polo, 208
Maria Theresa of Austria, 3, 33, 39
marketplaces, 59, 63, 67, 69, 73, 76, 78, 91, 92, 94, 95, 104, 113, 119, 129, 130–32, 138, 168, 169, 171
Matilde of Canossa, 169
Maxwell, James Clerk, 216
measure: and abstraction, 30, 39–40, 70, 97, 139–40, 146, 190, 192, 198; ambiguity of, 24, 70, 73, 194; and architecture, 8, 58–59, 83–85 (see also *pietre di paragone*); authority of, 10, 63, 75–76, 82, 97, 101–2, 213–14; in the Bible, 46, 111–13, 162, 165, 192; and clothes, 134–37; as commercial tools, 44–45, 79, 86, 111, 202–4, 207, 213 (*see also* marketplaces); devotion for, 145–52, 154–55; disappearance of, 53–54, 129, 147, 165; discrepancies in, 18–20, 28–31, 95, 181, 209–10; as epistemological filters, 6–7, 9–10, 11–12, 142; faith in, 98, 111, 113–14, 147, 155, 163, 239n15; and the ground, 7, 50, 60, 65, 127, 129–30, 134, 135–36, 140–41, 162, 172, 199; healing, 155, 194–95; how to, 50–52, 59, 67, 69, 92–93, 94, 118; and the human body, 46–48, 154, 155, 194–200, 206; invisibility of, 39–40, 130, 144, 178; and the king's body, 47, 49, 165; labor of, 27, 63, 117–18, 128–29; liquidity of, 45–46; as makers of order, 73, 111, 139–44, 148, 192–93; as mirrors, 38–40, 198; as nightmares, 4, 99–100; as

INDEX 307

numbers, 29, 117, 181 (*see also* counting); and penitence, 112, 115–20, 136–37, 141; and remaking, 141; as ribbons, 148, 154, 155; rights on, 63, 76–78, 81–82, 86, 92; self-measure, 136–38; as social bonds, 39, 64, 86, 89, 103–4, 114, 137–38, 142; as tools of coercion, 89–90, 118, 142, 213; as weapons, 89–90, 213
Mégnié, Pierre-Bernard, 38–39
meter: brass standard of 1794, 33; platinum bar of 1799, 33–40. See also standards
metric system, 5; adoption by Italian states, 6, 11, 28–30, 45, 49; creation and launch in France, 5–6, 11, 28, 32–34, 38–40; and the *toise du Perou*, 37
metrology, 33, 47, 52, 222n6, 229n23
Mexico, introduction of the metric system, 6
Mickiewicz, Adam, 50
Milan (Italy), 11, 90, 116, 117–19, 143–44, 165, 177; Beccaria's reform, 3, 28, 30; introduction of the metric system, 37 (*see also* Oriani, Barnaba); premodern measures of, 24, 25, 29–30, 31, 69, 72, 77, 78, 82, 169, 223n4
Minturnae, 178
Misson, Maximilien, 17–19, 22, 31, 50, 280n39
Modena (Italy), 117, 183, 244n22, 262n17; cathedral of, 188–92; measures of, 60, 72, 73, 190–91
Monza (Italy), measures of, 85
Mouton, Gabriel, 11, 110
Muttoni, Francesco, 17–18

Napione, Gian Francesco Galeani, 52–53
Naples (Italy), 14, 17, 25, 28, 69, 165, 293n10; measures of, 15, 17–24, 302n32
Napoleon, 6, 8, 11, 25, 29, 49, 65, 213
Newton, Isaac, 37, 38
Nicholas IV (Girolamo d'Ascoli), 122
Nicholas V (Tommaso Parentucelli), 150–51
Nicholas of Reggio, 195
Novara (Italy), 117, 269n12
Nugent, Thomas, 50

oaths, 57, 80, 97, 116, 120
Obizzo Malaspina, 76

obviousness, xvii
Odysseus, 46
Olivi, Pierre de Jean, 100, 103–4
optics, 38, 52–53, 196–98, 210, 216, 297n15
Oriani, Barnaba, 28, 37, 227n2
Orsini, Latino Malabranca, 134–35
Ottaviano degli Ubaldini, 123–24, 128
Otto III (king), 168
Ovid, 161–63, 172–73, 214

Padua (Italy), 102–3, 129–30, 132, 177, 182, 207, 249n10, 270n14; measures of, 85–86, 117
Paestum (Italy), 17
Palamedes, 46
Palladio, Andrea, 17, 52, 150
Paltanieri, Simone, 123–24
Paoli, Paolo Antonio, 14–17, 19, 50–52
Paris (France), 33, 154, 216; Académie des sciences, 5, 36–37, 39; first international conference of 1799–1800, 5–6, 33, 38, 45; 1960 international conference of weights and measures, 216; premetric measures of, 14, 22, 226n21, 227–28n13; *toise*, 10, 28, 37–38, 41, 44, 49, 227–28n13
Parma (Italy), 29, 49, 102, 103, 109, 115, 117, 119, 134, 135, 138, 269–70n12; measures of, 72, 73, 82
Paschal II (Ranierius), 66
Paucton, Alexis-Jean-Pierre, 22, 41–48
Paul the Deacon, 164–65
Pavia (Italy), 167, 182, 184; measures of, 287n23
Pegolotti, Francesco Balducci, 207–8
Penitents (*fratres penitentie* or *fratres de penitentia*), 109, 115–19, 136–37, 141, 143, 171
Perrault, Charles, 69
Peru (old term for Ecuador): mission in, 36–39, 231n26; *toise* of *Perou* (*see* Paris [France]: *toise*)
Perugia (Italy), 67, 102; measures of, 65–66, 93; *petra iustitiae*, 67
Peter Leopold, 3
Peter Lombard, 111
Peto, Luca, 83
Petrarch, 140, 177
Pheidon, 46

Piacenza (Italy), 76, 90, 270n14, 282n10, 292n4; measures of, 60
pietre di paragone, 65–73, 77, 103, 110, 129, 136; on churches, 58–59, 60, 65–67, 73, 85, 191; on communal palaces, 60, 83–86, 72, 115; definition, 67–69, 75; destruction of, 73, 149; incised in stone, 60, 66–69, 70, 164–65; near doors and gates, 169–71; as negative impressions, 129, 149–50
Pilio of Medicina, 78, 125, 252n8, 272n17
Pisa (Italy), 95, 112, 142, 201, 206, 252n1; cathedral of, 185–86, 190; measures of, 81, 89, 93–94, 97, 168, 190, 202–4
Pistoia (Italy), 72, 77, 119–20, 123, 124; church of San Jacopo, 92, 120; measures of, 78, 168
platinum, 33–34, 36–40
Pliny the Younger, 46
podestà, 57–58, 98, 116, 119, 135, 138, 247n24; in charge of measurements, 72, 78, 79, 85, 89, 90, 92, 93, 97, 115, 130, 132, 241n33; election, 79
Pomponius Mela, 42
Prato (Italy), 87, 140
precision, xii, 5–6, 11, 15, 20, 30, 33–34, 39, 76, 95, 118, 132–33, 213–14, 292n37
punishment of violators, 69, 82, 90–92, 98, 100, 268n7
Pythagoras, 47, 96

Ravenna (Italy), 124, 165; measures of, 65
Raynal, Guillaume Thomas, 38
regalia, 76–77, 81, 167. *See also* measure: rights on
Reggiani, Napoleone, 29–30
Reggio Emilia (Italy), 57–59, 72, 119, 134, 138, 164, 171; measures of, 58–59, 75, 92, 93, 239n5
Remigio dei Girolami, 137
Ricci, Amico, 52
Richer, Jean, 11
Ripoll (Catalonia), monastery of, 176
Rome (Italy), 29, 37, 150, 156, 165, 182, 206; church of Santa Maria Maggiore, 186; church of St John Lateran, 150–52; premetric measures, 22, 27, 29, 31, 42, 44, 83–85, 97, 126, 178, 181, 183–84, 190, 206 (see also *actus*); temple of Castor and Pollux, 209; Trajan Column, 11
Ruskin, John, 239n15

Sacchetti, Franco, 98–100, 101, 221n3
Salerno (Italy): cathedral of, 294n19; *passus* of, 169
Salimbene de Adam, 103, 135, 138, 206, 271n7, 273n24
sameness, 22, 97, 211; failure in producing, 24, 29, 176, 208; promise of, 111, 155, 163, 204, 215; as repetition of the identical, 40, 97–98, 111, 136
San Galgano (Italy), monastery of, 119
San Gimignano (Italy), 134, 206
Sansedoni, Ambrogio, 153, 154
Sbarra Franciotti, Ferrante, 17
scale, xv, 9–10, 22, 53, 123, 146; and history, 64, 155, 173, 183, 207, 208, 212; in maps and plans, 15, 17, 22, 45, 53; vs. size, 12–13, 53
scales, 73, 82, 83, 88, 92, 112, 144
Scarpellini, Feliciano, 27–28, 31, 51–52, 83
Schlegel, Friedrich, 53
Schopenhauer, Arthur, 53
Selvatico, Pietro, 52
Seroux d'Agincourt, Jean Baptiste, 53
Sicardo of Cremona, 171
Siena (Italy), 81–82, 91, 119, 132, 134, 140–41, 252n2; cathedral, 112; church of San Domenico, 153–54; measures of, 65, 89, 104, 248–49n3; palazzo comunale, 88–90; Piazza del Campo, 104
Silvester II (Gerbert d'Aurillac), 70, 176–77, 186–88, 190, 198
size, 22, 53, 95, 214. *See also* scale
standards: destruction of, 28–29, 65, 72–73, 287–88n23; divisions of, 5, 11, 27, 32–34, 93, 203; incised in stone, 168 (see also *pietre di paragone*); inspectors and checks of, 59, 67, 69, 72, 78, 81–83, 90–94, 101–2, 104–5, 109, 115–17, 119, 120, 130, 141, 144, 160, 168, 241n31; line-standards vs. end-standards, 34; as metal bars, 24, 72 (*see also* meter); names of, 65–66, 136; physicality of, 29–30, 37, 40, 75, 103;

INDEX 309

safekeeping of, 33, 40, 72, 169; stamping of, 33, 67, 82, 85, 92; as strings, 125; temporality of, 70, 75–76, 184, 212–13; truthfulness of, 31, 39, 163; variance of, 29–30; violators of, xi–xiii, 70 (*see also* punishment of violators); weathering, 75; in wood, 67, 69, 95, 119
stima (estimate), 58, 95–96, 252n2
Strabo, 42, 46, 240n23
sumptuary laws, 135–37, 272–73n20
surveying, 4, 10, 20, 28, 30, 57, 124–25, 144, 156–58, 164, 172–73, 175, 176–84. See also *Corpus agrimensorum romanorum*; *terminatio*

Talleyrand, Charles-Maurice de, 32
tavole di ragguaglio (conversion charts), 25–31, 39, 42–44, 53, 60; Italy's official *tavole* of 1877, 29, 227n11; Napoleonic *tavole* of 1801, 37 (*see also* Oriani, Barnaba)
terminatio, 129–32, 136, 269n9, 269–70n12.
See also Bologna (Italy): *libri terminorum*; Brescia (Italy): *Liber viis*
Tertullian, 195
Todi (Italy), measures of, 60, 85, 239n5
toise du Perou. See Paris (France): *toise*
tourism, 8, 14, 17–20, 50, 52
towers, 60, 94; height of, 50, 123, 140, 142–43, 197–98, 201
Trachtenberg, Marvin, xii, 275n22, 293n9
Treviso (Italy), 140, 177, 291n30, 297n10; measures of, 169
triangulation, 30, 37, 196, 197–98
Tunis (Tunisia), 207–8; measures of, 302n32
Turin (Italy), 45, 165; measures of, 227n3

Udine (Italy), 125
Ugonio, Pompeo, 151
Uguccione da Pisa, 96–97, 194

Ulloa, Antonio de, 37
United States of America, xiv, 6, 38, 213, 232n36
Urban II (Odo of Châtillon), 185

Varro, 176, 178
Venice (Italy), 41, 92, 101; Academy of fine arts, 52; measures of, 82–83, 85–86, 89, 90, 92, 207–8, 247–48n26, 302n32; Peace of, 77; Rialto, 92; San Giacomo di Rialto, 67
Verona (Italy), 177, 182, 186, 206, 291n30; cathedral, 168; church of San Zeno, 187; measures of, 60, 85, 87, 91–92, 126, 247–48n26; measuring the distance between San Fermo and Santa Maria della Scala, 124–25
Vicenza (Italy), 140–41, 269n12; measures of, 17
Vienna (Austria), 36; congress of, 6, 45, 49
Vienne (France), council of, 104
Virgil, 14, 17, 82
Visconti: Milanese family, 82, 90–91; Pisan family, 81
Visconti, Ferdinando, 20–22, 24, 30, 226n22, 226n24
Viterbo (Italy): measures of, 72; monastery of Saint Rose, 154
Vitruvius, 47, 189, 234n24
Vitruvius Rufus, 175
Voltaire (François-Marie Arouet), 42, 212
Volterra (Italy), 90, 95, 140; *canna* of, 170

Wenzel, Anton, 17

Ximenes, Leonardo, 24, 29, 37

Zurich (Switzerland), 186